International Workshop
on *Gelidium*

Developments in Hydrobiology 68

Series editor

H. J. Dumont

International Workshop on *Gelidium*

Proceedings of the International Workshop on *Gelidium*
held in Santander, Spain, September 3–8, 1990

Under the Sponsorship of
Diputación Regional de Cantabria
International Seaweed Association

Edited by
J. A. Juanes, B. Santelices and J. L. McLachlan

Reprinted from Hydrobiologia, vol. 221 (1991)

Springer Science+Business Media, B.V.

Library of Congress Cataloging-in-Publication Data

International Workshop on Gelidium (1991 : Santander, Spain)
    International Workshop on Gelidium : proceedings of the
International Workshop on Gelidium held in Santander, Spain,
September 3-8, 1991, under the sponsorship of Diputación Regional de
Cantabria, International Seaweed Association / edited by J.A.
Juanes, B. Santelices, and J.L. McLachlan.
        p.   cm. -- (Developments in hydrobiology ; 68)
    "Reprinted from Hydrobiologia, 221."
    Includes bibliographical references and index.
    ISBN 978-0-7923-1372-4   ISBN 978-94-011-3610-5 (eBook)
    DOI 10.1007/978-94-011-3610-5
    1. Gelidium--Congresses.   2. Gelidium--Biotechnology--Congresses.
I. Juanes. J. A. (José A.)   II. Santelices G., Bernabé.
III. McLachlan, J. L. (Jack Lamont), 1930-   . IV. Cantabria
(Spain). Diputación Regional. V. International Seaweed
Association. VI. Hydrobiologia. VII. Title. VIII. Series.
QK569.G4I57 1991
589.4'1--dc20   91-25037

ISBN 978-0-7923-1372-4

Printed on acid-free paper

International Workshop on *Gelidium*

held in Santander, Spain, September 3–8, 1990

ORGANIZATIONAL COMMITTEE

Co-chairman:	F. Torrontegui
	J. L. McLachlan
Secretary:	S. González
Members:	J. A. Juanes
	C. Fernández

SCIENTIFIC COMMITTEE

Co-chairman:	C. Fernández
	B. Santelices
Secretary:	J. A. Juanes
Members:	J. L. McLachlan
	R. Armisen

COMMITTEE OF HONOUR

Excmo. Sr. D. Juan Hormaechea Cazón
Excmo. Sr. D. Carlos Romero Herrera
Excmo. Sr. D. Aldolfo Pajares Compostizo
Excmo. Sr. D. Antonio Pallarés Sánchez
Ilmo. Sr. D. Manuel Huerta Castillo
Ilmo. Sr. D. Vicente de la Hera Llorente
Excmo. Sr. D. José Loira Rúa
Excmo. Sr. D. José María Ureña Francés
Excmo. Sr. D. Ernest Lluch Martín
Ilmo. Sr. D. Rafael Jaen Vergara
Ilmo. Sr. D. Rafael Robles Pariente

SPECIAL SPONSORS

Comisión Interministerial de Ciencia y Tecnología (CICYT)
Universidad de Cantabria
Instituto Español de Oceanografía

OTHER SPONSORS

Caja Cantabria
Banco Santander
Banco Hispano Americano
Caja Rural de Burgos
Banco Español de Crédito
Tina Menor, S/A
SAYOMAR
Marinalg International
Gomas Marinas S/A
Hispanagar S/A
Industrias Roko S/A
Algas Marinas S/A
Setexam

Contents

Hydrobiologia **221**: ix–xii, 1991.
J. A. Juanes, B. Santelices & J. L. McLachlan (eds), International Workshop on Gelidium.

Opening addresses

Honourable Guests, Ladies and Gentlemen:

Welcome to Cantabria; welcome to Santander:

My name is Fernando Torrontegui Mirones, and I am Head of the Servicio de Actividades Perqueras of the Diputación Regional de Cantabria and Co-chairman of the Organizing Committee. On their behalf, I would like to welcome you and express our hope that your stay with us will be a pleasant and profitable one for all. As we proceed with the official opening of this International Workshop on *Gelidium*, it is usual to acknowledge those people and organizations who have collaborated in the holding of this workshop, especially the International Seaweed Association and its former Chairman, Dr J. L. McLachlan, whose collaboration, together with that of the Diputación Regional de Cantabria has been decisive in the organization of this workshop. Once again, welcome to this International Workshop on *Gelidium* and sincere thanks for being with us here today in Santander.

FERNANDO TORRONTEGUI
Co-chairman
Organizing Committee

In recent years it has been shown that *Gelidium* has become a resource that fulfills an important role in our region. At this very time, the collecting and drying of this seaweed can be observed along our coastal area and in the adjacent fields.

When we were approached by the Diputación Regional de Cantabria and the International Seaweed Association, who proposed our integration into the preparatory activities for this workshop, our participation was immediately confirmed, given the importance of this endeavour.

It is well known that the Diputación Regional de Cantabria is giving increased consideration to *Gelidium* resources. My University is thus happy to make its small contribution to this workshop, and also to openly express its intention to study, and to include, topics of regional interest in its activities.

The University of Cantabria hopes to create, in the future, an Institute of Marine Sciences in Santander, the constitution of which will be founded on several bases. The work that has been carried out by the laboratory of the Instituto Español de Oceanografía of Santander has already provided substantial resources and information on this subject. The incorporation of the Escuela Superior de la Marina Civil into the University, the work in physical oceanography and, lastly, studies on the ecology of coastal areas and estuaries that the University has been carrying out have sown the seeds that will, in the future, become the Institute of Marine Sciences.

I hope that this activity represents a step forward in the knowledge of this important resource and will, on the other hand, motivate our University to continue to increase its research capacity in this field of Marine Science.

I would like to wish all of you a pleasant and scientifically productive stay. We hope that with this meeting we can absorb your expertise for the good of our region and so be able to administer this important resource that is present on our coast.

On behalf of both the University and the Diputación Regional de Cantabria, we thank you for your contributions which will, without doubt, benefit our region, others in the north of Spain and the people who are trying to make advances in the information available on *Gelidium*.

JOSÉ MARÍA UREÑA FRANCÉS
Rector Magnífico
Universidad de Cantabria

Honourable Guests, Ladies and Gentlemen:

Welcome to this first International Workshop on *Gelidium* which is being held in our region, Cantabria. In the name of the Diputación Regional de Cantabria, of our Department and personally, I express our satisfaction with the selection of this region for the holding of this Workshop.

It is easy to understand Cantabria's interest in everything related to seaweeds and, specifically, to *Gelidium*, bearing in mind that around our shores there are large beds of *Gelidium* of high commercial value. These beds exist not only in our autonomous community, but also around the borders of the Basque Country and Asturias, along more than 600 km of coastline. The annual collection of *Gelidium* in Cantabria reaches approximately 2,300 tonnes dry weight and employs no less than 600 people in harvesting the weed washed up on the shore; this does not take into account those people involved in related activities. Considering these data and the growing interest worldwide in this valuable marine resource and its derivatives, the decision made by us, and the organizations here represented, is both positive and normal. It has made the holding of this event a reality, allowing us to spend a few days studying *Gelidium* in our region, in Spain and in the world, and to review its present situation. For this reason it is necessary to expand studies that we, ourselves, started some years ago, and which have allowed us to take measures for planning in this sector in 1986.

It is obviously necessary to continue these studies in greater depth and to acquire additional information for the development and commercial exploitation of *Gelidium*, while, at the same time, considering the relative fragility of these natural populations and the social problems that may arise in some cases. We hope to authorize new classes of farming, when requisite scientific information becomes available, and also to authorize and promote marine cultivation activities.

We are continuing to expand with all the means at our disposal to meet these objectives. Specifically, we have the Santander Coastal Centre at the Instituto Español de Oceanografía that carried out the first studies and allowed us to organize this sector in 1986. Research and development will undoubtedly be intensified in the future with the soon-to-be-launched Seaweed Cultivation Unit in Cantabria as a pilot centre in association with the above-mentioned Institute. We hope that this will start working in the near future.

It is especially important to emphasize at this point that the studies will first be undertaken on a technical level. This will be followed by extensive meetings and administrative agreements to be reached by the Fishing Authorities of the communities of Galicia, Asturias, the Basque Country and Cantabria, with the participation and agreement of the corresponding central government department, Departamento Secretaría General de Pesca Marítima.

In this context, we cannot ignore the studies that have been undertaken through the existing agreement between the Department that I represent, Diputación Regional de Cantabria, and the University of Cantabria. These studies are pertinent and completely up to date. This is an attempt to expand the existing knowledge, to try and develop additional activities, and to find new outlets for this product.

For all of these reasons, the Diputación Regional de Cantabria, after receiving a proposal from the International Seaweed Association, decided to organize this Workshop. The aims are to discuss and compile current information available on this group of gelidiod seaweeds, and to promote new research that will lead to maximum utilization and better applications of these resources and the products extracted from them.

I, therefore, believe that this workshop which we are opening today is the ideal framework for the exchange of information between participants and provides a unique opportunity to discuss, scientifically, the exploitation of the seaweed resources of our region. I feel that, because of this, we can more than justify the holding of this Workshop, and this has been emphasized by the previous speakers at this opening ceremony.

I should like to finish by thanking all of the attending participants and those organizations which have made their presence at this Workshop possible. Naturally, I, myself, address those representing organizations and entries from my region and county. Above all, however, I wish to thank those who are visiting us from abroad, those who have made the effort to travel to Cantabria to develop this workshop from outside our region or country. We have representatives from no less than 13 countries: Canada, Chile, China, Denmark, France, England, Japan, Morocco, Norway, New Zealand, Portugal, South Africa and U.S.A. I wish all of you a pleasant and scientifically fruitful stay among us, and enjoyment of our beautiful countryside, all of which, I hope, will persuade you to visit us again.

Many thanks to all of you.

On behalf of the Department of the Diputación de Cantabria, I now declare the First International Workshop on *Gelidium* open!

VICENTE DE LA HERA LLORENTE
Consejero
Ganadería, Agricultura y Pesca
de la Diputación Regional de
Cantabria

Hydrobiologia **221**: xiii–xiv, 1991.
J. A. Juanes, B. Santelices & J. L. McLachlan (eds), International Workshop on Gelidium.

Foreword

Seaweed extractives that are important commercial phycocolloids have numerous and varied applications. The industry is largely a post-World War II development that has grown extremely rapidly over the past several decades, and it is presently a sophisticated, multi-million dollar enterprise. Resources yielding phycocolloids are obtained from many areas of the world, and a diversity of seaweed species are used in the international seaweed industry. As phycocolloids are natural products, the continuing availability of adequate resources is a primary factor, on which the continuance and future success of the seaweed-extractive industry is dependent. Resource availability has especially hindered the development of the agar sector of the industry, which has been unable to keep pace with demand for this material. Agar, as a consequence, has been, and remains, the most valuable of the phycocolloids. This is especially true of extracts of *Gelidium*, and related species, which yield a high-grade form of agar. Agars have extensive industrial applications, and these could most likely be significantly expanded, if additional quantities were available. The agar industry has, however, been characterized by both erratic and limited supplies of natural resources. There are certainly means of increasing and augmenting the current resource base, although this clearly requires a better and fuller understanding of these agarophytes and their extractives. It is, moreover, especially important to integrate the scientific data-base with social and economic concerns in the development of this broad-based industry, and such integrations must be given appropriate consideration in any discussion of the seaweed industry.

It is not surprising that a concept of an international workshop on *Gelidium* was developed. The basic idea for this workshop was formulated several years ago, and a proposal was put forth by Dr Consolación Fernández and her Spanish colleagues to Sr. Fernando Torrontegui and his associates in the Regional Government of Cantabria. This proposal was received favourably, and it was agreed that the Regional Government of Cantabria, jointly with the International Seaweed Association, would sponsor a workshop on *Gelidium*, to be held in Santander. Publication of this volume is the final outcome of the devoted efforts of a number of people over the past few years.

In the early 1970's, a small symposium was convened to discuss *Chondrus crispus*, an important seaweed yielding a commercially-valuable extract. This effort was modestly successful, and most of the information available at that time on Irish moss was reviewed and brought together in a publication. The 'little red book' continues to serve a useful purpose, even though there have been extensive advancements in our understanding of *Chondrus crispus* since 1970. Similarly, there was an update and review published on the important alginophyte, *Macrocystis*. This effort, too, has provided an ongoing and invaluable service to those interested in these seaweeds and their utilization.

An attempt has been made to achieve similar results for the present workshop, and, hopefully, these will also have a long-lasting value to the *Gelidium* industry and to others who are simply interested in these seaweeds. Those invited to participate in this workshop were drawn widely from the international community. This reflects, in part, the worldwide distribution of commercial species of *Gelidium*. The participants came not only from Europe, but also from the Americas, Asia, Africa and Australia – this truly represented an international workshop. Species of *Gelidium* are economically important seaweeds in many of the participants' home countries; nevertheless, the major global importance of *Gelidium* is in Iberia, and particularly in Cantabria, where a thriving industry is centered.

There was a broad range of topics discussed during the week of the workshop, including papers on the biological and economical importance of *Gelidium*, world resources, ecology, exploitation, physiology, cultivation and genetics and cell culture, together with an overview of the *Gelidium* industry in Spain.

Discussions were not limited to the invited participants; indeed, those in attendance were encouraged to enter into the wide-ranging and vigorous exchange of ideas.

The International Seaweed Association has, by its very nature, a vested interest in *Gelidium*, and it was an honour for this organization to co-sponsor this workshop. The International Seaweed Association is keenly interested in these small, focussed meetings, and the Association would like to see more of them undertaken in future on a variety of topics. It was patently obvious at the First International Seaweed Symposium that the extent and availability of seaweed resources were the primary issues confronting the industry. During the immediate past seaweed symposium (ISS XIII), nearly forty years later, it is both interesting to note and significant that similar concerns and views were being expressed. Resource availability, thus, continues to claim attention in the development of the seaweed industry; this is certainly directly applicable to *Gelidium* and other agarophytes. The present workshop had the opportunity to address some of the problems specifically related to *Gelidium*, and it is anticipated that subsequent workshops will be convened to take advantage of the progress that has been made here; the International Seaweed Association will perhaps become equally involved with these.

The co-sponsorship of the International Seaweed Association was largely in kind. The real substance of support came from the Regional Government of Cantabria, supplemented by that from a number of other organizations. Many individuals, in addition, contributed significantly to the success of this symposium. The efforts of Sr. D. Fernando Torrontegui, Sr. D. Santiago González and Sr. D. José A. Juanes must, in particular, be acknowledged, while significant organizational support was received from the Universidad de Cantabria and the Departamento de Ciencias y Tecnicas del Agua y del Media Ambiente; this is greatly appreciated. In the preparation of this volume for publication, we have been most fortunate to have had the valuable editorial assistance of Ms Mary J. Sakurai and Mr F. James Hendra, and technical help was provided by the Department of Biology, St. Francis Xavier University.

J. L. McLachlan
Co-chairman
Organizing Committee

Hydrobiologia **221**: xv–xviii, 1991.
J. A. Juanes, B. Santelices & J. L. McLachlan (eds), International Workshop on Gelidium.

List of participants

Afonso-Carrillo, Juilo
Departamento Biología Vegetal
Facultad Farmacia
Universidad de La Laguna
La Laguna
38271 Tenerife, Islas Canarias
Spain

Augirre, Mercedes
Department of Pure and Applied
Biology
University of Leeds
Leeds LS2 9JT
England

Akatsuka, Isamu
Department of Life Science
Osaka Prefectural University
4-103, Hino 6498, Hino-si
Tokyo
Japan 191

Albertos, José
Gummagar S/A
Abanto y Ciérvana
48508 Vizcaya
Spain

Aler, Jorge
Industrias Roko S/A
Avenida de la Sardiñeira, 37
15007 La Coruña
Spain

Alvarez, Faustina
Agar de Asturias S/A
Bricia s/n, Posada de Lanes
33594 Lanes, Principado de Asturias
Spain

Anderson, Robert
Seaweed Unit
Sea Fisheries Research Institute
Private Bag X2
Roggebaai 8012
Republic of South Africa

Armisén, Rafael
Hispanagar S/A
Apartado 392
09080 Burgos
Spain

Berasategui, Estefanía
Departamento de Biología Vegetal y
Ecología
Facultad de Ciencias
Universidad del Pais Vasco
Apartado 644
48080 Bilbao
Spain

Betancort, José
Excelentísimo Cabildo Insular de
Gran Canaria
Jardín Botánico 'Viera y Clavijo'
Apartado 14 de Tafira Alta
35017 Las Palmas, Islas Canarias
Spain

Borja, Angel
Servicio de Investigación de Oceano-
gráfica
AZTI-SIO
Gobierno Vasco
Avenida Satrústegui 8
20008 San Sebastián
Spain

Carvallo, Paloma
Area Coordinacíon e Informes
Secretaría General de Pesca
Marítima
Ortega y Gasset, 57
28006 Madrid
Spain

Catoira, José L.
Xunta de Galicia
Consellería de Pesca
Casa del Mar, 5 Planta
La Coruña
Spain

Cendrero, Orestes
Laboratorio de Santander
Instituto Español de Oceanografia
Apartado 240
39080 Santander
Spain

Cotilí, José
Hispanagar S/A
Apartado 392
09080 Burgos
Spain

Courtney, William
Coast Biological Ltd.
P.O. Box 58103
East Tamami, Auckland
New Zealand

del Alamo, Concepción
Avenida Oviedo 31
39300 Torrelavega, Cantabria
Spain

Depolo, Miguel
Algas Marinas S/A
P.O. Box 9529
Fidel Oteiza 1956, p 14°
Santiago
Chile

Díaz, Begoña
Paseo Ocharan – Mazas 26
Castro Urdiales, Cantabria
Spain

Errico, Angelo
Gummagar S/A
Abanto y Ciérvana
48508 Vizcaya
Spain

Fei, Xin-geng
Institute of Oceanology
Academia Sinica
7 Nan-hai Road
Qingdao
People's Republic of China

Fernández, Consolación
Departamento Biología de
Organismos y Sistemas Ecología
Facultad Biología
Universidad de Oviedo
33005 Oviedo, Principado de Asturias
Spain

Fernández, Carlos
Laboratorio de Santander
Instituto Español de Oceanografia
Apartado 240
39080 Santander
Spain

Fredriksen, Stein
Department of Biology
Marine Botany
University of Oslo
P.O. Box 1069, Blindern
N-0316 Oslo-NTH
Norway

García, José L.
Industrias Roko S/A
Avenida de la Sardiñeira, 37
15007 La Coruña
Spain

García, Manuel
Departamento Biología Vegetal
Facultad Biología
Universidad Santiago de Com-
  postela
15760 La Coruña, Santiago
Spain

García-Barredo, Maite
Calleja Norte 3, p 2, 4°
39005 Santander
Spain

García-Reina, Guillermo
Departamento de Biología
Facultad Ciencias del Mar
Universidad de las Palmas de Gran
  Canaria
Apartado 550
35017 Las Palmas, Islas Canarias
Spain

Gil, Candelaria
Departamento de Biología Vegetal
Facultad Farmacia
Universidad de La Laguna
La Laguna
38271 Tenerife, Islas Canarias
Spain

Givernaud, Thierry
Laboratoie Algologie Fondamentale
  et Appliquée
39 rue Desmoueux
14000 Caen
France

Gómez, Juan L.
Departamento de Biología
Facultad Ciencias de Mar
Universidad de las Palmas de Gran
  Canaria

Apartado 550
35017 Las Palmas, Islas Canarias
Spain

González, Nieves
Excelentísimo Cabildo Insular de
  Gran Canaria
Jardín Botánico 'Viera y Clavijo'
Apartado 14 de Tafira Alta
35017 Las Palmas, Islas Canarias
Spain

González, Santiago
Servicio de Actividades Pesqueras
Pasaje de la Puntida 1–1°
39001 Santander
Spain

Gordon, Margaret
School of Biological Sciences
Victoria University of Wellington
P.O. Box 600
Wellington
New Zealand

Gorostiaga, José M.
Departamento de Biología Vegetal y
  Ecología
Facultad Ciencias
Universidad del Pais Vasco
Apartado 644
48080 Bilbao
Spain

Gutierrez, Luis
Departamento Biología de
  Organismos y Sistemas Ecología
Facultad Biología
Universidad de Oviedo
33005 Oviedo, Principado de Astu-
  rias
Spain

Hansen Danebo, Diana
FMC Corporation, Litex A/S
Risingevej 1
DK-2665 Vallensback Strand
Denmark

Haroun, Ricardo
Departamento Biología Vegetal
Facultad Farmacia
Universidad de La Laguna
La Laguna
38271 Tenerife, Islas Canarias
Spain

Henocq, Vincent
Pronatec S/A
CITTN – rue Chanzy
59260 Hellemmes
France

Holmsgaard, Jens-Eric
Holmsgaard Consult Ltd.
Fredensvej 37
DK-2920 Charlottenlund
Denmark

Ivanac, Ivan
Algas Marinas S/A
P.O. Box 9529
Fidel Oteiza 1956, p 14°
Santiago
Chile

Juanes, José
Departamento Ciencias y Tecnicas
  del Agua Ecología
Universidad de Cantabria
Avenida de los Castros s/n
39005 Santander
Spain

Kettani, Rachid
Marokagar S/A
Uurue Abou Baker Wahrani
B.P. 2121
Casablanca 05
Morocco

Lahaye, Marc
Institut National de la Recherche
  Agronomique
Laboratoire de Biochimie et Tech-
  nologie de Glucides
B.P. 527
44026 Nantes Cedex 03
France

Lanuza, Paloma
Museo Marítimo
Diputación Regional de Cantabria
Santander
Spain

Lebbar, Nabil
Setexamm
B.P. 210
Kenitra
Morocco

Leivar, Javier
Hispanagar S/A
Apartado 392
09080 Burgos
Spain

Lemos, Ana M.
Sicomol S/A
Boavista de Lavos
3080 Figueira da Foz
Portugal

Macler, Bruce
Water Management Division
Environmental Protection Agency,
Region IX
1235 Mission St.
San Francisco, CA 94103
USA

Matsuhiro, Betty
Departamento de Química
Facultad de Ciencias
Universidad de Santiago de Chile
Matucana 28-D, Casilla 5659
Santiago
Chile

McLachlan, Jack L.
National Research Council
1411 Oxford St.
Halifax NS
Canada B3H 3Z1

Melo, Ricardo
Departamento Biología Vegetal
Facultad de Ciências
Universidade de Lisboa
R. Ernesto de Vasconcelos
Bloco C-2, 4°
1700 Lisboa
Portugal

Modliszewski, James
FMC Corporation, Marine Colloids
Division
2000 Market St.
Philadelphia, PA 19103
USA

Moreno, Manuel
Hispanagar S/A
Apartado 392
09080 Burgos
Spain

Myslabodski, David E.
Institute for Biotechnology
University of Trondheim
N-7034 Trondheim-NTH
Norway

Neushul, Michael
Department of Biological Sciences
University of California at Santa
Barbara
Santa Barbara, CA 93106
USA

Niell, F. Xavier
Departamento de Ecología
Facultad de Ciencias
Universidad de Málaga
Campus Teatinos
29071 Málaga
Spain

Oliveira, José C.
Departamento Recursos Halieuticos
Instituto Nacional de Investigação
das Pescas e Algés
1400 Lisboa
Portugal

Patwary, Moshin
National Research Council
1411 Oxford St.
Halifax NS
Canada B3H 3Z1

Peterson, Harlan
Perny Inc.
P.O. Box 721
Ridgewood, NJ 07451
USA

Puente, Araceli
Fernández Vallejo 31 C
39316 Torrelavega, Cantabria
Spain

Requena, Salvador
Sucromatic S.L.
Avenida Dr. Waksman, 21-Pta. 28
46006 Valencia
Spain

Revenga, Silvia
Area Coordinación e Informes
Secretaría General de Pesca
Marítima

Ortega y Gasset, 57
28006 Madrid
Spain

Rico, José M.
Departamento Biología de Organis-
mos y Sistemas Ecología
Facultad Biología
Universidad de Oviedo
33005 Oviedo, Principado de
Asturias
Spain

Robertson, Bruce
Department of Botany
Institute for Coastal Research
University of Port Elizabeth
P.O. Box 1600
Port Elizabeth 6000
Republic of South Africa

Robledo, Danie
Departemento de Biología
Facultad Ciencias del Mar
Universidad de las Palmas de
Gran Canaria
Apartado 550
35017 Las Palmas, Islas Canarias
Spain

Rochas, Cyrille
Centre de Recherches sur les Macro-
molécules Vegétales
Domaine Universitaire, B.P. 53 X
38041 Grenoble Cedex
France

Rodrígues-Cabello, Cristina
Servicio de Actividades Pesqueras
Consejería de Ganadería, Agricultu-
ra y Perca
Pasaje de la Puntida 1
1° Santander
Spain

Salinas, Juan M.
Departamento de Cultivo de Algas
Instituto Español de Oceanografia
Apartado 240
39080 Santander
Spain

San Martín, Maite
Pérez Galdós 14
39006 Santander
Spain

San Martín, Jorge
Algas de Cantabria
S. Cooperativa Ltda.
V. Velasco 3A, 3°B
39011 Santander
Spain

Santelices, Bernabé
Departamento de Ecología
Facultad Ciencias Biológicas
Pontifica Universidad Católica de
Chile
Casilla 114-D
Santiago
Chile

Santos, Rui
Departamento Estudos de Impacte
Industrial
Laboratório Nacional Engenharia e
Tecnologia Industrial
Estrada do Paco do Lumiar
1600 Lisboa
Portugal

Seip, William F.
Becton Dickinson Microbiology Sys-
tems
Department of Research and Devel-
opment
250 Schilling Circle
Cockeyvill, MD
USA

Simões, José A.
Sicomol S/A

Boavista de Lavos
3080 Figueira da Foz
Portugal

Soler, Emilio
Departamento de Biología
Facultad Ciencias del Mar
Universidad de las Palmas de Gran
Canaria
Apartado 550
35017 Las Palmas, Islas Canarias
Spain

Sosa, Pedro
Departamento de Biología
Facultad Ciencias del Mar
Universidad de las Palmas de Gran
Canaria
Apartado 550
35017 Las Palmas, Islas Canarias
Spain

Sousa-Pinto, Isabel
Instuto Botânico
Universidade do Porto
R. do Campo Alegre, 1190
4100 Porto
Portugal

Torres, María
Departamento de Ecología
Facultad de Ciencias
Universidad de Málaga
Campus Teatinos
29071 Málaga
Spain

Torrontegui, Fernando
Servicio de Actividades Pesqueras
Pasaje de la Puntida 1–1°
39001 Santander
Spain

van der Meer, John
National Research Council
1411 Oxford Street
Halifax NS
Canada B3H 3Z1

Vega, José L.
Gomas Marinas S/A
Juan Canalejo, 83
15003 La Coruña
Spain

Villardefrancos, Emilio
Industrias Roko S/A
Avenida de la Sardiñeira, 37
15007 La Coruña
Spain

Vizcaíno, Alberto
Centro de Experimentación Pes-
quera
Consejería de Ganadería
Agricultura y Pesca
Oviedo, Principado de Asturias
Spain

Hydrobiologia **221**: 1–17, 1991.
J. A. Juanes, B. Santelices & J. L. McLachlan (eds), International Workshop on Gelidium.
© *1991 Kluwer Academic Publishers.*

1

Intrageneric differences in cystocarp structure in *Gelidium* and *Pterocladia*

B. Santelices
Departamento de Ecología, Facultad de Ciencias Biológicas, Pontificia Universidad Católica de Chile, Casilla 114-D, Santiago, Chile

Key words: cystocarps, *Gelidium*, morphological variability, *Pterocladia*, taxonomy

Abstract

Unilocular cystocarps, with ostioles opening to one frond surface, have traditionally distinguished *Pterocladia* from *Gelidium*, described as having bilocular cystocarps, with ostioles opening to both surfaces; however, unequally developed locules have been described in *Pterocladia* and differences in cystocarpic architecture between *Pterocladia capillacea* and the type species of the genus, *P. lucida*, have been recently found. As heterogeneity in cystocarp architecture raises questions of basic intergeneric distinction, a survey of reproductive morphology of species in both genera is presented in this study. Six morphologically-different types of cystocarps are distinguished among the five species of *Pterocladia* and the seven species of *Gelidium* examined.

Introduction

The taxonomic segregation of *Gelidium* and *Pterocladia* is the single most important, yet unsolved problem in the taxonomy of the Gelidiales. These are two of the most diverse genera in the Order, with several ecologically- and economically-important species (Santelices, 1988). Taxonomic clarification should permit a better understanding of structural and qualitative differences among agars produced by members of this Order.

Up to six morphological characters have been proposed to separate *Gelidium* from *Pterocladia*. Critical evaluations (Santelices, 1990) indicate, however, that these characters have limited taxonomic value. Cystocarp structure, the basic generic distinction, is described to be unilocular in *Pterocladia*, with ostioles opening to only one surface, while cystocarps in *Gelidium* are bilocular, opening to both frond surfaces (Bornet & Thuret,

1876). Fan (1961) found, however, two unequally developed locules opening to only one surface in two species of *Pterocladia*; therefore, this segregation character is a relative rather than an absolute difference.

As fertile female thalli are infrequent in field collections of *Gelidium* and *Pterocladia*, vegetative segregation characters have been suggested. Generic segregation based on hyphal distribution in the thallus and on the shape of medullary cells was proposed by Okamura (1934). Neither characters can be used because the internal structure of some species of *Gelidium* and *Pterocladia* changes during the life of the frond (Dixon, 1958). Basal bending at the point of branching of indeterminate laterals, supposed to be exclusive for *Gelidium* (Stewart, 1976), also occurs in species of *Pterocladia*. Disposition of surface cortical cells, especially at the most basal parts of axes, allow generic segregation (Akatsuka, 1981) although a few exceptions also exist (Rodríguez & Santelices,

1988); likewise, apical architecture allows segregation among some of the species of both genera (Rodríguez & Santelices, 1987), while many small-sized species of *Pterocladia* have *Gelidium*-type apices (Rodríguez & Santelices, 1988). Although cystocarpic structure, apical architecture and arrangement of external cortical cells in the basal portions of axes are, therefore, of some taxonomic value, none of these characters allows complete generic segregation, as there are species showing exceptions to each of these features.

Recent studies (Santelices, 1991) have revealed differences in cystocarpic architecture among some species of *Pterocladia*. Heterogeneity in cystocarpic structure, either in the genera *Pterocladia* or *Gelidium*, raises questions about the basic intergeneric distinction proposed by Bornet and Thuret (1876); consequently, a survey of reproductive morphologies among species in these two genera has been undertaken. The present study reports the first results of this survey, distinguishing six different types of cystocarps among five species of *Pterocladia* and seven species of *Gelidium*. As in a previous study (Santelices, 1991), the primary focus is on late-cystocarpic development, with immediate post-fertilization events included where pertinent.

Materials and methods

The representative specimens of the five species of *Pterocladia* and seven species of *Gelidium* included in this study are listed in Table 1. Fertile female gametophytes were examined under a stereomicroscope for cystocarps. A sample of 10 to 20 cystocarps of different sizes, and presumably of different ages, was obtained from each specimen. Cystocarps were gradually rehydrated, avoiding tissue damage, and fixed in a 10% formaldehyde solution in seawater. Fixed cystocarps were embedded in gelatin and cut, 30–40 μm thick to avoid destruction of the sporogenous tissues, using a Leitz freezing microtome. Sections were stained with Methylene Blue and photomicrographs were taken with a Nikon Biophot microscope.

Results

Six morphologically-different types of cystocarps can be recognized among the species presently referred to the genera *Gelidium* and *Pterocladia*. A description characterizing the most distinctive features of each type follows:

The Pterocladia lucida-type

The margins of the fertile fronds are ruffled (Fig. 1), and in longitudinal section they bear carpogonial branches on only one surface. Functional carpogonia are large, with elongated trichogynes and prominent supporting cells (Fig. 2). In my specimens, the supporting cell was often the second basal cell of a cell row of the third order. The most basal cell of that cell row produced the nutritive filaments.

Some separation of cortical tissues between the cell rows of third and fourth orders on the fertile side was evident at the time of carpogonial formation (Fig. 1). After fertilization, and simultaneous with growth and expansion of the gonimoblast, the cystocarpic wall of the fertile surface elongates. This generates a cystocarpic cavity which has a placenta of fertile tissue at its base (Fig. 3). As the growth of the gonimoblast progresses from the center towards the margins of the fronds, the cystocarpic cavity enlarges. Elongation is always preceded by cell separation (Fig. 4). Elongation of internal cortical cells and the cystocarpic locule occur simultaneously.

After separation of the two frond surfaces, third-order cell rows remaining on the floor of the cystocarp differentiate nutritive filaments (Fig. 5). These filaments later fuse with gonimoblast filaments, generating a network of irregularly-shaped fertile cells above the elongated cells of the second-order cell row. Some of these fertile cells had pit connections with cells of the second-order cell row (Fig. 6).

Carposporangia are cut off successively, by transverse divisions, forming chain-like rows in a mucilaginous envelope. Young cystocarps exhibit short chains with elongated, ovoid carpospores

Table 1. Cystocarpic specimens used in this study.

Species	Locality	Determined by	Collections Numbers
P. bulbosa Loomis	Salt Pond Pavillion, Port Allen, Kauai, Hawaii	B. Santelices	SS/UC 6669
P. caerulescens (Kützing) Santelices	Kuhio Beach, Oahu, Hawaii	B. Santelices	SS/UC 6670
P. capillacea (Gmelin) Bornet & Turet	Praia Grande Arraval del Cabo (Cabo Frio), Brazil	Yocie Yoneshigue-Valentin	Y.Y. 2185, 2187, 2188
	Ensenada de Lomo, Cabo Frio, Brazil	Yocie Yoneshigue-Valentin	Y.Y. 2519, 2521
	Saco de Inglés, Brazil	Yocie Yoneshigue-Valentin	Y.Y. 2652, 2653, 2656, 2657, 2658
	E.S. Mateus, Cabo Frio, Brazil	Yocie Yoneshigue-Valentin	Y.Y. 2473
P. lucida (R. Brown) J. Agardh	Wharariki Beach, New Zealand	N.M. Adams	D.M. A3717
	Otaki Beach, New Zealand	N.M. Adams	D.M. A3707
	Alderman Islands, New Zealand	W. Nelson	D.M. A13767
	Mataikona, E. Wairarape, New Zealand	N.M. Adams	D.M. A4313a, A736
	Wha Wha (N of Mercury Bay) New Zealand	U.V.D.	D.M. A2136
P. macnabbiana Dawson	Bahia Culebra, Costa Rica	E.Y. Dawson	US 1777 (ex BF)
G. amansii (Lamouroux) Lamouroux	Enoshima, Kanagawa Prefecture, Japan	T. Yoshida	SS/UC 6671 (ex herb. Okamura)
G. coulteri Harvey	Pta. Pinos, California	I.A. Abbott	SS/UC 6672
G. galapagensis Taylor	Golfo de Fonseca, El Salvador	E.Y. Dawson	US 3650 (ex BF)
G. musciformis (Taylor) Santelices	Playitas, near La Union Golfo de Fonseca, El Salvador	E.Y. Dawson	US 3488 (ex BF)
G. pteridifolium Norris, Hommersand & Fredericq	Dwesa Reserve, Transkei	B. Santelices	SS/UC 6673
G. pusillum var. *pulvinatum* (J. Agardh) Feldmann	Maalea Bay, Maui, Hawaii	B. Santelices	SS/UC 6674

US (ex BF) = Algal collection, U.S. National Herbarium, Smithsonian Institution (ex Beaudette Foundation).
Y.Y. = Private collection of Yocie Yoneshigue-Valentin
D.M. = Herb. Dominium Museum, Wellington, New Zealand
SS/UC = Sala de Sistemática, P. Universidad Católica de Chile

4

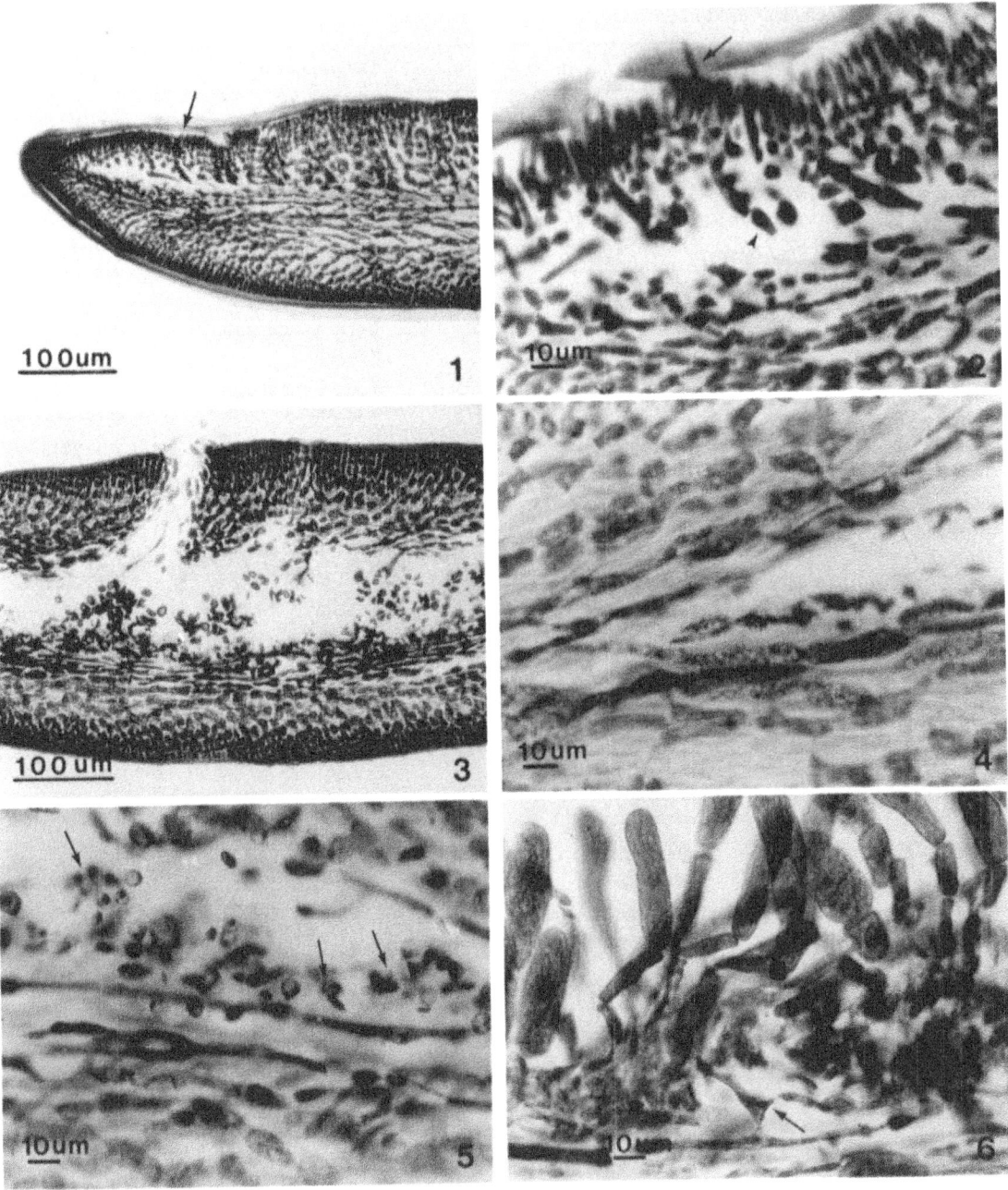

Figures 1–6. Pterocladia lucida

Fig. 1. Longitudinal section of a fertile tip showing ruffled margins and carpogonia (arrow).

Fig. 2. Carpogonium with large trichogyne (arrow) and prominent supporting cell (arrow head).

Fig. 3. Longitudinal section of a young cystocarp with a single cavity and a placenta of fertile tissue at the base of the cavity.

Fig. 4. Cell separation in the border of the expanding cystocarpic cavity. The expansion of the gonimoblast has not reached this part of the cystocarp.

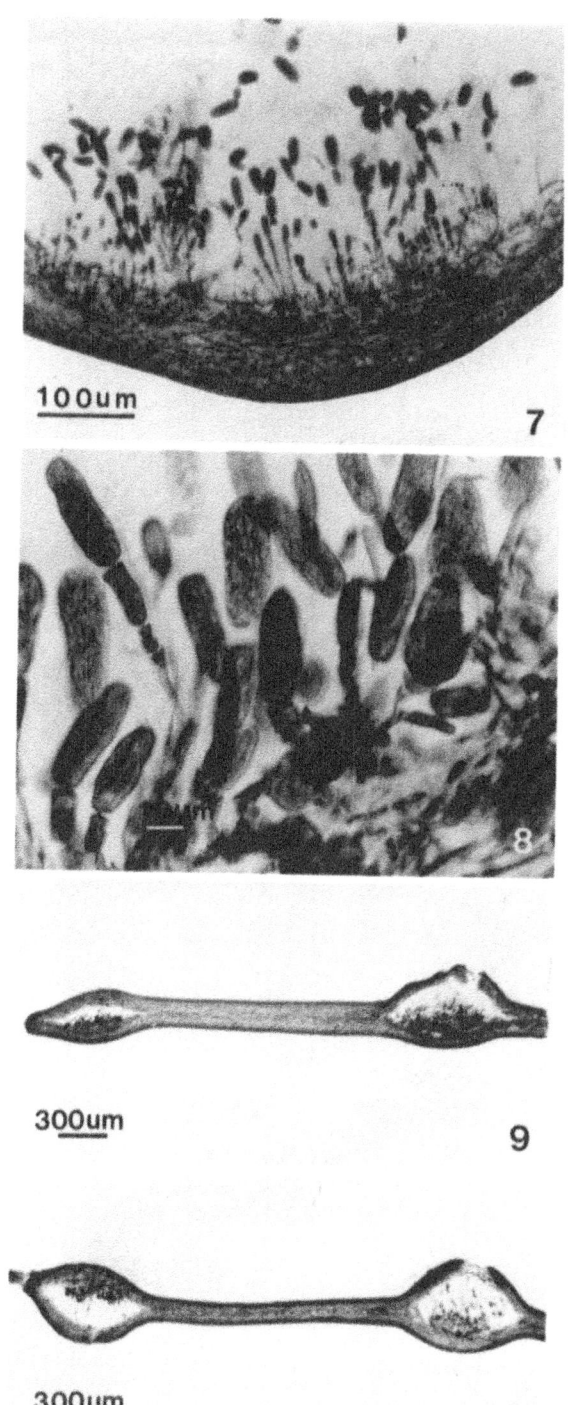

100um

7

8

300um

9

300um

10

(Fig. 7). Older cystocarps have longer, strongly-clavate chain-like sporangia (Fig. 8). The basal spores in these rows are short and cylindrical, while the terminal are ovoid or clavate. Elongated cortical cells extending from the bottom to the cortex of the cystocarp are more common in older, than in young cystocarps.

Although carpogonia formation in *Pterocladia lucida* occurs only on one of the frond surfaces, the specific side does not seem predeterminated as successive cystocarps can open to the same or to different surfaces (Figs. 9, 10).

The Pterocladia capillacea-type

Pterocladia capillacea and *P. caerulescens* exhibit essentially similar patters of cystocarpic development. The fertile pinnules have a notched apex, ruffled margins and a longitudinal furrow extending down the middle of the pinnule. Carpogonial branches are seen, in longitudinal section, developing towards both surfaces of the fertile ramuli (Fig. 11).

Functional carpogonia exhibit long trichogynes, while the supporting cell is less prominent than in *Pterocladia lucida* (Fig. 11). Nutritive filaments are not evident at the time of carpogonial formation. They differentiate, probably after carpogonial elongation, from the basal cell of the third-order cell row (Fig. 12) and are produced on both frond surfaces (Fig. 13).

Figures 7–10. Pterocladia lucida

Fig. 7. Carpospore production in chain-like rows.

Fig. 8. Details of transversely divided carposporangia in strongly clavate, chain-like rows.

Fig. 9. Successive cystocarps formed on the same frond surface.

Fig. 10. Successive cystocarps formed on different frond surfaces.

◄ *Fig. 5.* Nutritive cells (arrow) reaching the cystocarpic border after separation of cortical cells.

Fig. 6. Placenta of fertile cells at the base of the cystocarp. Some of these cells exhibit connections (arrow) with the elongated cells of the cell rows of the second order.

6

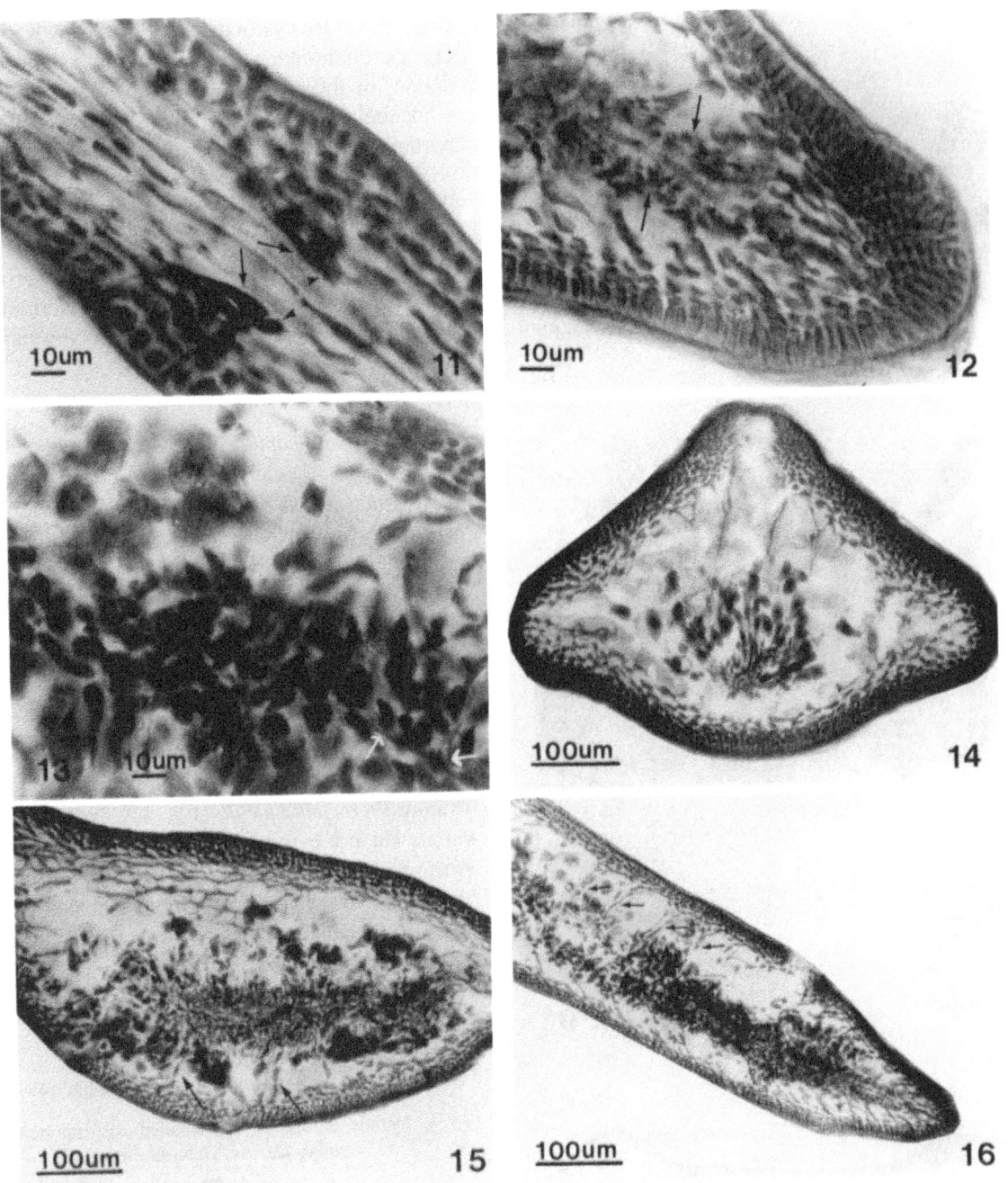

Figures 11–16. Pterocladia caerulescens
Fig. 11. Longitudinal section of fertile tip showing supporting cells (arrow heads) and carpogonial (arrows) directed towards both frond surfaces.

Fig. 12. Nutritive cells (arrows) produced towards both frond surfaces.

Fig. 13. Nutritive filaments surrounding the elongated cells of the cell row of second order (arrows).

Fig. 14. Transverse section through a young cystocarp showing cystocarpic cavity, with a placenta of fertile tissue at the base, a few inner cortical filaments and carpospores.

After fertilization and gonimoblast expansion, the cystocarpic wall of one of the two surfaces elongates, generating a cystocarpic cavity with a placenta of fertile tissue at the base of the cystocarp (Fig. 14). Median sections of these cystocarps (Fig. 15) reveal a few, elongated internal cortical filaments extending from the inner cortex at the bottom of the cystocarps to the placenta. As growth of the cystocarp progresses, the placenta detaches from the bottom of the cystocarp at about the cell row of third order on the opposite side. Carposporangia are then produced on both sides of the septum (Figs. 15, 17) with a gradual development of paired placenta. Elongated internal cortical filaments extend between the internal walls of the cystocarpic cavities and the cell row of second order (Figs. 17, 18). In some species, such as *P. caerulescens*, they form a three-dimensional network (Fig. 16), seemingly maintaining the structural integrity of the expanding cystocarp.

The fusion network generated by the fusion of gonimoblast filaments and nutritive filaments develops around the elongated cells of the cell row of second order. Earlier gonimoblast divisions producing carposporangia are either transverse or oblique (Fig. 19), and initials may form short stacks of rounded and flat cells resembling a pile of coins. Later divisions are predominantly oblique. Initials grow radially, conferring an arboriform appearance to the sporogenous tissue (Figs. 20, 21, 22). Chain-like rows are evident among immature spores, and are lost in more mature carpospores.

Even though the gonimoblast filaments of *Pterocladia caerulescens* mature and produce spores on both sides of the frond, one locule generally develops more strongly than the other and the cystocarpic wall is more extensively elevated on that side. One or more ostioles are formed in the elevated side; by contrast, mature cystocarps of *P. capillacea* may exhibit two well-developed, although asymmetric, locules, with the pericarp protruding equally on both frond surfaces and having one or more ostioles in each surface. Thus, the early cystocarpic development in these two *Pterocladia* species, although essentially similar, can be slightly different in later stages.

The Pterocladia bulbosa-type

The fertile female pinnules in this species (Figs. 23, 24) are slightly less acute than vegetative apices, have slightly ruffled margins and lack an apical notch. In longitudinal section, they exhibit carpogonia towards one (Fig. 23) or both frond surfaces (Fig. 24).

Functional carpogonia reach large sizes, although their supporting cells are less prominent than in other species of *Pterocladia*. Nutritive filaments, composed of subcuboidal, small cells (Fig. 25) originate from the basal cell of a third-order cell row towards both frond surfaces.

After fertilization, the cystocarpic wall extends only on one side. As in *Pterocladia lucida*, cell separation seems to precede gonimoblast expansion, and the cystocarp remains unilocular with a placenta of fertile tissue at the floor of the cystocarpic cavity even in the larger, and presumably older, cystocarps studied (Figs. 26, 27). Although the fertilization network develops mostly on one side of the frond, in close proximity to the elongated cells of the cell row of second order (Fig. 28), the larger cystocarps show proliferation of the fertilization network on both frond surfaces (Fig. 29) similar to *P. capillacea* and *P. caerulescens*; however, in *P. bulbosa* spore production occurs only on one side.

Although gonimoblast divisions in *P. bulbosa*

◄ *Fig. 15.* Longitudinal section of a more mature cystocarp. The placenta has detached from the bottom of the cystocarp and carposporangia are produced on both sides of the septum. A few elongated cortical filaments (arrow) extend from the cortex to the placenta.

Fig. 16. Mature cystocarp of *P. caerulescens* exhibiting a centrally-placed placenta, spores produced to both frond surfaces and a matrix of elongated, parallel, inner cortical filaments (arrows).

8

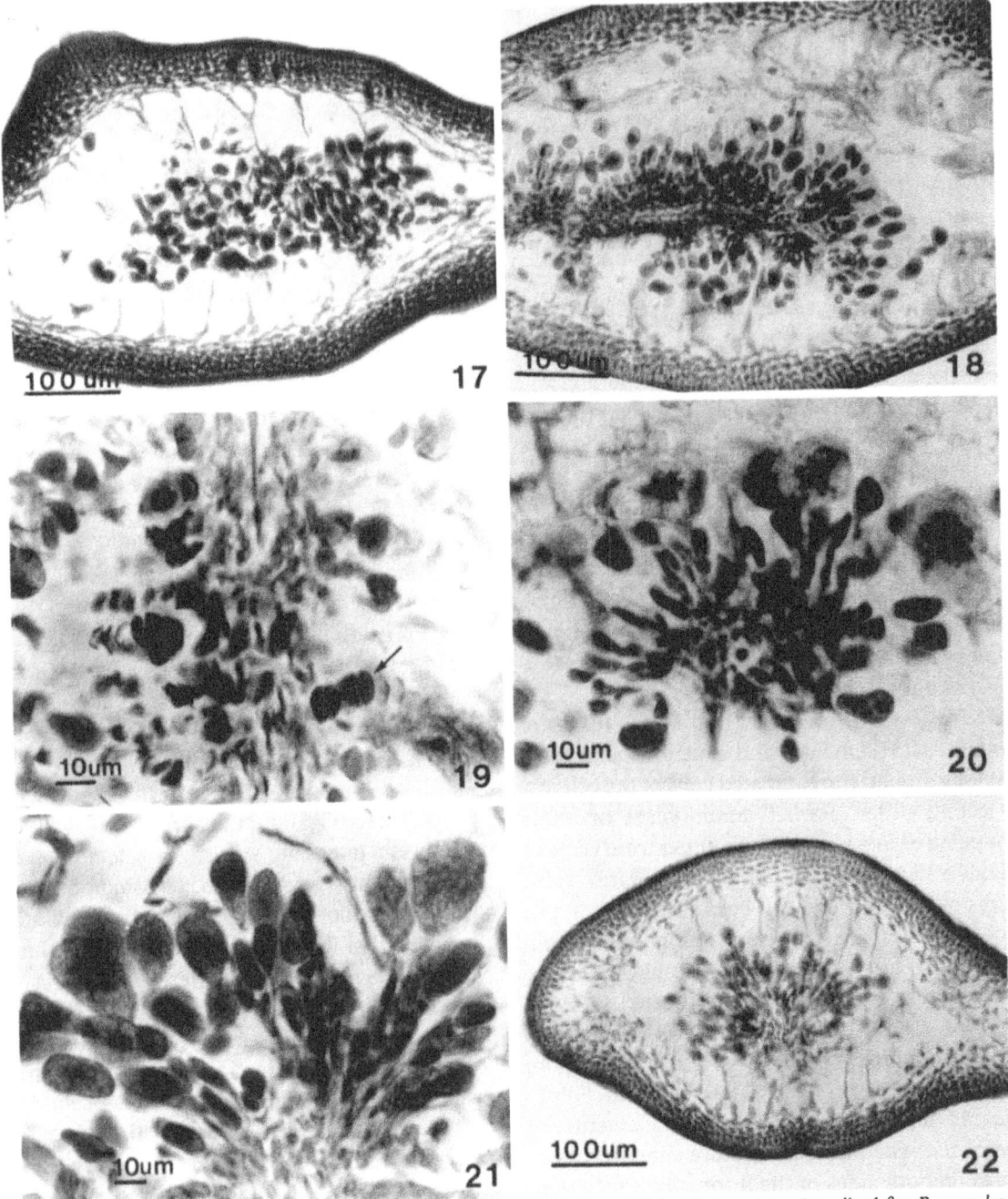

Fig. 17. Longitudinal section of a cystocarp of *P. capillacea* in similar state of development as described for *P. caerulescens* in Fig. 15.

Fig. 18. Mature cystocarp of *P. capillacea* exhibiting a centrally-placed placenta, spores produced towards both frond surfaces and a few elongated, parallel, inner cortical filaments.

Fig. 19. Transverse gonimoblast divisions in *P. caerulescens*. Initials from short stacks of flat cells, similar to a pile of coins (arrows).

Fig. 20. Carpospore production in *P. capillacea*. Mature spores are rounded and grow radially.

Fig. 21. Carpospore production in *P. caerulescens*. Mature spores are rounded and grow radially.

Fig. 22. Longitudinal sections of a cystocarp of *P. capillacea* exhibiting abundance of elongated inner cortical filaments and sporogenous tissue of arboriform appearance elevated from the bottom of the cystocarp and situated in center of the cystocarpic locule.

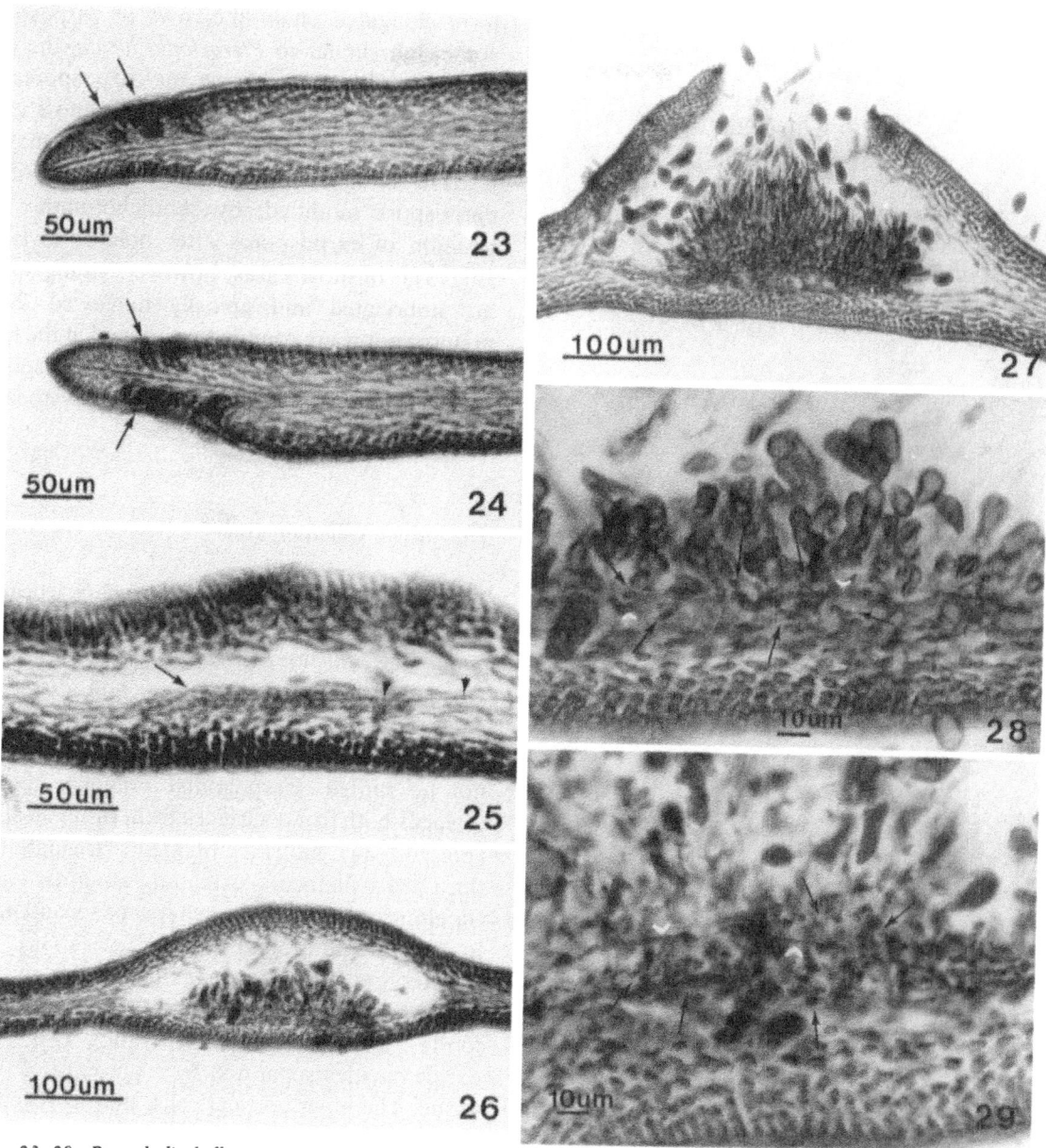

Figures 23–29. Pterocladia bulbosa

Figs. 23–24. Longitudinal sections of fertile female pinnules exhibiting carpogonial branches (arrows) to one (Fig. 23) or to both (Fig. 24) frond surfaces.

Fig. 25. Nutritive filaments (arrow) on both frond surfaces around the elongated cells of the cell row of second order (arrow head). Note splitting of the frond just above the fertile area.

Figs. 26–27. Longitudinal sections through cystocarps at different stages of maturation exhibiting a single cavity and a placenta of fertile tissue located at the base of the cystocarp.

Figs. 28–29. Different degrees of development of fertile tissue (arrow) around the elongated cells of the cell row of second order (arrow head). Some of these fertile cells exhibit cellular connections with the second-order cells.

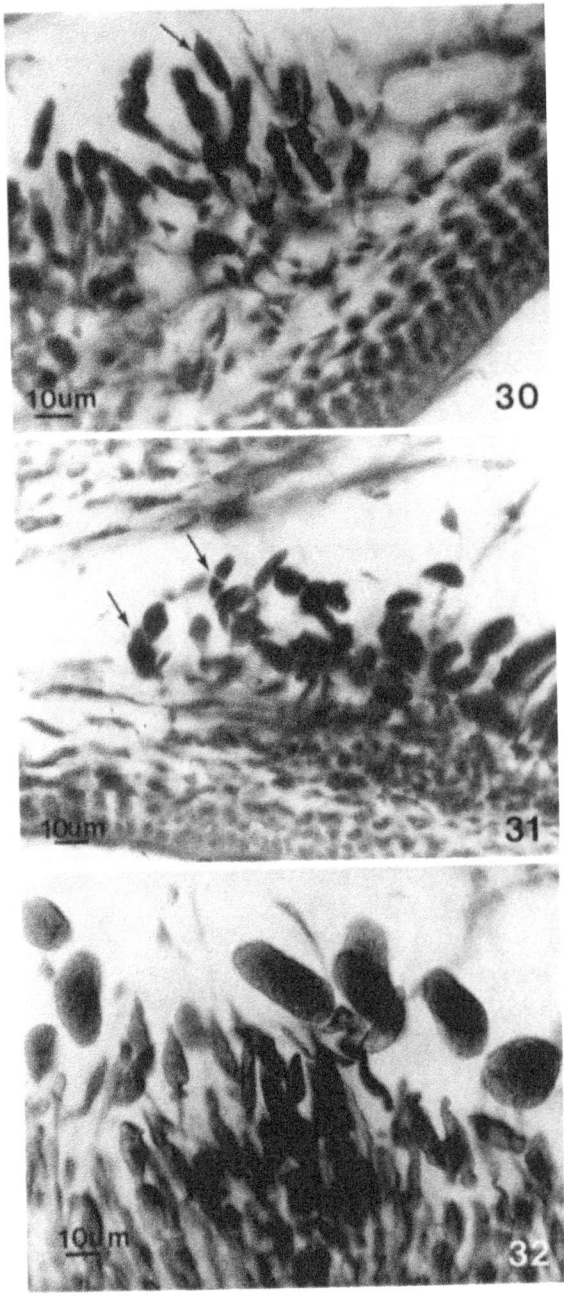

Figures 30–32. Pterocladia bulbosa

Fig. 30. Cell-row like carposporangia showing imbricated spores sagittate-shaped with narrowed apices (arrow).

Fig. 31. Transverse and subdichotomous divisions of some carpospores (arrows).

Fig. 32. Rounded and radially-expanded mature carpospores.

form elongated chain-like rows of carpospores somewhat similar to *Pterocladia lucida*, the divisions are generally oblique, the carpospores are imbricated, apically-narrowed and almost sagittate (Fig. 30). In a few cases, divisions may be transverse, or the uniseriate arrangement of the carpospore modified, by subdichotomous disposition of carpospores after oblique divisions (Fig. 31). In most cases, however, young spores are imbricated and apically narrowed. More mature spores are rounded, produced at the tip of 3-5 cell-long, chain-like rows and disposed, slightly-expanded, radially, in the cystocarpic cavity (Fig. 32).

The classic Gelidium-type

This is the commonest type of cystocarp in the specimens studied. It occurred in *Gelidium amansii*, *G. chilense*, *G. galapagensis*, *G. pusillum* var. *pulvinatum*, *G. pteridifolium* and *Pterocladia macnabbiana*.

The margins of the fertile pinnules may or may not be ruffled. Carpogonial branches develop towards both frond surfaces from the central axis (Figs. 33, 34). Nutritive filaments originate from third-order filaments, extending on either side of the elongated cells of the cell row of second order, forming a core around them (Figs. 35, 36).

After growth and expansion of the gonimoblast, the cystocarpic wall on both surfaces of the fertile ramuli elongates, generating two cystocarpic cavities separated by a placenta of fertile tissue. The inner cortical cells, that in vegetative ramuli ran parallel to the central axis, elongate during the expansion of the cystocarps (Fig. 37). They form a matrix of elongated, parallel, multicellular filaments that extend between the elongated cells of the cell row of second order and the cortical cells. The most internal cell of these filaments (Fig. 38) is a short, irregularly-triangular element that connects with the second-order cells on one side and with two elongated cortical cells on the other side.

The fusion network develops on both sides of the frond, and carposporangia are cut off by obli-

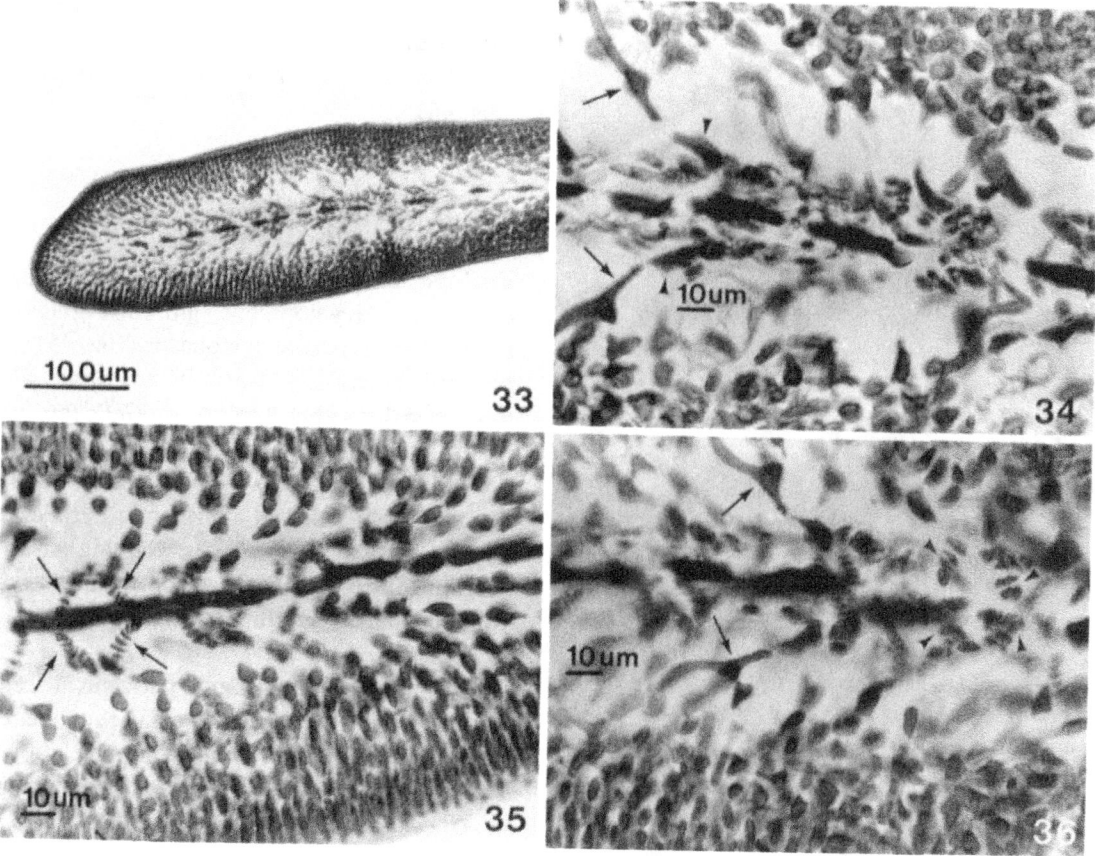

Figures 33–35. Gelidium pteridifolium

Fig. 33. Longitudinal sections through a fertile tip.

Fig. 34. Intercallary carpogonia (arrows) with a prominent basal cell (arrow heads).

Fig. 35. Nutritive cells (arrow) differentiated prior to the formation of carpogonial filaments, forming a core around the elongated cells of the cell rows of second order.

Fig. 36. Intercallary carpogonia (arrows) and nutritive filaments (arrow heads) in a fertile tip of *G. pusillum* f. *pulvinatum.*

que divisions. Mixtures of carposporangia of varying sizes coexist in young cystocarps, and some interspecific variation occurs in the manner carposporangia are produced; for example, in *Gelidium galapagensis* a few transverse and pseudodichotomous divisions occur (Fig. 39), while in most other species, divisions are oblique. In *G. galapagensis*, carpospores arise singly and directly from gonimoblast initials, while in *G. pteridifolium*, spores arise forming clusters (Fig. 40) and in *G. amansii*, spore-clusters are produced at the tip of elongated processes (Fig. 41).

The mature cystocarp is biconvex (Fig. 42), with a paired placenta around the axial and periaxial cells (rows of cells of second order), producing carpospores towards both sides of the fronds and with a matrix of parallel, elongated, inner-cortical filaments extending from the placenta to the cortex. Ostiolar openings develop on opposite sides (Fig. 42). The two cystocarpic locules generally develop at a similar rate and

12

are approximately symmetric; however, asymmetric cystocarps can be found in some species, such as *G. pusillum* (Fig. 43) and *P. macnabbiana* (Fig. 44).

The Gelidium coulteri type

The specimens examined lacked early stages of cystocarpic development, although longitudinal sections of immature cystocarps (Fig. 45) show a pattern essentially similar to that already described for other species of *Gelidium*. Cystocarps are biconvex, with a paired placenta around the medial septum, producing carpospores towards both sides of the frond. Parallel, elongated, cortical filaments extending from the septum to the cortex are absent from all but the youngest parts (e.g. the borders) of the expanding cystocarps.

The inner cortical and pericentral cells of *G. coulteri* are not as elongated as equivalent cells in the species of *Gelidium* described above (*cf.* Fig. 46; Figs. 37, 38). While the cystocarpic walls of *G. coulteri* expand, some elongation is also shown by the inner cortical cells and a few filaments are seen running from the placenta to the wall (Fig. 47). As in other species of *Gelidium*, the innermost cell of these filaments is also irregularly triangular and connects with the elongated cells of the cell row of the second order on one side and with two elongated cortical cells on the opposite side. As the expansion of the cystocarp progresses, the connection between this triangular cell and the elongated inner cortical cell is lost (Fig. 48). The innermost, irregularly-triangular cell becomes part of the fusion network (Fig. 49) which eventually produces spore initials (Fig. 50). The function of this innermost cortical cell is, therefore, significantly different from that of equivalent cells in other species of *Gelidium* and *Pterocladia*. The regular involvement of this cell in the fertilization network explains the lack of inner cortical filaments in all but the youngest parts of the cystocarps and sets this type of cystocarp apart from others described above.

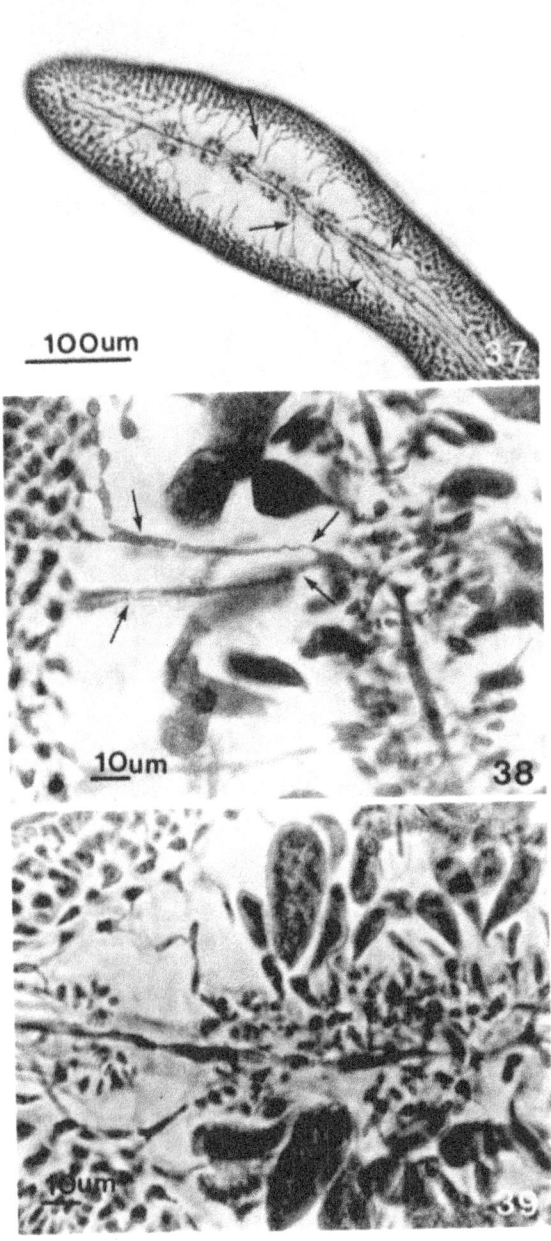

Figures 37–39. Gelidium galapagense

Fig. 37. Longitudinal section of a fertile tip. Note the change in disposition of elongated inner cortical cells (arrows) as the cystocarp develops.

Fig. 38. Detail of an elongated inner cortical filament (arrows).

Fig. 39. Transverse and oblique divisions during carpospore production.

13

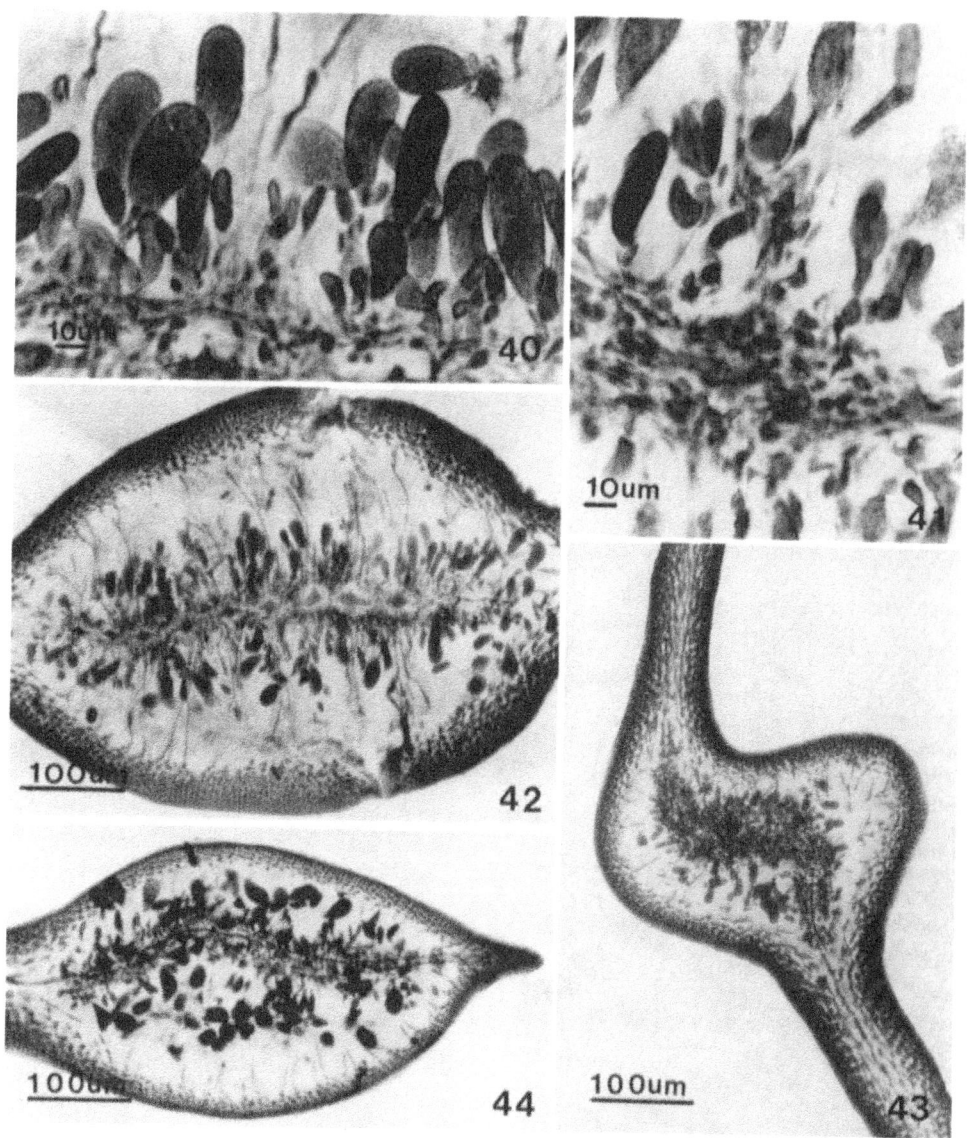

Fig. 40. Clusters of spores of *Gelidium pteridifolium*.

Fig. 41. Clusters of spores of *Gelidium amansii* at the tip of elongated gonimoblast branches.

Fig. 42. Transverse section of a mature cystocarp of *G. pteridifolium*, with ovoid carpospores and elongated cortical filaments running from the placenta to the cortex.

Figs. 43–44. Cystocarps of *Gelidium pusillum* f. *pulvinatum* (Fig. 43) and *Pterocladia macnabbiana* (Fig. 44) with asymmetric cystocarpic developments.

The Gelidium musciformis-type

In longitudinal section, the fertile ramuli exhibit carpogonial branches developing towards both surfaces (Figs. 51, 52). Infrequently, two carpogonia are formed by the same supporting cell, which is elongated and about one-half the size of the carpogonial cell. Nutritive filaments differen-

14

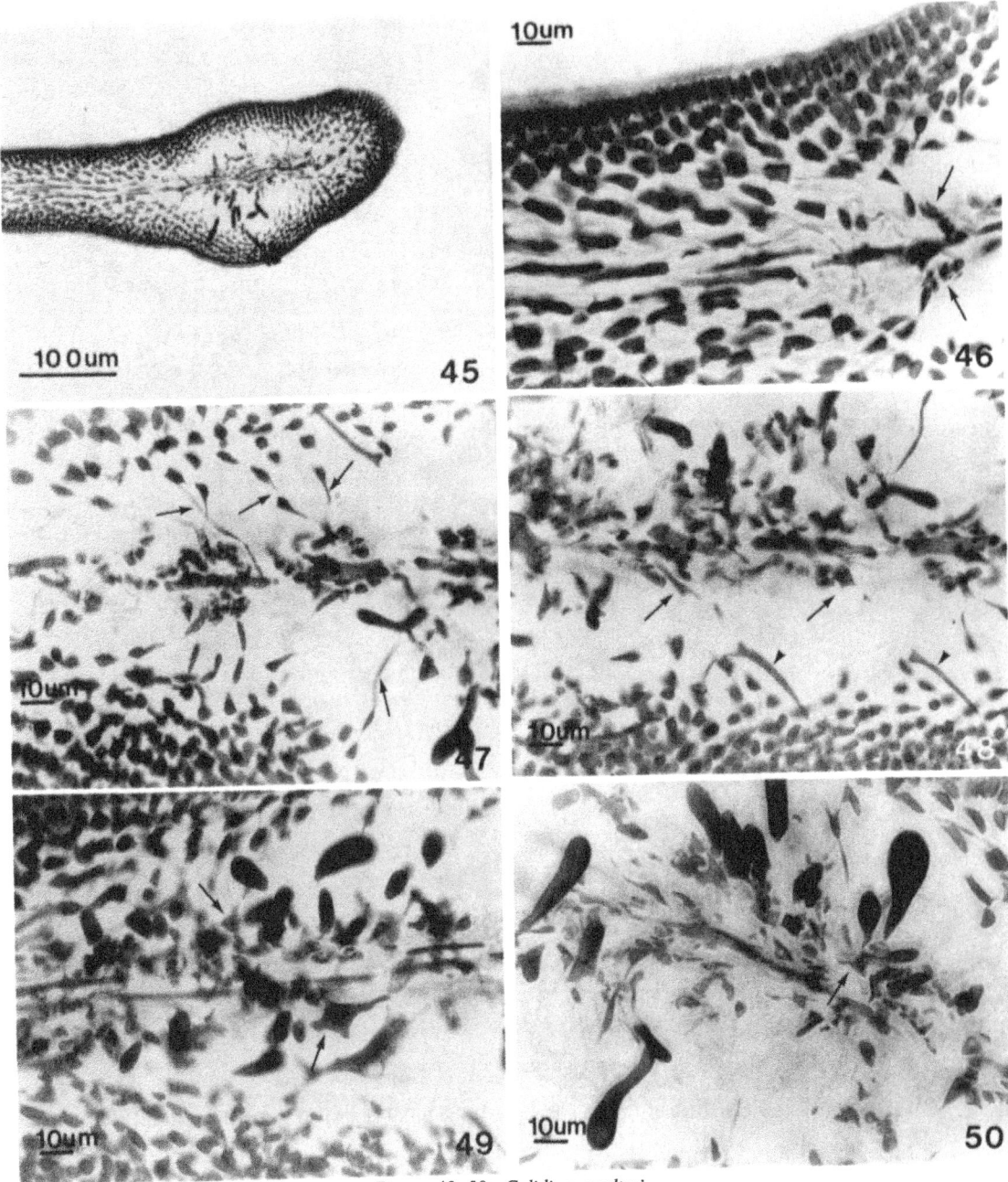

Figures 45–50. Gelidium coulteri

Fig. 45. Longitudinal section of a cystocarp. Note the scarcity of elongated inner cortical filaments.

Fig. 46. Nutritive filaments (arrow).

Fig. 47. Elongate inner cortical filaments (arrows) in the youngest part of a cystocarp.

Fig. 48. The inner cortical filaments later become discontinuous. The basal, irregularly-triangular cell (arrow) becomes a part of the fertilization network. Remains of elongated cortical filaments (arrows heads) are seen near the cortex.

Figs. 49–50. The basal, irregularly-triangular cell of the cortical filament can grow and connect with other cells in the fertile network (arrow in Fig. 49). Spore initials can be seen originating from these cells in some mature parts of the cystocarp (arrow in Fig. 50).

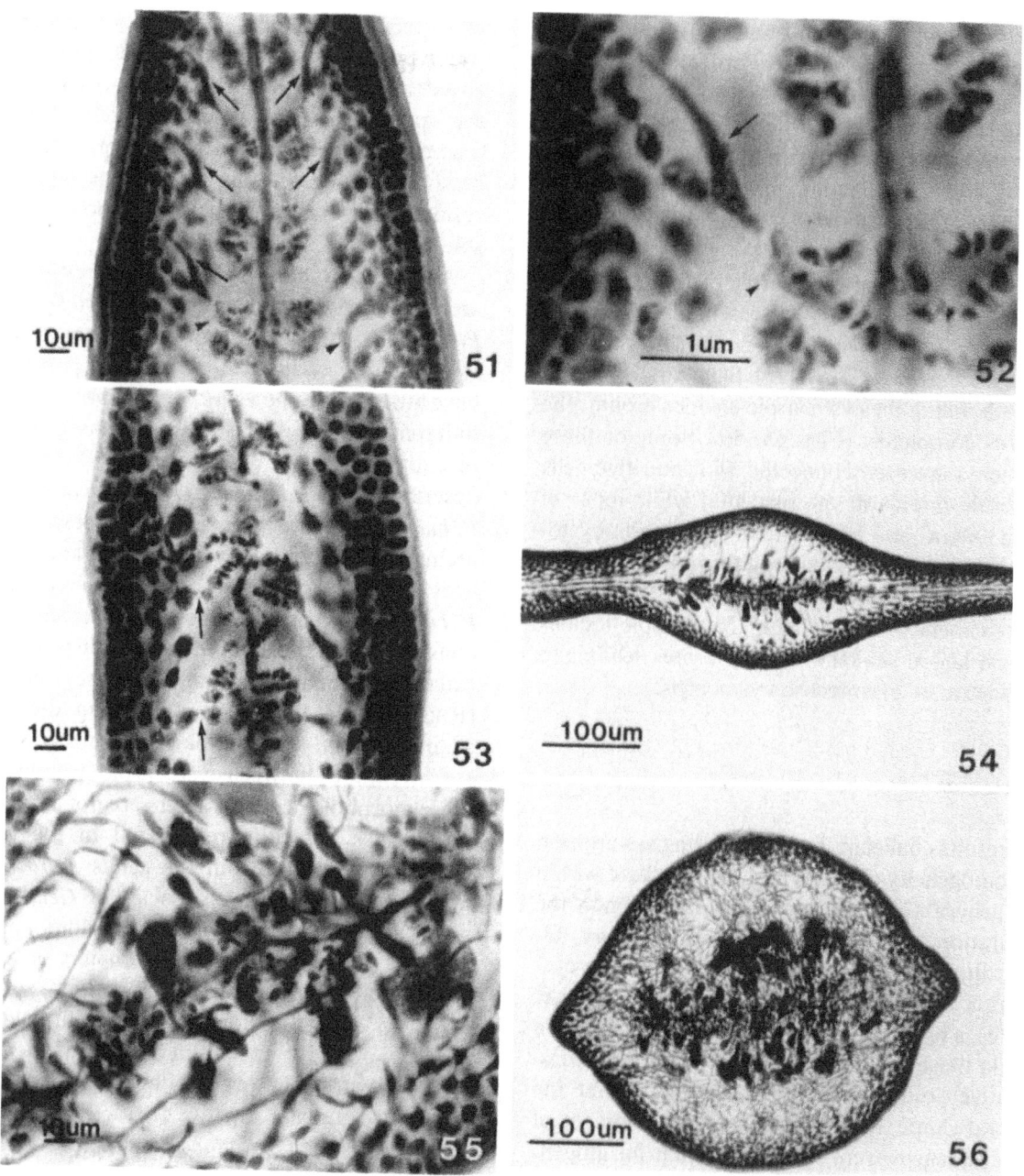

Figs. 51–52. Gelidium musciformis. Longitudinal sections through a fertile tip showing carpogonia (arrow), elongated basal cells (arrow head) and nutritive filaments.

Fig. 53. Connections between nutritive filaments and cortical cells (arrow).

Fig. 54. Longitudinal section of a young cystocarp showing carpospores and a few elongated cortical filaments.

Fig. 55. Proliferation of inner cortical filaments in the cystocarpic cavity.

Fig. 56. Transverse section through a mature cystocarp of *G. musciformis* showing proliferation of inner cortical filaments that fill most of the cystocarpic cavity.

16

tiate before carpogonia formation, from the most basal cell of a cell row of third order, and, in a few cases, they seem to originate also from the supporting cell (Fig. 52). Fusions between nutritive cells and cortical cells are frequent (Fig. 53), and soon after fertilization, the fusion network becomes connected with cortical cells from all sides.

In early stages of cystocarpic development, the inner cortical filaments are essentially similar to other species of *Gelidium* (Fig. 54); however, as the cystocarp develops, these filaments grow and branch, filling the cystocarpic cavities around the spores completely (Figs. 55, 56). Some of these filaments are seen connected with nutritive cells or fertile initials in the placenta, while most of them extend and branch freely inside the cystocarpic cavity. The function of this matrix of cortical filaments formed in later stages of cystocarp development is not evident. Cystocarpic locules can expand at similar or different rates, leading to symmetric or asymmetric cystocarps.

Discussion

My results challenge the longstanding assumption of homogeneity of cystocarpic architecture within the genera *Gelidium* and *Pterocladia*. Since the separation of *Pterocladia* from *Gelidium* (J. Agardh, 1851) and the description of mature cystocarps of *P. capillacea* by Bornet and Thuret (1876), a homogeneity of cystocarpic architecture within these genera has been assumed. In a comparative study, Fan (1961) concluded that the general shape, the structure and development of the cystocarps were strikingly similar for all genera included at that time in the Order Gelidiales. The exceptions were species of *Pterocladia* for which developmental stages were considered similar, though with differences in locules and ostioles; consequently, later taxonomic studies, including recent reviews of the genera in the Order (Akatsuka, 1986) have been largely limited to confirming the number of locules in the cystocarpic cavity so as to distinguish *Gelidium* from *Pterocladia*. My present results suggest important

interspecies differences in cystocarpic architecture in both genera. Such differences involve early, as well as late, post-fertilization events and include the manner in which carpogonial and nutritive filaments are produced, the direction of gonimoblast divisions giving rise to carposporangia, the resulting shape of immature carpospores and the role of internal cortical filaments.

As there is clear heterogeneity in cystocarpic architecture within the genera *Gelidium* and *Pterocladia*, this should stimulate re-assessment of vegetative characters judged, at present, to have little taxonomic value; for example, the three different types of cystocarps among species recently recognized in *Pterocladia*, which I have described for *P. lucida*, *P. capillacea* and *P. bulbosa*, parallel three different types of apical architecture reported by Rodríguez and Santelices (1988) for these species. Segregating *P. bulbosa* from other species of *Pterocladia* is also consistent with the different type of disposition of external cortical cells at the bases of axes (Rodríguez & Santelices, 1988) and the spore arrangement in stichidia-like branchlets.

Correlations between cystocarpic structure and other morphological characters are not as clear in the species presently recognized in the genus *Gelidium* as are those in the genus *Pterocladia*; however, the number of species of *Gelidium* is much larger than *Pterocladia* and additional differences either in vegetative characters or cystocarpic architecture might appear in future studies. As fertile female thalli are infrequent in field collections of these species, correlations between cystocarp architecture and either vegetative structure or asexual reproductive characters are necessary before changes in the delimitation of both genera can be proposed.

Acknowledgements

I thank I.A. Abbott, S. Fredericq, and J. McLachlan for reviewing and commenting on the manuscript. Cystocarpic specimens for this study were kindly provided by Drs I.A. Abbott, Yocie Yoneshige-Valentin, Wendy Nelson, James

Norris and T. Yoshida. My gratitude to all of them for these valuable materials. The patience of Mrs Verónica Flores and the quality of her histological and photographic work are appreciated. I am pleased to acknowledge the financial support of Fondo Nacional de Ciencias (Grant Fondecyt, 803–90) and Red Latinoamericana de Botánica (Grant N° 90-7) that allowed me to undertake this study.

References

Agardh, J. G., 1851. Species, Genera et Ordines Algarum. Lund. (2)1.vii + 336 + addenda and index.

Akatsuka, I., 1981. Comparative morphology of outermost cortical cells in the Gelidiaceae (Rhodophyta) of Japan. Nova Hedwigia 35: 453–463.

Akatsuka, I., 1986. Surface cell morphology and its relationship to other generic characters in non-parasitic Gelidiaceae (Rhodophyta). Bot. mar. 29: 59–68.

Bornet, E. & G. Thuret, 1876. Notes algologiques. Fasc. 1xx + (2)pp., pls. 1–50. G. Masson, Paris.

Dixon, P. S., 1958. The structure and the development of the thallus in the British species of Gelidium and Pterocladia. Ann. Bot. N.S. 22: 353–368.

Fan, K-C., 1961. Morphological studies of the Gelidiales. Univ. Calif. Publ. Bot. 32: 315–368.

Okamura, K., 1934. On Gelidium and Pterocladia of Japan. J. Imp. Fish. Inst. 29: 47–67.

Rodríguez, D. & B. Santelices, 1987. Patterns of apical structure in the genera Gelidium and Pterocladia (Gelidiaceae, Rhodophyta). Hydrobiologia 151/152: 199–203.

Rodríguez, D. & B. Santelices, 1988. Separation of Gelidium and Pterocladia on vegetative characters. In Abbott I. A. (ed), Taxonomy of Economic Seaweeds with Reference to some Pacific and Caribbean Species. Vol II. California Sea Grant College Program, La Jolla, California: 115–125.

Santelices, B., 1988. Synopsis of biological data on the seaweed genera Gelidium and Pterocladia (Rhodophyta). FAO Fisheries Synopsis 145: 1–55.

Santelices, B., 1990. New and old problems in the taxonomy of the Gelidiales (Rhodophyta). Hydrobiologia 204/205: 125–135.

Santelices, B., 1991. Variations in cystocarp structure in Pterocladia (Gelidiales: Rhodophyta). Pac. Sci. 45 (in press).

Stewart, J. G., 1976. Gelidiaceae. In Abbott I. A., Hollenberg G. J., Marine Algae of California. Stanford University Press, California: 340–352.

Hydrobiologia **221**: 19–29, 1991.
J. A. Juanes, B. Santelices & J. L. McLachlan (eds), International Workshop on Gelidium.
© 1991 *Kluwer Academic Publishers.*

Worldwide distribution of commercial resources of seaweeds including *Gelidium*

Dennis J. McHugh
Department of Chemistry, University College, University of New South Wales, Australian Defence Force Academy, Campbell ACT 2600, Australia

Key words: agar, alginate, carrageenan, commercial resources, food, *Gelidium*, seaweeds

Abstract

The commercial exploitation of seaweeds for use as food and for the production of agar, alginate and carrageenan is outlined. The quantities of seaweed harvested for each purpose are tabulated and discussed. Seaweeds for food are derived chiefly from China, Japan and Korea, with almost 94% obtained by cultivation. Alginophytes are collected in 15 countries but six of these account for more than 80% of the total harvest; all are from natural stocks except for a large quantity of *Laminaria* cultivated in China. Natural carrageenophytes, from 12 countries, now account for only 20% of the total harvest; the remainder is cultivated *Eucheuma* species, 99% of which is produced in only two countries, the Philippines and Indonesia. Of the four categories of commercial resources of seaweeds considered, agarophytes are spread more evenly over a greater number of countries; they come from 20 countries and only five of these are minor contributors to the total. *Gelidium* species are particularly important because of the high quality agar they yield; their distribution and location are discussed.

Introduction

In the past 30 years, there have been several surveys of the worldwide resources of a variety of seaweed species. Many of the resulting statistics are approximate, because accurate surveys are difficult, time-consuming and expensive and can usually be justified only when commercial exploitation is being considered; nevertheless, these surveys have been useful in guiding commercial interests to likely locations for harvesting. Even for many of the commercial seaweed beds, estimates for the total available resource are variable, not only because of the limitations of physical measurements, but also because the resource itself varies with changing environmental factors from year to year.

Commercial resources of seaweeds are sometimes only a minor proportion of the total worldwide standing crop. Access and abundance are two of the factors which determine the commercial viability of a resource, and they have been significant determinants of how the industry has developed. Access to some seaweeds has been improved through cultivation; for example, the quantity of *Eucheuma* produced by cultivation now exceeds the accessible natural standing stocks. Other important factors which determine commercial viability include costs of harvesting (for labour and/or equipment), drying, transportation, chemicals, water supply and environmental protection measures.

Illustrative examples can be drawn from the brown seaweeds suitable for alginate extraction.

The Kerguelen Islands in the Indian Ocean are reputed to be the richest resource of *Macrocystis* and *Durvillaea* in the world, yet their distance (4800 km) from both South Africa and Australia makes transportation costs prohibitive. Large quantities of brown seaweeds are found on the Falkland Islands (Malvinas), about 800 km from Argentina, but the climate and lack of fuel make drying difficult. An alginate factory could be established there, although its size would be determined by the minimum size of harvesting vessel that could be operated. Such a vessel would have to be large enough to withstand the rough seas and weather conditions which can prevail in this area. It would need to be used to its full potential to justify the capital cost, so any factory would have to be large enough to cope with the resulting quantity of raw material. The necessary chemicals, solvents and fuels would have to be shipped from Argentina; sufficient water storage would be required. The combination of all these factors would make the factory uneconomical to operate. Large natural resources are, therefore, not necessarily of commercial value.

This survey is confined to seaweeds of commercial interest, and to quantities and locations actually harvested or cultivated. It encompasses seaweeds used in the food, alginate, carrageenan and agar industries. Resources of *Gelidium* species, a valuable source of high-quality agar and agarose, are considered in more detail.

The statistics in all the tables are expressed as 'dry weight', a term used by industry for the mass of the seaweed after it has been dried by natural means. This dried seaweed usually contains about 20% moisture, but this can vary, depending on the post-harvest treatment and the type of seaweed. The main exception is *Eucheuma*, a source of carrageenan, which buyers prefer to contain 35% moisture, for shipping convenience.

Seaweeds for food

Seaweeds used in food are most popular in China, Japan and Korea, although they are used in other Asian countries, and in countries where there are ethnic Asian communities. The main types are *Laminaria* species (kombu), *Undaria pinnatifida* (Harv.) Suringar (wakame), *Hizikia fusiformis* (Harv.) Okamura (hiziki) and *Porphyra* species (nori). The Japanese names are shown in

Table 1. Seaweeds for food – quantity harvested in dry tonnes.

	Laminaria	Undaria	Hizikia	Porphyra
Argentina				5 (w)
Chile				10 (w)
China	150000	500		13000
		200 (w)		100 (w)
France				10 (w)
Japan	10000	20000		44300
	22000 (w)	1300 (w)		
Korea R (Sth)	2300	50700	6200	11600
	300 (w)	1200 (w)	3400 (w)	100 (w)
Korea DPR	110000	7500		
New Zealand				5 (w)
Totals	294600	81400	9600	69130
% of Table total	64.8%	17.9%	2.1%	15.2%

(w): harvested from wild resource. Other quantities are from cultivation.

Source: Quantities shown in the Table are the author's estimates, based on published government statistics and/or information received from representatives of industrial, academic and government organisations.

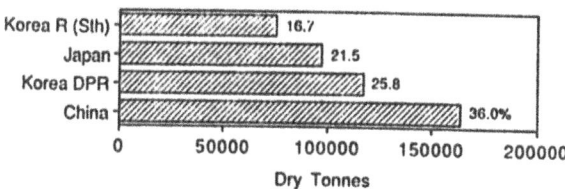

Fig. 1. Seaweeds for food. Total contribution by each country, in dry tonnes and as a percentage of the total for all countries.

brackets; other vernacular names are used in China and Korea. Table 1 shows the quantities of these seaweeds that are produced by cultivation, or harvested from the wild, by the major producing countries. The total production of seaweeds used as food by each country is shown in Fig. 1; each country's contribution by percentage is also shown.

China and Japan are large consumers of *Laminaria*, while South Koreans have a preference for *Undaria*. The Chinese annually cultivate about 200 000 tonnes (dry weight) of *Laminaria japonica* Aerschoug of which about 75% is used in food; Japanese cultivation is on a smaller scale, and the harvest of wild material exceeds the yield from cultivation (Table 1). In South Korea, even less *Laminaria* is cultivated and very little wild material is collected. However, very large quantities of *Laminaria* are cultivated on both the east and west coasts of North Korea, where it is used mostly as food, generally without being dried; some is also sold to the USSR, loaded wet onto ships at sea.

In both Japan and South Korea, most of the *Undaria* comes from cultivation, with South Korea production being about double that of Japan. When relative populations are considered, South Korea produces about six times more than Japan per capita. Most of the South Korean production is in the south-west corner of the south coast, while the northern areas around Hokkaido are the main source in Japan; North Korean cultivation is on a smaller scale. The low *Undaria* production in China is a reflection of the greater interest in *Laminaria* cultivation.

Hizikia is both cultivated and wild in South Korea with about half as much material being collected from the wild as from cultivated beds.

There are no separate statistics for *Hizikia* in Japan; it is probably included in the 'Other' category for both cultivated and wild seaweeds in the official Fishery statistics. Appreciable quantities are imported from South Korea – 4 700 tonnes in 1989.

The cultivation of *Porphyra* is well-developed (Mumford & Miura, 1988), especially in Japan, where the production is far greater than in South Korea and China; by comparison, the quantities collected from the wild in all three countries are almost negligible. Small quantities of *Porphyra* (5–10 tonnes) are collected from the wild in France, New Zealand, Argentina and Chile.

There are minor uses of other seaweeds as food but the quantities are not recorded here; for example, for centuries in China and Japan, *Gelidium* has been collected from the shore by local people who boil it in water to make an edible jelly. *Gracilaria* species are collected from the wild and used fresh as a salad vegetable in Hawaii and some Asian countries. *Caulerpa lentillifera* finds a similar use because of its delicate taste and soft, succulent texture. It is commercially cultivated in ponds, particularly on Mactan Island on the outskirts of Cebu, Philippines (Trono, 1986). In Chile, small quantities of wild *Durvillaea*, locally-called cochayuyo, are used in food.

Prices for seaweeds sold as food are always much greater than for those used for extraction of colloids. The wholesale price in Japan for brown seaweeds of food quality can vary from US$ 7500–10 000 per dry tonne (Nisizawa, 1987) while international prices for brown seaweeds used in alginate extraction range from US$ 150–500 per dry tonne (McHugh, 1987). Red seaweeds may command even higher prices. In Japan, good-quality *Porphyra* (nori) sells wholesale at US$ 24 per kg, while the retail price can be at least double this (Nisizawa, 1987); in Nova Scotia, *Palmaria* (dulse) sells for Can$ 22 per kg (J.L. McLachlan, *in litt.*).

Seaweeds for alginate production

The types of seaweed used for the extraction of alginates are shown in Table 2. Such a wide range

Table 2. Seaweeds for alginate production – quantity harvested in dry tonnes.

	Laminaria	Ascophyllum	Other
Argentina			500 – *Macrocystis*
Australia			4000 – *Durvillaea*
Canada		7400	
Chile			16500 – *Lessonia*
			2100 – *Macrocystis*
			1000 – *Durvillaea*
China	50000 – *japonica*		
France	13000 – *digitata*	1000	1000 – *Fucus*
	1000 – *hyperborea*		
Iceland	250 – *digitata*	4000	
Ireland	1000 – *hyperborea*	12000	
Mexico			4000 – *Macrocystis*
Namibia	20 – *schinzii*		25 – *Ecklonia*
Norway	25000 – *hyperborea*	8500	
Scotland	1000 – *hyperborea*	3000	
South Africa	800 – *schinzii*		1600 – *Ecklonia*
Spain	2000 – various species		300 – *Fucus*
USA			10000 – *Macrocystis*
Totals			16600 – *Macrocystis* (9.7%)*
			16500 – *Lessonia* (9.6%)*
			5000 – *Durvillaea* (2.9%)*
			2925 – Other
	94070	35900	41025
* % of Table total	55.0%	21.0%	24.0%

Source: Quantities shown in the Table are the author's estimates, based on published government statistics and/or information received from representatives of industrial, academic and government organisations.

of species is used in alginate production for two principal reasons: there is a strong demand for any good source of alginate which can be harvested economically; secondly, the properties of alginates vary from one species to another and sometimes within the parts of one plant (for example, the fronds and stipes of *Laminaria hyperborea* (Gunn.) Fosile). By having a variety of sources, a manufacturer can blend his products to meet the requirements for any particular use; for example, some applications rely on the formation of gels. The addition of calcium ions to an alginate solution will cause a gel to form; the gel may be soft or firm depending on the alginate used. Alginic acid is a linear polymer based on two different monomers, mannuronic acid and guluronic acid. The properties of alginates extracted from different seaweeds will vary according to the ratio of mannuronic acid to guluronic acid (the M/G ratio) in the polymer. A high proportion of guluronic acid (a low M/G ratio) in the polymer gives an alginate which forms firm gels; the gels are much weaker if mannuronic acid predominates. The stipes of *Laminaria hyperborea* yield an alginate with a low M/G ratio (about 0.4), and this alginate is finding increasing demand for the preparation of immobilised cells and enzymes in biotechnology. Alginates with a range of gel strengths can be made by blending alginate from *L. hyperborea* with others that have higher M/G ratios.

Species of *Laminaria* are the largest resource for alginate production (Table 2). Commercial resources of *L. hyperborea* (16.4% of the total alginophyte resource) are utilised off the coasts of Norway, Ireland, Scotland and France. In Norway, mechanical harvesting, using a drag rake, can yield from a half- to a tonne of wet seaweed per minute; usually the whole plant is used, without drying, for alginate extraction. In Ireland and Scotland, the fronds are removed from cast plants and the stipes ('rods') are air-dried on the stony shore, piled like cordwood. France collects only a small proportion (3.6%) of the estimated available crop of *L. hyperborea* and is looking for suitable harvesting methods. *Laminaria digitata* (L.) Lamouroux is harvested predominantly in France; a 'twisting pole' with hooks on the end is used to pull plants up and off the sea bed. Other species in Table 2 include *L. schinzii* Fosile for South Africa, unspecified species for Spain (Gallardo *et al.*, 1990) and that portion of the Chinese *L. japonica* which is not used for food, although it is not clear whether all of this 50 000 tonnes (29% of the total alginophyte resource) goes into alginate manufacture. Some of the *L. japonica* cultivated in North Korea is used for alginate manufacture, but this has not been included in the table; as there is only one alginate factory, the quantity used would be comparatively small. Information on harvests from USSR is not readily available but, after a recent visit, McLachlan and Ragan (1990) estimate an alginate production of about 200 tonnes per annum in the far east and north. This would require a harvest of about 1000 tonnes dry-weight of the *Laminaria* species used (*L. saccharina* (L.) Lamouroux and *L. digitata*); the harvest comes from both natural stocks and material cultivated on a two-year cycle.

Ascophyllum nodosum (L.) Le Jolis, the next largest resource (Table 2), gives darker alkaline extracts than most *Laminaria* species; it contains large amounts of phenolic compounds which are oxidised and polymerised to brown products during extraction. It is an intertidal plant, and the largest single resource, in Ireland (33% of the *Ascophyllum*) is harvested by hand. The two smallest resources, in Scotland (8%) and France (3%), are also hand-harvested, but mechanical methods are used in Norway (24%) and Iceland (11%); in Nova Scotia (21%), both mechanical- and hand-harvesting are done.

Lessonia is cast onto the shores of the northern desert region of Chile; some material is also harvested by hand. It was adversely affected by the warm waters of the El Niño phenomenon in 1982–83, and the current yield is only two-thirds of that obtained before the beds were devastated.

The main resources of *Macrocystis* are found on the west coasts of USA and Baja California. The values in Table 2 represent average annual yields; the standing crop can be seriously affected by storms and the El Niño phenomenon. The harvest fell to 800 dry tonnes in 1983, after the 1982–83 El Niño. The only major commercial resource from South America is in Chile; small and variable annual quantities (100–500 tonnes) are available from Argentina.

Small but important contributions to the variety of available alginates are made by *Ecklonia*, *Fucus* and *Durvillaea*. Dried *Durvillaea* has an unusually high alginate content of 40–45%. Unsuccessful attempts have been made to base an alginate industry on the large resource of *Nereocystis lutekeana* (Mert.) Postels & Ruprecht on the west coast of Canada. Accurate surveys of the best areas show that at least 440 000 wet tonnes of seaweed are available. Only about 200 wet tonnes of this (not shown in Table 2) are currently harvested annually.

All weights in Table 2 are in dry tonnes, but some of the resources are processed without drying; in France and Norway, the *Laminaria* species are processed wet. All of the *Ascophyllum* in Scotland and at least 50% of it in Norway and Canada are wet processed, as is all of the *Macrocystis* from USA and Mexico. The total commercial resources available from each country are shown in Figure 2; the only cultivated resource is that from China.

Seaweeds for carrageenan production

The seaweed supply for carrageenan production is now dominated by the *Eucheuma* species, which are cultivated mainly in the Philippines and Indonesia (Table 3). Development of cultivation, beginning in the 1970s, has allowed considerable

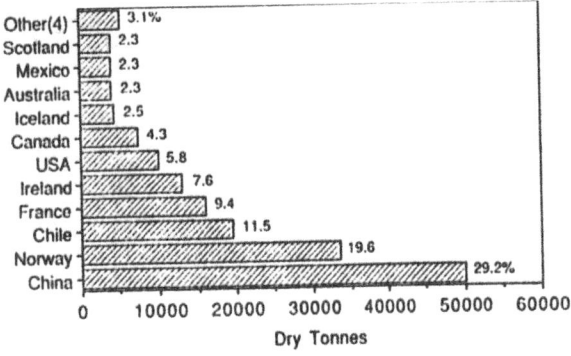

Fig. 2. Seaweeds for alginate production. Total contribution by each country, in dry tonnes and as a percentage of the total for all countries. Other (4) = total contribution by four other countries (listed in Table 2).

the natural supplies of *Chondrus crispus* Stackh. from Canada and France and species of *Gigartina/Iridaea*, mainly from South America, Mexico and southern Europe; these now constitute only about 20% of the total raw material. New uses for carrageenan have recently placed increased pressure on the seaweed supply and the resulting shortage has caused a rise in price of about 100% from 1987 prices. Carrageenan production for the next year is expected to expand by about 20% and virtually all the necessary seaweed resource must come from cultivation.

There are three main types of carrageenan: *lambda*, *kappa* and *iota*; each has differences in properties and uses. *Chondrus*, as harvested, yields a mixture of *lambda* and *kappa*. The *Gigartina* from southern Europe is a good source of *lambda* (Stanley, 1987) while the product from

expansion of the carrageenan industry. Previously, output of carrageenan was restricted by

Table 3. Seaweeds for carrageenan production – quantity harvested in dry tonnes.

	Eucheuma	*Chondrus*	*Gigartina*	Other
Argentina			80	
Brazil				1100 (*Hypnea*)
Canada		4900		
Chile			2900	4400 (*Iridaea*)
China	300 (gelat)[a]			
Denmark				400 (*Furcellaria*)
Fiji	100 (cot)[b]			
France		800		
Indonesia	8000 (cot)			
	6500 (spin)[c]			
Kiribati	100 (cot)			
Korea R (Sth)		150		
Mexico			300	
Morocco			200	
Peru			20	
Philippines	50000 (cot)			
	1000 (spin)			
Portugal		220	200	
Spain		300	600	
Totals	66000	6370	4300	5900
% of Table total	79.9%	7.7%	5.2%	7.2%

[a] *Eucheuma gelatinae.*

[b] *Eucheuma cottonii.*

[c] *Eucheuma spinosum.*

Source: Quantities shown in the Table are the author's estimates, based on published government statistics and/or information received from representatives of industrial, academic and government organisations.

the South American seaweeds is probably best described as *iota*; the extracted colloids do not necessarily match the exact composition and structures which chemists have assigned to *lambda*, *kappa* and *iota*. *Eucheuma cottonii* Lueben van Bosse yields almost ideal *kappa*-carrageenan, while *Eucheuma spinosum* (L.) J. Agardh. yields almost ideal *iota*-carrageenan (recently some *Eucheuma* species have been transferred to *Kappaphycus*).

The greatest demand is for *kappa*-carrageenan, and *E. cottonii* accounts for about 90% of the cultivated *Eucheuma* species. *E. spinosum* is the more difficult of the two to grow in the Philippines; however when cultivation was introduced to Bali, Indonesia, it was found that both species grow equally well, so the bulk of *E. spinosum* now comes from Bali. The only wild *Eucheuma* in Table 3 is 300 tonnes from China, where the species is *Eucheuma gelatinae* (Esp.) J. Agardh.

In considering the Chilean resources in Table 3, it should be noted that in the past some *Iridaea* species have been listed as *Gigartina*. The values here are derived from recent Chilean government statistics which show three species of *Gigartina* and a single, separate classification of '*Iridaea*'.

Among the minor producers of *Eucheuma*, Fiji had raised production to about 300 tonnes in 1987, but this decreased after the military coup. Efforts are now being made to revive interest among potential farmers and a large company farm is under development. Kiribati has had difficulty in marketing its seaweed because of its isolation, but the increase in price and strong demand will allow it to expand production. China has experimental farms on Hainan Island (latitude 17–18 °N), so production could increase considerably in the future.

The largest suppliers of carrageenophytes are the Philippines and Indonesia (Fig. 3), mainly from cultivated material, followed by Chile and Canada, sources of the more traditional wild seaweeds.

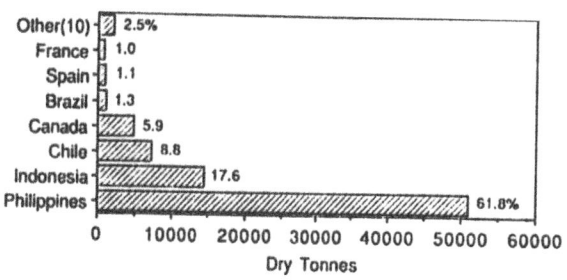

Fig. 3. Seaweeds for carrageenan production. Total contribution by each country, in dry tonnes and as a percentage of the total for all countries. Other (10) = total contribution by ten other countries (listed in Table 3).

Seaweeds for agar production – *Gelidium* species

The principal sources of agar are species of *Gelidium* and *Gracilaria*, with minor contributions from *Pterocladia* and *Gelidiella* (Table 4). *Gelidium* and *Pterocladia* are usually regarded as giving the best quality agar, and they command a higher price. Species of *Gelidium* on the coast of Japan were the original source of agar, but these became depleted by industrial expansion and pollution.

Three main grades of agar are recognised: bacteriological agar, sugar-reactive agar and food-grade agar. *Gelidium* and *Pterocladia* are the sources of bacteriological agar because their extracts best meet the requirements of gel strength (about 600 g cm^{-2}) and temperature hysteresis (the difference between melting temperature and gelling temperature); the desirable characteristics are, gelling temperature of 32–36 °C, melting temperature of 85–86 °C (Armisen & Galatas, 1987). Sugar-reactive agar is obtained largely from some *Gracilaria* species found in the eastern Pacific (Santelices & Doty, 1989). This type of agar retains its gel strength with the addition of sugar (at least up to 75 g per 100 mL of 1% agar gel) and the gel becomes elastic. This is in contrast to agar from *Gelidium*, which loses gel strength with the addition of even small amounts of sugar while the gel remains brittle (Moss & Doty, 1987). Agar which cannot meet the specifications for the above two grades is sold as food grade, this being the least expensive of the three grades.

Each country's harvest and its contribution to

Table 4. Seaweeds for agar production – quantity harvested in dry tonnes.

	Gelidium	approx. % of total *Gelidium* harvest	*Gracilaria*	Other
Argentina			2500	
Australia	10			
Brazil			1000	
Chile	430	2	6800	
			6020 – cult.	
China	200	1	2000 – cult.	50 (G)[a]
France	300	1.5		
India			600	300 (G)
Indonesia	1400	6.5	3450	
Japan	3100	14.5		
Korea R (Sth)	2900	13.5	50	
Korea DPR	270	1		
Madagascar	200	1		
Mexico	2200	10		
Morocco	4000	18.5	200	
Namibia			1200	
New Zealand				250 (P)[b]
Portugal	2000	9.5	120	700 (P)
South Africa	190	1		
Spain	4300[c]	20		
Taiwan			1600 – cult.	
Totals	21500	100	25540	1300
% of Table total	44%		53%	3%

[a] *Gelidiella*.

[b] *Pterocladia*.

[c] Based on 60% of actual harvest of 7100 t; estimate approx. 40% of harvest is not *Gelidium*.

Source: Quantities shown in the Table are the author's estimates, based on published government statistics and/or information received from representatives of industrial, academic and government organisations.

the total *Gelidium* harvest are shown in Table 4. The most prolific geographical area is the Iberian Peninsula and Morocco. The largest harvest appears to come from Spain (7100 tonnes in 1989, although a more average year would yield 5500–6000 tonnes); however, much of the Spanish *Gelidium* is badly contaminated (up to 50%) with other seaweeds so that in an average year's harvest, the actual amount of *Gelidium* would be closer to 3000 tonnes. This can be compared to the harvest from neighbouring Portugal, where the 2000 tonnes collected is 'clean' *Gelidium*. The Spanish harvest area (yielding 20% of the world *Gelidium* harvest) runs along the entire north coast and the adjoining part of the west coast to the border with Portugal; from the border to Porto in Portugal, *Chondrus crispus* is collected rather than *Gelidium*. In Portugal (9.5%), the main areas for *Gelidium* collection commence almost halfway down the west coast, about 70 km of coast extending north from Peniche; other areas are south of Lisbon, near Cabo Espichel, around Vila Nova, and on the southern coast near Lagos. In Morocco, the main harvest areas (18.5%) lie between Larache in the north and the 30° parallel of latitude, south of Agadir. Also used are three other areas on the former Western Sahara coast; a little south of Cap Juby near the former border, around Cap Boujdour, and about 50 km of coastline just south of the 25° parallel of latitude.

Japan remains a major producer of *Gelidium* (14.5%). The Izu Islands, part of Tokyo Prefecture, and the Izu Peninsula have long been major sources. Significant amounts also come from the east and west ends of the island of Shikoku, from the prefectures of Hyogo and Wakayama which border the Seto Inland Sea on each side of Osaka, and from Nagasaki prefecture on Kyushu (Japan, 1990). The Japanese agar industry has the capacity to use all the local *Gelidium* and for some years also imported *Gelidium* and large quantities of *Gracilaria*; however, the increasing costs of processing seaweeds in Japan, coupled with the rising price of *Gracilaria*, has led some companies to establish extraction plants in other countries, such as Chile, where costs are lower and seaweed supplies are closer. In South Korea (13.5%), *Gelidium* is reputed to be present around most of the coastline but commercial harvests are concentrated in the area between Pusan and Pohang, with the agar processors based in Pusan. The only data available for North Korea is the importation of 270 tonnes ($\sim 1\%$) by Japan from that country.

Commercial *Gelidium* harvests for Mexico (10% of the world *Gelidium* harvest) are made on the west coast of Baja California in selected areas from Punta Descanso, B.C., which is north of Ensenada, to Bahia Tortugas, B.C.S., south of the 28° parallel of latitude. The quantity of *Gelidium* from Indonesia has fluctuated widely in the past, but with the increased demand and price, substantial quantities are now being regularly harvested (6.5%). Collection is spread over a wide area; the southern coast of Java yields about 30% of the total, the islands between Java and Timor about 25%, Sumatra about 15%, and the remainder comes from several areas to the north and east of Timor. In Chile, *Gelidium* accounts for only 3% of the total agarophyte output of the country, all of which is exported; it is harvested near Cabo Tablas, north of Valparaiso, and at two locations about midway between Valparaiso and Concepcion.

Seaweeds for agar production – Other species

Gracilaria species had been known to give good yields of agar with poor gel strength; treatment of the seaweed with alkali was found to lower the yield but increase the gel strength (Funaki & Kojima, 1951). With the industrialization of this discovery, *Gracilaria* became a useful source of agar; the conditions for the alkaline treatment can vary according to the origin of the *Gracilaria* (Okazaki. 1971), but they range between 60–90 °C for 1 to 3 h using 4–7% sodium hydroxide.

The polymer molecules in agar usually form themselves into helices; the interaction of these helices causes gel formation. The agar from untreated *Gracilaria* has been found to contain units of L-galactose 6-sulfate in its polymer molecules; its presence causes kinks (irregularities) in the helices. Treatment with alkali converts the L-galactose 6-sulfate into 3,6-anhydro-L-galactose and the shape of this new molecule removes the kinks from the helices. This allows the helices to align more closely with each other as the gel forms, giving a higher gel strength.

Gracilaria production is dominated by Chile, with about 50% of the world total; almost half of this comes from recently-introduced cultivation. The Chinese production (8%) is all from cultivation, in the southern provinces of Guangdong and Hainan Island. Taiwan produces 1600 tonnes, cultivated in ponds, but 85% of this is used for abalone farms so the Taiwanese agar industry imports raw material. There are other significant producers: Indonesia (13%), Argentina (10%), Namibia (5%), Brazil (4%). Other minor producers are shown in Table 4.

The contributions to agar production of *Gelidium* and *Gracilaria* species were about equal five years ago, but successful cultivation has led to an increased availability of *Gracilaria*; its use now exceeds that of *Gelidium* (Table 4). Commercial sources of *Pterocladia* are limited to natural stocks in the Azores and New Zealand. The contribution of *Gelidiella*, from India and China, to world agar production is very small.

McLachlan and Ragan (1990) report that, in

the USSR, production of galactans from red seaweeds has been as high as 1000 tonnes per year. The largest factory is at Odessa, and it is dependent on *Phyllophora nervosa* (DC) Greville from the Black Sea, a resource which has been considerably depleted in recent years by industrialisation and pollution; some aquaculture of this species is being undertaken in the Black Sea. In the north and far east, agar is produced from *Ahnfeltia* species. Only small quantities of agar are produced in the north (Archangel), using *A. plicata* (Huds) Fries. In the far east (Vladivostok), *A. tobuchiensis* (Kanno + Martsub.) Makami is used, with a production of about 400 dry tonnes per year from Stark Field and the Kuril Islands (J. L. McLachlan, *in litt.*); also being evaluated is *Gracilaria* from Vietnam, where farming in ponds has been expanding since 1987.

Each country's total contribution of agarophytes is shown in Fig. 4: Chile leads with its combination of wild and cultivated *Gracilaria*: then follows a group of five countries each producing 3000–5000 tonnes, all predominantly *Gelidium* except for Indonesia, where *Gracilaria* production is higher, probably because of the introduction of cultivation in some areas. From Fig. 4, it can be seen that the supply of agarophytes is spread more evenly over a much wider range of countries than that of the other types of seaweed considered here (Figs. 1–3).

The current world demand for seaweed foods is matched by the supply and any future increase in demand could be adequately met by extending cultivation. The natural stocks of alginophytes are meeting world demand but there is little surplus; natural disasters such as the El Niño, or strong storms on the southers Californian coast, have lead to difficulties in meeting demand. Larger quantities of *Ascophyllum* could be harvested from natural stands in Nova Scotia and New Brunswick, Canada; present methods of cultivating brown seaweeds give a product which is affordable for food but normally too expensive to use for alginate extraction. New uses for carrageenan have led to increased demand for carrageenophytes in the past two years, and the consequent shortage has increased prices, particularly for *Eucheuma*. Natural stocks of traditional species are limited, and the industry is dependent on increased cultivation of *Eucheuma*, which is now being undertaken. The applications of agarose in biotechnology are increasing, and the demand for bacteriological agar continues to grow. There has been difficulty in meeting the resultant demand for high quality agarophytes, with most natural *Gelidium* beds being well exploited and the quality of the available natural *Gracilaria* declining. Mixed results have been obtained from the cultivation of *Gracilaria*, careful choice of species, strain and growth conditions being necessary to obtain a good quality agar. The commercial cultivation of *Gelidium* has proved to be difficult, but current research and development is promising and offers the best prospects for those hoping to meet future demands for a high-quality agar.

Acknowledgements

The author is indebted to the many people from industrial, government and academic organisations who contributed the data that have made this paper possible.

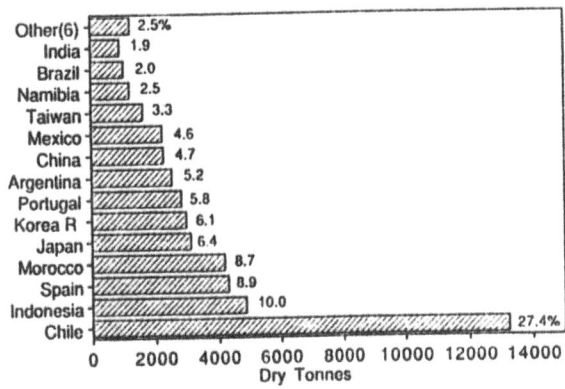

Fig. 4. Seaweeds for agar production. Total contribution by each country, in dry tonnes and as a percentage of the total for all countries. Other (6) = total contribution by six other countries (listed in Table 4).

References

Armisen, R. & F. Galatas, 1987. Production, properties and uses of agar. In McHugh D.J. (ed.), Production, properties and uses of agar. In McHugh D.J. (ed.), Production and Utilization of Products from Commercial Seaweeds. FAO Fish. Tech. Pap. 288: 1–57.

Funaki, K. & Y. Kojima, 1951. Studies on the preparation of agar-agar from *Gracilaria confervoides*. 1. Bull. Jap. Soc. Sci. Fish. 16: 401–404.

Gallardo, T., M. A. Cobelas & A. A. de Meneses, 1990. Current state of seaweed resources in Spain. Hydrobiologia 204/205: 287–292.

Japan, 1990. Fisheries and Aquaculture Statistics Yearbook 1988. Statistics and Information Department, Japan Ministery of Agriculture, Forestry and Fisheries, Tokyo.

McHugh, D. J., 1987. Production, properties and uses of alginates. In McHugh D. J. (ed.), Production and Utilization of Products from Commercial Seaweeds. FAO Fish. Tech. Pap. 288: 58–115.

McLachlan, J. L. & M. A. Ragan, 1990. Visit to USSR marine phycology laboratories. Appl. Phycol. For. 7(2): 3–6.

Moss, J. R. & M. S. Doty, 1987. Establishing a seaweed industry in Hawaii: an initial assessment. Hawaii State Department of Land and Natural Resources, Honolulu. 73 pp.

Mumford, T. F. & A. Miura, 1988. *Porphyra* as food: cultivation and economics. In Lembi C. A., Waaland J. R. (eds), Algae and Human Affairs. Cambridge University Press, Cambridge, 87–117.

Nisizawa, K., 1987. Preparation and marketing of seaweeds as foods. In McHugh D. J. (ed.), Production and Utilization of Products from Commercial Seaweeds. FAO Fish. Tech. Pap. 288: 147–189.

Okazaki, A., 1971. Seaweeds and Their Uses in Japan. Tokai University Press, Tokyo.

Santelices, B. & M. S. Doty, 1989. A review of *Gracilaria* farming. Aquaculture 78: 95–133.

Stanley, N., 1987. Production, properties and uses of carrageenan. In McHugh D. J. (ed.), Production and Utilization of Products from Commercial Seaweeds. FAO Fish. Tech. Pap. 288: 116–146.

Trono, G. C., 1986. Seaweed culture in the Asia-Pacific region. RAPA Publication 1987/8. Regional Office for Asia and the Pacific, Food & Agriculture Organisation of UN, Bangkok. 41 pp.

Hydrobiologia **221**: 31–44, 1991.
J. A. Juanes, B. Santelices & J. L. McLachlan (eds), International Workshop on Gelidium.
© 1991 *Kluwer Academic Publishers.*

Production ecology of *Gelidium*

B. Santelices

Departamento de Ecología, Facultad de Ciencias Biológicas, Pontificia Universidad Católica de Chile, Casilla 114-D, Santiago, Chile

Key words: *Gelidium*, production ecology, stock distribution, yields

Abstract

Available data on determinants of production in species of *Gelidium* suggest several general patterns. Species diversity is higher in tropical latitudes, whereas in temperate latitudes the size of the fronds is larger, the species are ecologically dominant and commercially viable. Typically, the species occur on rocky substrate, often on coralline crusts, associated with rapid water movement and arranged in successive belts that can extend down to 25 m depth. Yields vary among species, to a maximum of 2.0 kg m^{-2} y^{-1}. Growth and production in many species can best be explained by complex interactions between irradiance and nutrients. Temperature can interact synergistically with irradiance, while water movement compensates for nutrient limitations. Increased water movement or the addition of nutrients can prevent, to an extent, bleaching by high light and high temperature. Available data suggest the existence of at least eight biological factors affecting predicted productivity of *Gelidium* crops: morphology, age of the frond, thallus part, reproductive state, seasonality, crop density, life history phase and geographic and ecological origin of the species. At least four events can remove or destroy *Gelidium* crops: extreme low tides, storms, grazing and careless harvesting. Only the last-named factor has been analyzed over more extensive experimental periods.

Introduction

Nearly 100 species of *Gelidium* have been described, and many are widely distributed in tropical and temperate waters. Several species are conspicuous in intertidal and subtidal communities, and some become locally dominant because of their size, density, cover or biomass. About 10 species are consumed directly by indigenous people, over 20 have been reported to be grazed on by invertebrates and fishes and 50 species are industrially or domestically used in agar and agarose production (Santelices, 1988).

Ecological understanding of this group of algae is both academically and economically important. Understanding population dynamics and species interactions in these isomorphic taxa are necessary before theories on reproductive ecology and life-history strategies can be formulated. Environmental control of growth and agar production is central to the cultivation and management of useful species. As all commercial crops of *Gelidium* are presently gathered from wild populations, scientifically-based resource management programs are needed to maintain maximum sustainable yields.

The present review considers recent developments in the ecology of the species of *Gelidium* with emphasis on the environmental control of production. Although the subject was reviewed by Santelices (1974, 1988), several recent studies (e.g., Carter & Anderson, 1986; Macler, 1986,

1988; Borja, 1987, 1988; Macler & West, 1987; Carter & Simons, 1987; Juanes & Fernández, 1988; Frederiksen & Rueness, 1989; Rueness & Frederiksen, 1989) have added new information on these species, either supporting or falsifying previous hypotheses. I have emphasized the search for general patterns of environmental responses in species of *Gelidium*. Although several such patterns have emerged, they are accepted tentatively because comparative information has not always been gathered using similar methods; however, by approaching the problem from this perspective, I supplement and update previous studies and suggest areas for future research.

Geographic distribution of stocks and yields

Species of *Gelidium* are components of most floras, while the majority of species and records are from warm water or tropical regions (e.g. Fig. 1). The maximum poleward distribution of this genus seems to be *G. crinale* (Turn.) Lamouroux in the Falkland Islands (Hariot,

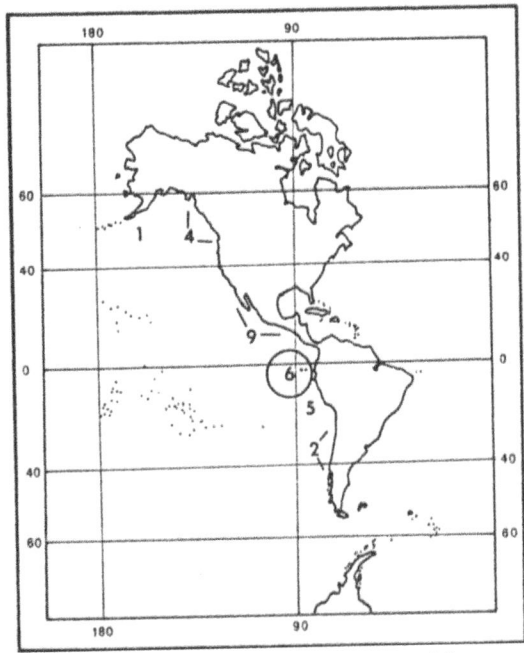

Fig. 1. Number of species of *Gelidium* reported for several geographic areas along the Pacific coast of America.

1889) and *G. latifolium* (Grev.) Bornet & Thuret and *G. pusillum* (Stackh.) Le Jolis in the southwest coast of Norway (Rueness & Tanager, 1984; Rueness & Frederiksen, 1989). Despite this essentially warm-temperate geographic distribution, most economically important crops of *Gelidium* are harvested in areas with summer water temperatures less than 25 °C (Fig. 2). Although the total production has declined in recent years (Akatsuka, 1986), the annual total world production, estimated from several sources (Santelices, 1988), ranges from 15 to 20×10^3 tonnes of dry matter. Between one-third and one-half of these resources come from intertidal and subtidal beds in Spain and Portugal, with a combined annual production of 6 to 9×10^3 tonnes. Japan and Korea have traditionally been important sources of *Gelidium*, providing 5 to 7×10^3 tonnes annually. All these productive areas occur where seasonal temperatures range between 10 and 25 °C. In many areas where surface summer temperatures reach 25 °C, commercial crops of *Gelidium* occur primarily in locales with cold upwelling waters (e.g. Baja California; Dawson *et al.*, 1960; Guzmán del Proó & de la Campa de Guzmán, 1978). The only warm-water commercial crops of *Gelidium* are from Indonesia and India, where water temperatures exceed 25 °C. The Indian crop is composed of the widespread, ecologically tolerant *Gelidium pusillum* combined with stocks of the tropical species *Pterocladia heteroplastos* (Børg.) Rao & Kaliaperumal (Kaliaperumal & Umamaheswara Rao, 1981). The Indonesian materials include significant quantities of *Gelidiella acerosa* (Forsskål) Feldman & Hamel (= *Gelidium rigidum* Greville; Soegiarto, 1978).

The reasons for limited biomass accumulation of species of *Gelidium* in tropical and subtropical surface waters remain unknown. These areas are generally richer in forms and species of *Gelidium* than temperate areas (Dawson, 1952, 1954; Santelices, 1977), and it is not unusual to find that a number of tropical species occur partially sympatrically or closely replace each other along an environmental gradient (e.g. water movement; Santelices, 1978); however, the individuals sel-

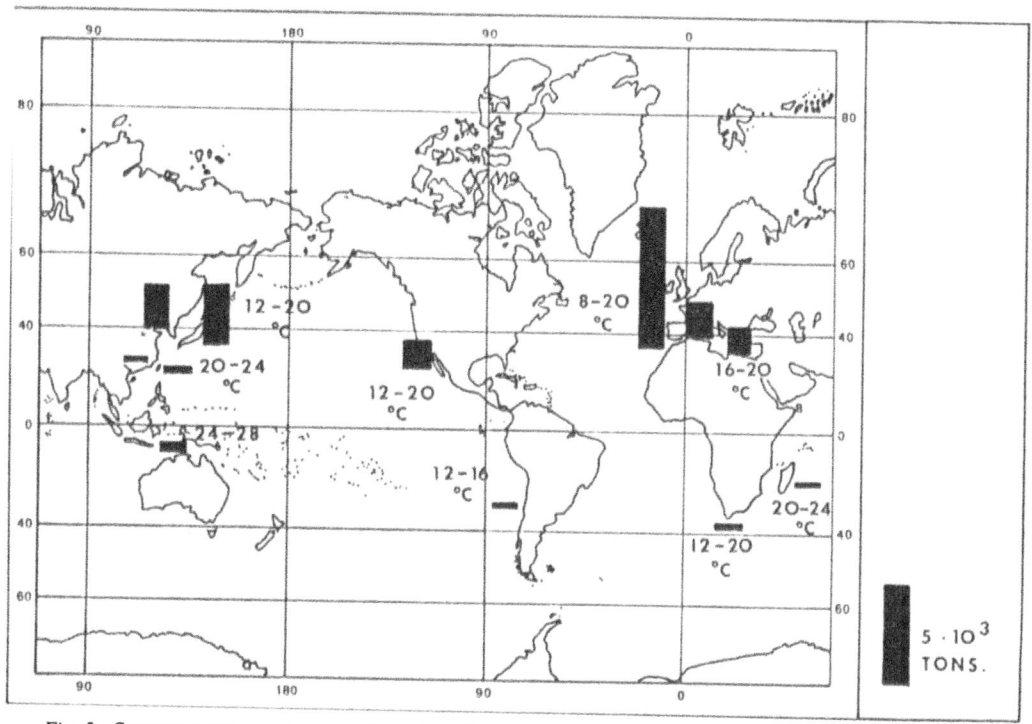

Fig. 2. Seawater temperatures and stocks of species of *Gelidium* in commercially harvested areas.

dom reach the size, abundance or frequency exhibited by harvestable monocultures of *Gelidium* in temperate waters.

Grazing by herbivorous fish, competitive interactions with other tropical algal species, morphological divergences between tropical and temperate species of *Gelidium* or environmental limitations in the tropics owing to high light intensities and reduced nutrient levels are all possible, yet untested, explanations to account for limited biomass accumulations of species of *Gelidium* in tropical waters.

Standing stocks in *Gelidium* beds vary from a few hundred grams to a maximum of about $2 \, kg \, m^{-2}$ (Table 1). The beds with larger stock values are those of *G. robustum* (Gard.) Hollenberg & Abbott in Pacific Mexico (to $2.06 \, kg \, m^{-2}$). Most *Gelidium* fields, however, approach values of $0.15–0.2 \, kg \, m^{-2}$ recorded for *G. coulteri* Harvey from the Californian coast (Hansen, 1980). With the exception of *G. pusillum* from India, the commercial fields of *Gelidium* (Table 1) generally exhibit standing stock values

above $0.3 \, kg \, m^{-2}$ and plant lengths above 100 mm. This is probably related to the harvesting practices used. Most fields are harvested by hand picking of intertidal or subtidal plants, and only the Japanese have used instruments for harvesting *Gelidium* (Yamada, 1976).

Growth in species of *Gelidium* results from activity of an apical cell, and yield measurements after experimental harvesting are, therefore, quantitative expressions of the rates at which cortical cells differentiate and growth. The measurement of field-elongation rates was the most popular method used in the past to calculate growth. The results (Santelices, 1988) suggested homogeneous and slow elongation rates, $90–100 \, mm \, y^{-1}$, regardless of species or latitude. Stewart (1984) called attention to these rates, as compared with those of other types of algae, and suggested that this was an intrinsic feature, unlikely to be significantly increased by external physical factors.

Growth and yields in species of *Gelidium* are also influenced by branching patterns and by the field density of erect axes. Hence, more recent

Table 1. Standing stocks, yields and plant length values of 4 commercial species of *Gelidium.*

Species	Location	Standing stock kg m^{-2}(dry)	Plant length (mm)	Yield mm y^{-1}	Yield kg m^{-2}y^{-1}
G. amansii	Japan (Izu Isls.)	0.5–1.5 (12) (17)	Up to 200 (17)	10 (16)	0.1–0.6 (11) (17)
G. pristoides	South Africa (5)	0.3–0.45 (4)	Up to 200 (4)	54–133 (6)	0.3–1.1 (4) (5)
G. robustum	California, Baja California (8)	1.08–2.06	Up to 350 (7)	90–100 (1) (15)	0.5–1.0 (15)
G. sesquipedale	N. Spain	0.2–0.6 (9) (2) (3)	Up to 250 (14)	90–100 (13) (14)	0.45–0.6 (14)

(1) Barilotti & Silverthorne, 1972; (2) Borja, 1987; (3) Borja, 1988; (4) Carter & Anderson, 1985; (5) Carter & Anderson, 1986; (6) Carter & Simons, 1987; (7) Guzmán del Proó & de la Campa de Guzmán, 1969; (8) Guzmán del Proó & de la Campa de Guzmán, 1978; (9) Juanes & Fernández, 1988; (10) Kaliaperumal & Umamaheswara Rao, 1981; (11) Yamada, 1972; (12) Okasaki, 1971; (13) Seoane Camba, 1966; (14) Seoane Camba, 1969; (15) Silverthorne, 1977; (16) Suto, 1974; (17) Yamada, 1976.

studies have included biomass increments after experimental harvesting, together with field-elongation rates (Table 1). The results indicate maximum production of 1.1 kg m^{-2} y^{-1}, and interspecific differences in productivity are evident. On an annual basis, the fields of *G. robustum* and *G. pristoides* (Turn.) Kützing in South Africa are twice as productive as those of *G. amansii* (Lamour.) Lamouroux in Japan. In addition, results in Table 1 indicate no clear relationship between standing stocks and yields. For example, although the fields of *G. pristoides* have smaller standing stocks, compared with other commercial species, they exhibit fast regeneration capacities; by contrast, fields of *G. amansii* can have relatively high stocks with slow regrowth.

The seasonality exhibited by the regrowth process weakens the hypothesis that homogeneous and slow growth rates are intrinsic features of these taxa. In *G. pristoides* from South Africa, recovery of individual plants to control (unharvested) levels took 3–4 mo if harvested in spring-summer, and 4–5 mo, if harvested in winter (Anderson *et al.*, 1989). The growth rate of the sublittoral *G. robustum* from Baja California (Guzmán del Proó & De la Campa de Guzmán, 1969; Barilotti & Silverthorne, 1972) was maximal during summer (August–September). The

cover in populations of *G. amansii* from Japan started to increase during or after winter (Saito & Atobe, 1970) and reached their maxima before summer. Maximum production of the *G. latifolium* fields of northern Spain occurs during spring (Juanes & Fernández, 1988). Seasonal changes in field regrowth suggest that, in these species, growth rates and productivity are also regulated by changes in external physical factors; in fact, by culturing in optimal range for all environmental conditions, vegetative *G. coulteri* was maintained for over a year at experimental growth rates >10% day^{-1} (Macler & West, 1987) and *Gelidium latifolium* from Norway had maximum growth rates of 8.8% day^{-1} when incubated under nutrient-sufficient conditions and photon flux densities of 300 μmol m^{-2} s^{-1} (Frederiksen & Rueness, 1989).

Habitat characterization

An analysis of the ecological patterns of distribution of species of *Gelidium* in temperate and local latitudes indicates several common features that characterize abiotic environmental habitat parameters of productive *Gelidium* beds (Santelices, 1974, 1988 for specific data). The populations frequently occur in habitats with rapid water movement, as belts near the lower limits of spring tides, extending to variable extents into the intertidal and the subtidal zones. Some species have restricted vertical distributions, whereas others form monocultures which can extend down to 10- or 20-m depth. Frequently, species of *Gelidium* occur as successive belts along a vertical elevation (Fig. 3), suggesting close similarity among species with respect to environmental tolerance. Interspecific competition is likely to be important in these situations, as Montalva & Santelices (1981) have shown for *G. chilense* (Mont.) Santelices &

Montalva and *G. lingulatum* Kützing in central Chile.

Species of *Gelidium* seem to prefer specific types of substratum. Precipitous rocky surfaces, with steep slopes, are the most frequent attachment surfaces. Very few species (e.g. *G. pusillum*, *G. crinale*) are found on muddy or sandy bottoms. When living on rocks, most species occur on calcareous substrate, especially crustose coralline algae. Several hypotheses have been advanced (Santelices, 1974, 1988) to account for this association. Spores or fragments of *Gelidium* may germinate or attach more readily to calcareous sub-

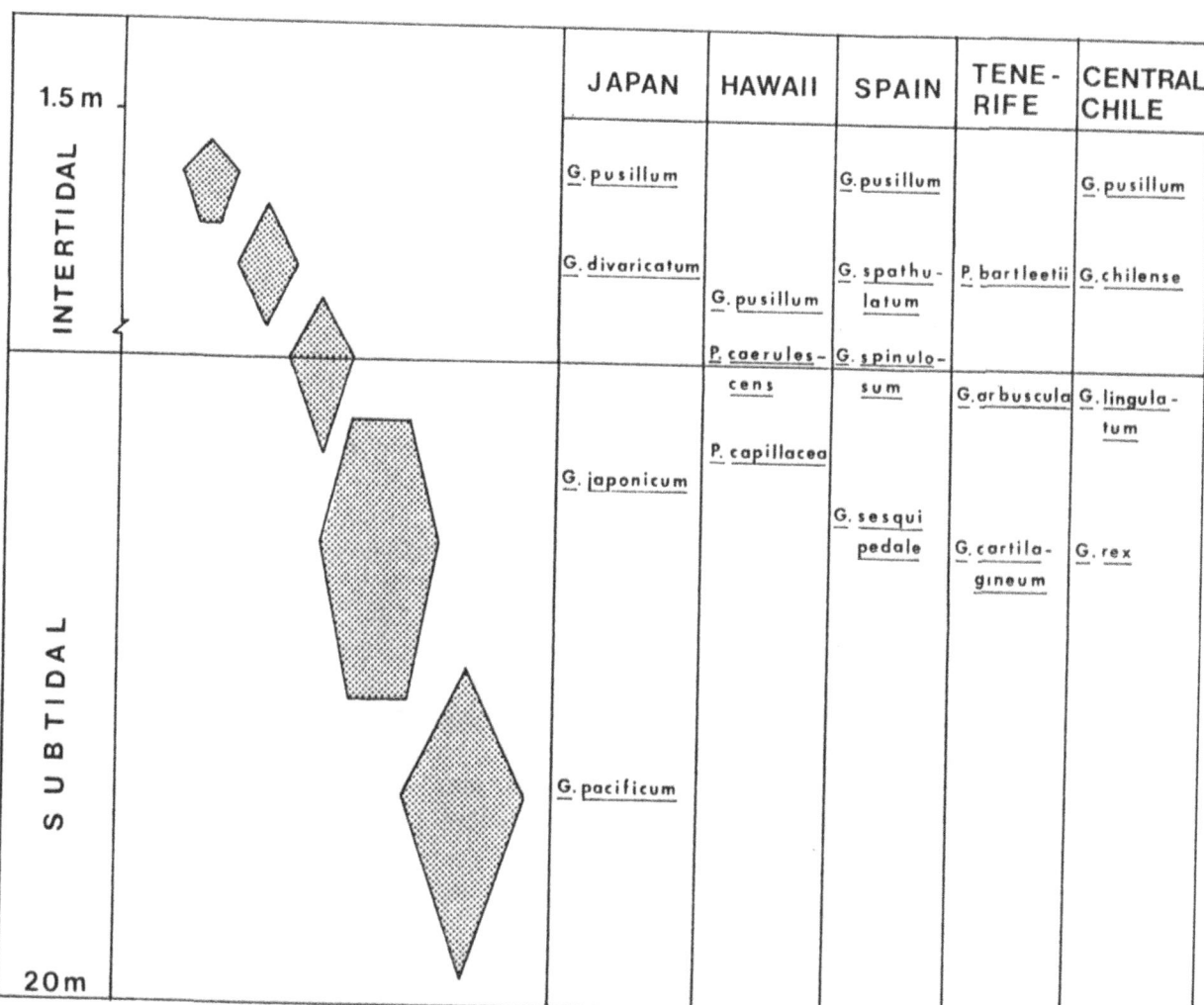

Fig. 3. Diagramatic representation of ecological distribution of species of *Gelidium* forming successive intertidal and subtidal belts in different latitudes.

strata, or crustose coralline algae may protect the creeping axes of the species of *Gelidium*. Dixon (1958) suggested that creeping axes of British intertidal populations of *Gelidium* could survive summer light only in rock crevices, below a cover of crustose coralline algae. Regeneration and growth of erect fronds from creeping axes are common in species of this genus, and vegetative propagation is a frequent method of colonization. The crustose algal cover could act as a light screen in some habitats or as a protection against grazers in others. Possibly, crustose-coralline algae and Gelidiaceae have a similar or closely-related optima for some ecological factors, such as water movement; therefore, both groups would tend to occur in the same habitat, perhaps competing for substrate and light.

Characteristically, species of *Gelidium* are dominant – or its individuals become larger – in shaded habitats receiving 25–50% of incident light (Santelices, 1974, 1988). Bleaching is common among summer populations in temperate latitudes, upper intertidal individuals during daytime low tides, and tropical populations exposed to full-light intensities.

Species of *Gelidium* are more productive in nutrient-rich waters, including upwelling (Guzmán del Proó & De la Campa de Guzmán, 1978), artificially-fertilized commercial crops (Yamada, 1972, 1976) or areas of domestic pollution (Littler & Murray, 1975; Hirose, 1979); however, these species are by no means restricted to those areas; they persist, although at low productivities in nutrient-poor waters, as in Japanese populations of *Gelidium amansii*, where the addition of artificial fertilizers to these beds in late summer increased the length, weight and nitrogen content of the thalli and changed their colour from yellowish to dark-purple. This indicates that, prior to fertilizer application, these populations had been living under nutrient-limited conditions (Yamada, 1976).

Field measurements of salinity tolerance have indicated that some species of *Gelidium* can occur under conditions of reduced (e.g. *G. pusillum*; Lawson, 1957; *G. corneum* (Huds.) Lamouroux; Conover, 1964) or increased (*G. crinale*; *G. cor-*

neum; Conover, 1964) salinity. Experimental laboratory studies (reviewed by Santelices, 1988; Macler, 1988) have shown that the range of salinity tolerance of various species of *Gelidium* normally exceeds salinity variations of marine waters; nevertheless, with the exception of *G. crinale* and *G. pusillum*, the species of *Gelidium* are normally absent from estuarine environments, perhaps owing more to reduced water movement than to reduced salinity.

Factors affecting productivity

I have distinguished three groups of factors affecting productivity of natural *Gelidium* beds: abiotic factors controlling growth, organismic determinants of production, and factors that remove the crop. Although factors in each group can interact with others in the same or in different groups, they are, for clarity, discussed separately.

Abiotic control of growth

Growth of species of *Gelidium*, as with other red algae, is regulated by complex interactions among irradiance, temperature, nutrients and water movement (Santelices, 1978). Light intensity and water movement can affect growth directly or through interactive effects. When one factor is limiting, the effects of other factors on growth cannot be fully expressed. Nutrients can compensate for inadequate water movement, with maximum growth rates, when water motion is comparatively low. In turn, an enhanced diffusion, resulting from high water movement or addition of nutrients, results in more effective uses of higher levels of irradiance and temperatures, resulting in faster growth and higher pigment concentrations.

Although several laboratory studies with species of *Gelidium* have been successful in defining physiological optima for growth, many such results are from single-factor experiments and must be interpreted with caution; a few general patterns emerge, however. Under field or laboratory conditions without nutrient enrichment, species of *Gelidium* are more light sensitive than other seaweeds with similar habitats; for example, *Gelidium caulacantheum* J. Agardh had a greater photosynthetic rate at 800 ft-c than at 2000 ft-c, the saturation level shown for other species with similar habitat (Chapman, 1966). The saturation intensity for *G. amansii* was below 5 klux, a low value compared with saturating values of about 21 klux for other intertidal species in the same locality (Ogata & Matsui, 1965). Bleaching of field thalli, followed by necrosis with exposure to high light intensities, has been commonly described in field observations of commercial crops (data in Santelices, 1988).

The situation changes, however, when nutrients, especially nitrate, are replete; for example, under N-sufficient conditions, *Gelidium latifolium* increased growth with increasing photon flux densities, to the maximum experimental value of $300 \, \mu mol \, m^{-2} \, s^{-1}$ (Frederiksen & Rueness, 1989). Significant differences in growth rates were recorded between the highest (300) and the lowest $(20 \, \mu mol \, m^{-2} \, s^{-1})$ photon flux densities, and pigmentation was affected by both photon flux densities and nitrogen availability. Phycobiliproteins and chlorophyll *a* decreased with increasing photon flux density, with the lowest pigment content in plants grown under N-deficient conditions. Phycobiliproteins are probably used as a nitrogen source when nitrogen becomes a limiting factor (Frederiksen & Rueness, 1989). High contents of phycobiliproteins may have the ability to absorb maximum light at low irradiances and to protect against photo-oxidation of chlorophyll *a* during periods of high irradiance.

The above findings explain some field situations, such as colour changes obtained by Yamada (1976) upon application of artificial fertilizers to wild crops of *Gelidium* beds at the end of the summer in Japan. These data also support early predictions (Santelices, 1978) of interacting effects of light intensity and nutrients. It should be noted, however, that this interaction could be even more complex, as Macler (1986) has demonstrated that the photosynthetic ^{14}C fixation by *G. coulteri* and the flow of photosynthate into various end products depend on the nitrogen sta-

tus of the tissue. Plants given luxury levels of nitrogen showed ^{14}C fixation rates several times that of plants starved for nitrogen.

Studies of temperature effects on species of *Gelidium* have often been related to seasonal growth exhibited by the species in temperate waters (Fig. 4). Thus, on reefs of Sagami province, Matsuura (1958) found that the increase in growth rate of four species of *Gelidium* was coincident with a rise in temperature from 13 to 20 °C. In Japan, the growth of *Gelidium amansii* correlates with temperature increments of 13 to 20 °C (Yamada, 1976). The species disappears or becomes reduced at temperatures close to 25 °C. In Baja California, the seasonal increase in growth rates of *G. robustum* was statistically related (Barilotti & Silverthorne, 1972) to increases in water temperature up to about 20 °C.

Laboratory studies have revealed, however, that the close correlation between increasing temperatures and growth occurs only up to temperatures of 15–20 °C. Beyond this level, most species so far studied exhibit a rather broad temperature optimum for production; for example, even though photosynthetic rates for *Gelidium amansii* increased with increasing temperatures up to 15 °C, no statistically significant differences were found between 15 and 30 °C (Yokohama, 1972). In *G. coulteri* from California, growth rates

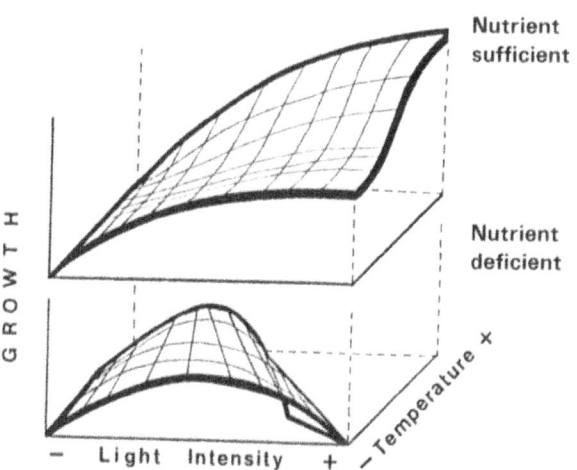

Fig. 4. Diagrammatic representation of the growth of *Gelidium* as a function of the interacting effects of light intensity, temperature and nutrients.

increased with increasing temperatures between 5 and 15 °C, although differences between 20 and 30 °C were not significant (Macler & West, 1987); likewise, while *G. latifolium* from Norway exhibited increasing growth rates between 10 and 15 °C (Rueness & Tanager, 1984), temperature increments from 14° to 28 °C had little effect (Frederikson & Rueness, 1989). These results point to the need for more critical analyses of the correlations between seasonal growth of species of *Gelidium* and temperatures, especially in temperate latitudes. Increasing temperatures would affect growth only during spring and summer. The decline in growth rates of most temperate species at the end of the summer is not likely the result of increased temperatures. Alternative explanations should be sought in increased solar radiation and the decreased nutrient concentrations usually occurring at that time of the year. Working with the related genus *Pterocladia*, Santelices (1978) found that under nitrogen-deficient conditions, the bleaching effects of light intensity were increased at high temperatures.

The relationship between water movement and growth of species of *Gelidium* has been generally described in field accounts of these species, although it has seldom been tested. The growth of 4 species of *Gelidium* cultured at various rotary speeds, yielded inconclusive results (Oliger & Santelices, 1981). As expected, the growth rate of some species (e.g. *G. chilense*) increased with increasing water movement up to a certain level, decreasing thereafter. Other species, however, either showed an exceptionally-broad range for growth (*G. lingulatum*) or were relatively unaffected by the water movement regimes (*G. rex* Santelices & Abbott). Besides increasing rates of diffusion, water movement may play other ecologically important roles in these species. Experimental field cultivation trials in central Chile (Santelices, 1987) have shown that whenever *G. rex* and *G. lingulatum* are moved into waters calmer than their respective habitats, they become covered by epiphytes. Similar problems have affected experimental field farms near Goleta, California, USA (Wheeler *et al.*, 1981).

Organismic determinant of production

Primary productivity of marine macrophytes is affected by the abiotic environment and by interpopulation and interspecific differences in some biological attributes (Littler & Arnold, 1980). Specific studies on the growth and production effects of these factors have not been completed for any species of *Gelidium*; however, the information available suggests that these differences might have overall importance.

Morphology: Most species of *Gelidium* have axes with variable degrees of branching. Some species exhibit erect axes entirely devoid of branches, while others display up to 3 or 4 orders of branches. Since erect axes and branches grow by activity of an apical cell, thalli growing under similar conditions, with many lateral branches, would increase their biomass faster than simple or sparingly branched plants (Barilotti, 1980). This explained, for example, the faster growth rates of *G. nudifrons* Gardner, compared with *G. robustum*, in experimental farms in California (Wheeler *et al.*, 1981), and the differences in yield per unit of initial stocking density between *G. lingulatum* and *G. rex* in central Chile (Santelices, 1987). Even though the branching pattern can be a specific character, it can be modified within a given species by diverse factors, including seasonal variations in light and temperature, length of emersion during low tides, reproductive periodicity and grazing (Santelices, 1988; Frederiksen & Rueness, 1989); thus, intraspecific and interspecific variations in production are expected.

Age of thallus: Mature plants are expected to have greater light requirements than sporeling stages, because of increased proportions of non-photosynthetic internal tissues. In agreement with this prediction, Correa *et al.* (1985) found the sporelings of *Gelidium lingulatum* and *G. chilense* saturated at $50 \mu E\,m^{-2}\,s^{-1}$ while mature thalli had maximum growth at $120 \mu E\,m^{-2}\,s^{-1}$ (Oliger & Santelices, 1981).

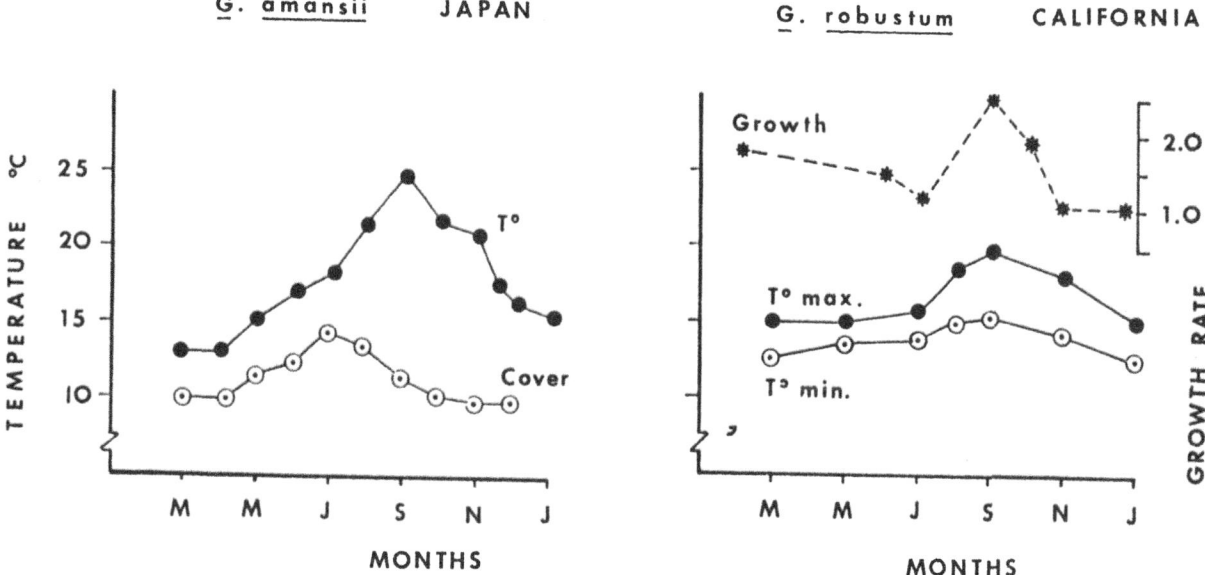

Fig. 5. Examples of correlations between seasonal growth and temperature in species of *Gelidium*.

Thallus parts: Working with the taxonomically related genus *Pterocladia*, Felicini (1970) described polarity in the light response of regenerating thalli. The adventitious buds that regenerated from the abapical portions of excised thalli always developed into creeping axes; by contrast, buds regenerated from the apical portion of excised thalli could develop either into creeping axes or erect fronds, depending on the light intensity. More recently, D'Antonio & Gibor (1985) showed that branching frequency and the type of branch produced were affected by both photoperiod and irradiance. When *G. robustum* was grown under long days (16 : 8), branch initiation was facilitated by higher light levels (85 μE m^{-2} s^{-1}), and under reduced light (35 μE m^{-2} s^{-1}), development of rhizoidal cell clusters was stimulated. These unpigmented rhizoidal cells, more than 500 μm in length, attach the thallus to the substratum. At photoperiods of 12 : 12 h, no differences resulting from high light regimes were observed. It is an important finding that long days together with reduced light induce the formation of rhizoidal cell clusters which enhances the ability of germlings to adhere to substrata. Even though D'Antonio & Gibor (1985) found this response in

germlings starting from spores, it suggests the possibility that fragments of adult fronds may re-attach when grown under reduced light intensity. Several species of *Gelidium* (reviewed by Santelices, 1988) undergo periodic hydrodynamic fragmentation owing to increased water movements, and many such fragments might re-attach in the field, becoming creeping axes under the appropriate light and photoperiodic conditions.

Reproductive state: Profound biochemical and physiological changes occur within reproductive thalli, prior to the onset of gamete or spore release (Littler & Arnold, 1980). Such studies have not been performed with species of *Gelidium*; however, our present understanding of nutritive cells and gonimoblast development in these species, and in red algae in general, imply energy dependence of reproductive tissues on more photosynthetically active thallus parts. As the prominent apical cell in fertile female branches lies in a notch overtopped on both sides by the more rapidly growing thallus margin (Hommersand & Fredericq, 1988), this is additional evidence of retarded growth in sexually reproducing thalli of *Gelidium*. Marked differences in productivity could be, therefore, expected among thalli of a given species at different reproductive stages growing in similar environments.

Life history phases: Species of *Gelidium* have a 'Polysiphonia-type' life history with alternation of isomorphic generations. In laboratory studies with *G. pristoides* (Carter, 1985) and *G. coulteri* (Macler & West, 1987), no significant differences were found in growth responses to environmental changes between the sporophytic and gametophytic generations of either species. Notwithstanding, the literature is rife with observations (reviewed by Santelices, 1988) of conspicuous field differences in proportions of tetrasporangial and sexual thalli; often tetrasporangial thalli are more frequent or more abundant, by several orders of magnitude. These data suggest increased sensitivity to environmental conditions and reduced competitive ability of the sexual, haploid phase. If this is true, significant differences in productivity might be expected in different beds, depending on the predominant life history phase.

Seasonality: Primary producers often show seasonal patterns of production that can be measured, at least over a short term, under constant incubation conditions. Seasonal patterns of production are usually constant for a given species and highly variable when a multitude of species is assessed. Some have their highest productivities in summer, while others are more productive in spring or winter (Brinkhuis, 1977; Littler *et al.*, 1979). Seasonal variation in regenerative capacity has been recorded for at least two species of *Gelidium* [(*G. pusillum* in Spain; Seoane-Camba (1965); *G. robustum* in Baja California; Barilotti & Silverthorne (1972)]. These variations might result from changes in the abiotic environment rather than from intrinsic seasonal differences in the regeneration capacities of these species. It should be noted, however, that Littler *et al.* (1979) found seasonal differences in photosynthesis and respiration rates of *G. pusillum* and *G. robustum*.

Crowding and density: Clumping of thalli may increase intraspecific competition for light and nutrients. Experimental studies with *Gelidium pusillum* indicated that clumped forms of this species produced (and respired) at only half the rate of separate individuals (Littler & Arnold, 1980). This phenomenon was attributed to competition for CO_2 and to self-shading by overlapping thalli, even when surrounded by relatively large volumes of water. Although effects of crowding on the growth species of *Gelidium* have not been demonstrated, Frederiksen & Rueness (1989) have anticipated somewhat similar effects in suspended cultures. Cultured plants of *Gelidium* kept in suspension by aeration may become bushy and ball-like. This change in morphology involves decreased growth owing to self-shading and increased competition for nutrients among the growing branches.

Geographic and ecological origin: Within their geographical ranges, physiological optima of *Gelidium* species for growth with respect to environmental conditions seems to be a genetically fixed character, closely related to geographic and ecological origin. Thus, tropical species of *Gelidium* have higher temperature optima for growth than temperate taxa; likewise, uppermost intertidal species have higher temperature and higher quantum dose optima for growth than lower intertidal species (Oliger & Santelices, 1981). It remains to be tested if these differences result in differences in growth rates and productivity. Such differences have been suggested in previous studies, although many of these experiments seem to have been performed under nutrient-deficient conditions; therefore, these results probably reflect differences in tolerance limits rather than differences in productivity.

Removal of the crop

Several environmental events such as extreme low tides, storminess and grazing by invertebrates and fishes, can remove or destroy *Gelidium* crops and modify the population structure of *Gelidium* beds

(Santelices, 1988). We lack quantitative data, however, on long-term ecological effects of these events on productivity and on associated communities. Comparatively more experimental data are available on productivity affected by harvesting practices, and how these may modify the regenerating capacity of the creeping axes. Spore-mediated field recruitment seems to be irregular or strictly seasonal in several species of *Gelidium* (Santelices, 1974, 1988; Jernakoff, 1986), and many of these beds are maintained principally by the regenerative capacity of the basal-most parts of the plants.

Studies on subtidal populations of *Gelidium sesquipedale* (Turn.) Thuret from Spain (Seoane-Camba, 1966) and *G. robustum* from Baja California (Guzmán del Proó & la Campa de Guzmán, 1969), and intertidal individuals of *G. pristoides* from South Africa (Carter & Anderson, 1985) have shown that scraping the substratum to remove *Gelidium* results in the destruction of creeping axes. Often the scraped areas are invaded by other algae, and it may take several years before the pre-experimental biomass values of *Gelidium* are attained. In two subtidal populations, cutting the weed with scissors (shearing) resulted in faster regrowth than the pulling of seaweed from the substratum by hand (hand picking); supposedly, hand picking damages the holdfasts or creeping axes. These results, however, are not be general: in the intertidal fields of *G. pristoides* from South Africa, the rate of regrowth of sheared tufts did not differ significantly from the rate of regrowth of hand-picked rufts (cf. Anderson *et al.*, this volume). Hand-picking apparently leaves intact some of the older and longer fronds which may protect the regenerative holdfast region against desiccation.

Conclusions

Recent information has contributed significantly to our understanding of the production ecology of species of *Gelidium*. Time-space stocks and yield distribution patterns together with laboratory experiments have falsified ideas of intrinsic limi-

tations to growth in this group of algae. Crop management practices should preserve the resources, leading to increased productivity; likewise, the possibility of cultivating the more productive species is suggested. The importance of environmental and intrinsic biological attributes in the regulation of production is stressed in the present review.

The combination of field and experimental laboratory data has been useful to identify complex interactions of factors as the most likely explanation for time-space patterns of distribution and growth of these species. Even though interactions are complex, they provide a unified explanation for a diversity of biological phenomenon, and this should be most useful in future studies on cultivation of *Gelidium*.

The report of thallus polarity, made over 20 years ago, together with more recent data on photoperiodic and light intensity effect on regeneration patterns are significant. These point to the unique adaptation in these species, i.e. to reattachment after frond fragmentation, suggesting that this is a normal and frequent process in many of the species. Such regenerative propagation is obviously important with respect to present interpretations of reproductive processes, life history patterns and dispersal of species of *Gelidium*.

The possibility of stimulating rhizoidal production to enhance reattachment also has significant economic implications. Slow growth rates of sporelings has been one of the limitations of field farming of *Gelidium* which are outcompeted by weedy seaweeds (e.g. *Enteromorpha, Ulva, Ectocarpus*). Rhizoidal reattachment should initiate repopulation and field cultivation practices with larger frond fragments, conferring competitive advantages to *Gelidium* species. At least one commercial venture seems to be developing based on this phenomenon (J. M. Salinas, 1991).

In spite of the advances, several significant questions relating production ecology of *Gelidium* remain unanswered; for example, we lack adequate explanations for the general absence of commercial crops of *Gelidium* in tropical waters, and for the general occurrence of these species epiphytes on calcareous crusts. We do not have a comprehensive understanding of the many roles played by water movement in the biology of these organisms or on the influence of organismic variability in their production patterns. The effects of biological interactions such as competition, grazing and epiphytism on the growth of these species and longer-term studies on the ecological effects of several factors that destroy the crops remain to be studied.

Acknowledgements

This study was supported by grants Fondecyt-Chile (Grant N° 803/1990), Dirección de Investigación, Pontificia Universidad Católica de Chile (Grants DIUC E-20 & 8903-E) and from Fundación Andes (Convenio C-51284). My gratitude to J. McLachlan for reviewing and commenting on the manuscript.

References

Anderson, R. J., R. Simons & N. G. Jarman, 1989. Commercial seaweeds in southern Africa: A review of utilization and research. S. Afr. J. mar. Sci. 8: 277–299.

Anderson, R. J., R. H. Simons, N. G. Jarman & G. J. Levitt, 1991. *Gelidium pristoides* in South Africa. Hydrobiologia 221: 55–66.

Akatsuka, I., 1986. Japanese Gelidiales (Rhodophyta), especially *Gelidium*. Oceanogr. Mar. Biol. annu. Rev. 24: 171–263.

Barilotti, D. C., 1980. Genetic considerations and experimental design of outplanting studies. In Abott I. A., M. S. Foster, L. F. Eklund (eds); Pacific Seaweed Aquaculture. California Sea Grant College Program, Institute of Marine Resources, University of California: 10–18.

Barilotti, D. C. & W. Silverthorne, 1972. A resource management study of *Gelidium robustum*. Proc. int. Seaweed Symp. 7: 255–261.

Borja, A., 1987. Cartografía, evaluación de la biomasa y arribazones del alga *Gelidium sesquipedale* (Clem.) Born. et Thur. en la costa guipuzcoana. Inv. Pesq. 51: 199–224.

Borja, A., 1988. Cartografía y evaluación de la biomasa del alga *Gelidium sesquipedale* (Clem.) Born. et Thur. 1876 en la costa vizcaína (N. de España). Inv. Pesq. 52: 85–107.

Brinkhuis, B. H., 1977. Seasonal variations in salt-marsh macroalgae photosynthesis. II. *Fucus vesiculosus* and *Ulva lactuca*. Mar. Biol. 44: 177–186.

Carter, A. R., 1985. Reproductive morphology and phenology, and culture studies of *Gelidium pristoides* (Rhodo-

phyta) from Port Alfred in South Africa. Bot. mar. 28(7): 303–311.

Carter, A. R. & R. J. Anderson, 1985. Regrowth after experimental harvesting of the agarophyte *Gelidium pristoides* (Gelidiales: Rhodophyta) in the eastern Cape Province. S. Afr. J. mar. Sci. 3: 111–118.

Carter, A. R. & R. J. Anderson, 1986. Seasonal growth and agar contents in *Gelidium pristoides* (Gelidiales, Rhodophyta) from Port Alfred, South Africa. Bot. mar. 29: 117–123.

Carter, A. R. & R. H. Simons, 1987. Regrowth and production capacity of *Gelidium pristoides* (Gelidiales, Rhodophyta) under various harvesting regimes at Port Alfred, South Africa. Bot. mar. 30: 227–231.

Conover, J. T., 1964. The ecology, seasonal periodicity, and distribution of benthic plants in some Texas lagoons. Bot. mar. 7: 4–41.

Chapman, V. J., 1966. The physiological ecology of some New Zealand seaweeds. Proc. int. Seaweed Symp. 5: 29–54.

Correa, J., M. Avila & B. Santelices, 1985. Effects of some environmental factors on growth of sporelings in two species of *Gelidium* (Rhodophyta). Aquaculture 44: 221–227.

D'Antonio, C. M. & A. Gibor, 1985. A note on some influences of photon flux density on the morphology of germlings of *Gelidium robustum* (Gelidiales, Rhodophyta) in culture. Bot. mar. 28: 313–316.

Dawson, E. Y., 1952. Marine red algae of Pacific Mexico. I. Bangiales to Corallinaceae Subf. Corallinoideae. Allan Hancock Pac. Exp. 17: 1–239.

Dawson, E. Y., 1954. Marine plants in the vicinity of the Institut Oceanographique de Nha Trang, Viet Nam. Pac. Sci. 8: 372–469.

Dawson, E. Y., M. Neushul & R. D. Wildman, 1960. Seaweeds associated with kelp beds along southern California and north-western Mexico. Pac. Nat. 1: 1–81.

Dixon, P. S., 1958. The structure and development of the thallus in the British species of *Gelidium* and *Pterocladia*. Ann. Bot., Lond., N.S. 22: 353–368.

Felicini, G. P., 1970. Research on regeneration in *Pterocladia capillacea* cultured *in vitro*. II. Light intensity and development of the erect frond. G. Bot. Ital. 104: 35–47.

Frederiksen, S. & J. Rueness, 1989. Culture studies of *Gelidium latifolium* (Grev.) Born. et Thur. (Rhodophyta) from Norway. Growth and nitrogen storage in response to varying photon flux density, temperature and nitrogen availability. Bot. mar. 32: 539–546.

Guzmán del Proó, S. A. & S. de la Campa de Guzmán, 1969. Investigaciones sobre *Gelidium cartilagineum* en la costa occidental de Baja California, Mexico. Proc. int. Seaweed Symp. 6: 179–186.

Guzmán del Proó, S. A. & S. de la Campa de Guzmán, 1978. *Gelidium robustum* (Florideophyceae), an agarophyte of Baja California, Mexico. Proc. int. Seaweed Symp. 9: 303–308.

Hansen, J. E., 1980. Physiological considerations in the mariculture of red algae. In Abbott I. A., M. Foster & L. Eklund (eds), Pacific Seaweed Aquaculture, California Sea Grant Colege Program, University of California at San Diego: 80–92.

Hariot, P., 1889. Algues. Mission scientifique de Cap Horn, 1882–1883. Botanique, Paris 5: 3–109, pls. 1–9.

Hirose, H., 1979. Composition of benthic marine algae in relation to pollution in the Seto Island Sea, Japan. Proc. 9th Int. Seaweed Symp. 173–179.

Hommersand, M. & S. Fredericq, 1988. An investigation of cystocarp development in *Gelidium pteridifolium* with a revised description of the Gelidiales (Rhodophyta). Phycologia 27: 254–272.

Juanes, J. A. & C. Fernández, 1988. Ciclo anual y producción de *Gelidium latifolium* (Grev.) Thur. et Born. (1876), en la región de Cabo Peñas (Asturias, N. de España). Inv. Pesq. 52: 109–122.

Jernakoff, P., 1986. Experimental investigation of interactions between the perennial red algae *Gelidium pusillum* and barnacles on a New South Wales rocky shore. Mar. Ecol. Prog. Ser. 28: 259–263.

Kaliaperumal, N. & M. U. Rao, 1981. Studies on the standing crop and phycocolloid of *Gelidium pusillum* and *Pterocladia heteroplastos*. Indian J. Bot. 4(2): 91–95.

Lawson, G. W., 1957. Some features of the intertidal ecology of Sierra Leone. J. West Afr. Sci. Assoc. 3: 166–174.

Littler, M. M. & K. E. Arnold, 1980. Sources of variability in macro algal primary productivity. Sampling and interpretative problems. Aquat. Bot. 8(2): 141–156.

Littler, M. M. & S. N. Murray, 1975. Impact of sewage on the distribution, abundance and community structure of rocky intertidal macroorganisms. Mar. Biol. (Berl.) 30(4): 277–292.

Littler, M. M., S. N. Murray & K. E. Arnold, 1979. Seasonal variations in net photosynthetic performance and cover of intertidal macrophytes. Aquat. Bot. 7(1): 35–46.

Macler, B. A., 1986. Regulation of carbon flow by nitrogen and light in the red alga *Gelidium coulteri*. Plant Physiol. 82: 136–141.

Macler, B. A., 1988. Salinity effects on photosynthesis, carbon allocation, and nitrogen assimilation in the red alga, *Gelidium coulteri*. Plant Physiol. 88: 690–694.

Macler, B. A. & J. West, 1987. Life history and physiology of the red alga, *Gelidium coulteri*, in unialgal culture. Aquaculture 61: 281–293.

Matsuura, S., 1958. Observations in the annual growth cycle of marine algae on a reef at Manadzuru on the Pacific Coast of Japan. Bot. Mag. (Tokyo) 71: 93–109.

Montalva, S. & B. Santelices, 1981. Interspecific interference among species of *Gelidium* from Central Chile. J. exp. mar. Biol. Ecol. 53(1): 77–88.

Ogata, E. & T. Matsui, 1965. Photosynthesis in several marine plants of Japan as affected by salinity, drying and pH with attention to their growth habitats. Bot. mar. 8: 199–217.

Okasaki, A., 1971. Seaweeds and their uses in Japan. Tokai University Press, Tokyo, Japan, 165 pp.

Oliger P., Santelices B. (1981). Physiological ecology studies on Chilean Gelidiales. J. exp. mar. Biol. Ecol. 53(1): 65–76.

Rueness, J. & S. Frederiksen, 1989. Field and culture studies of *Gelidium latifolium* (Grev.) Born. & Thur. (Rhodophyta) from Norway. Sarsia 74: 177–185.

Rueness, J. & T. Tanager, 1984. Growth in culture of four red algae from Norway with potential for mariculture. Hydrobiologia 116/117: 303–307.

Saito, I. & S. Atobe, 1970. Phytosociological study of the intertidal marine algae. I. Usujiri Benten-Jima, Hokkaido. Bull. Fac. Fish. Hokkaido Univ. 21: 37–69.

Salinas, J. M., 1991. Spray system for re-attachment of *Gelidium sesquipedale* (Clen.) Born. et Thur. (Gelidiales: Rhodophyta). Hydrobiologia 221: 107–117.

Santelices, B., 1974. Gelidioid algae, a brief resumé of the pertinent literature. Tech. Rep. Mar. Agron. Sea Grant Program, Hawaii 1: 1–111.

Santelices, B., 1977. A taxonomic review of Hawaiian Gelidiales (Rhodophyta). Pac. Sci. 31: 61–84.

Santelices, B., 1978. Multiple interaction of factors in the distribution of some Hawaiian Gelidiales (Rhodophyta). Pac. Sci. 32: 119–147.

Santelices, B., 1986. The wild harvest and culture of the economically important species of *Gelidium* in Chile. In Doty M. S., J. F. Caddy & B. Santelices (eds), Case studies of seven commercial seaweed resources. FAO Fisheries Technical Paper 281, pp. 165–192.

Santelices, B., 1987. Métodos alternativos para la propagación y el cultivo de *Gelidium* en Chile Central. In Verreth J. A. J., M. Carillo, S. Zanuy & E. A. Huisman (eds), Investigación acuícola en América Latina. PUDOC Wageningen, pp. 349–366.

Santelices, B., 1988. Synopsis of biological data on the seaweed genera *Gelidium* and *Pterocladia* (Rhodophyta). FAO Fisheries Synopsis 145: 1–55.

Seoane-Camba, J., 1965. Estudios sobre las algas bentónicas en la costa sur de la Península Ibérica. Inv. Pesq. 29: 3–216.

Seoane-Camba, J., 1966. Algunos datos de interés en la recolección de *Gelidium sesquipedale*. Publ. Téc. Junta Estud. Pesca, Madrid 5: 437–455.

Seoane-Camba, J., 1969. Crecimiento, producción, desprendimiento de biomasa en *Gelidium sesquipedale* (Clem.) Thuret. Proc. int. Seaweed Symp. 6: 365–374.

Silverthorne, W., 1977. Optimal production from a seaweed resource. Bot. mar. 20: 75–98.

Soegiarto, A., 1978. Indonesian seaweed resources: Their utilization and management. Proc. int. Seaweed Symp. 9: 463–471.

Stewart, J., 1984. Vegetative growth-rates of *Pterocladia capillacea* (Gelidiaceae, Rhodophyta). Bot. mar. 27: 85–94.

Suto, S., 1970. Mariculture of seaweeds and its problems in Japan. NOAA Tech. Rep. NMFS Circ. (388): 7–16.

Wheeler, W. N., M. Neushul & B. W. W. Harger, 1981. Development of a coastal marine farm and its associated problems. Proc. int. Seaweed Symp. 10: 631–636.

Yamada, N., 1972. Manuring for *Gelidium*. Proc. int. Seaweed Symp. 7: 384–390.

Yamada, N., 1976. Current status and future prospects for harvesting and resource management of the agarophyte in Japan. J. Fish. Res. Bd Can. 33: 1024–1030.

Yokohama, Y., 1972. Photosynthesis-temperature relationships in several benthic marine algae. Proc. int. Seaweed Symp. 7: 286–291.

Hydrobiologia **221**: 45–54, 1991.
J. A. Juanes, B. Santelices & J. L. McLachlan (eds), International Workshop on Gelidium.
© 1991 *Kluwer Academic Publishers.*

Biological criteria for the exploitation of the commercially important species of *Gelidium* in Spain

José A. Juanes* & Angel Borja**
*Universidad de Cantabria; Departamento de Ciencias y Técnicas del Agua y Medio Ambiente; Avenide de los Castros s/n; 39005 Santander; Spain **AZTI-SIO; Avenida Satrústegui, 8; 20008 San Sebastián; Spain

Key words: casts, *Gelidium*, harvesting methods, management, production, Spain

Abstract

Exploitation of the commercially-important species of *Gelidium* in Spain (*G. sesquipedale* and *G. latifolium*) and development of the industry occurred after World War II, as a consequence of their use as resources for the extraction of agar. This resulted in the implementation of several harvesting methods, the most important of which is the gathering of cast seaweeds, both from the shore and the sea. From the very beginning, direct exploitation of these species (*i.e.* plucking) was controversial because of possible adverse ecological effects. Consequently, several biological and ecological studies of both species of *Gelidium* were begun. This included such aspects as growth, biomass production, productivity, reproduction, regeneration capacity and agar yield. Recently, the growing interest in using and conserving this resource has led to increased knowledge of the biology and ecology of these species. In this paper, we provide an overview of the research carried out by different Spanish groups and suggest guidelines for rational management of these resources.

Introduction

The use of *Gelidium* in Spain in agriculture (*i.e.* as fertilizer) is an old tradition, whereas the development of industrial exploitation dates back only to World War II (Cabrero Gomez, 1951, *in* Establier, 1964), when two of the species of *Gelidium* in Spain, *G. sesquipedale* (Clem.) Born. et Thur. and *G. latifolium* (Grev.) Thur. et Born. (*sensu* Dixon & Irvine, 1977), were used in the extraction of agar.

The exploitation of seaweeds was first regulated in Spain in 1945, when the initial agar industries were set up. From then on, intensification of harvesting progressively developed on the north Atlantic coast, between Cape Peñas (*ca.* 6° W) and the French border (*ca.* 1° 47′ W), an area where both species are most abundant (Alvarez de Meneses, 1972; Borja, 1987a; 1988; Fernández & Anadón, 1989) (Fig. 1).

These developments led to increased exploitation of cast seaweed, and to improvements in methods for direct exploitation of the beds (*i.e.* plucking, cutting machines). The possible destructive effects of these methods on the resources resulted in controversy and led to the first studies on *Gelidium* (Seoane-Camba, 1966, 1969; Cendrero & Ramos, 1967).

As the basic aspects of the dynamics of this resource were unknown, it was difficult to devise an acceptable exploitation program, and a conservative approach was, therefore, maintained in order to preserve the commercial beds from possible over-exploitation.

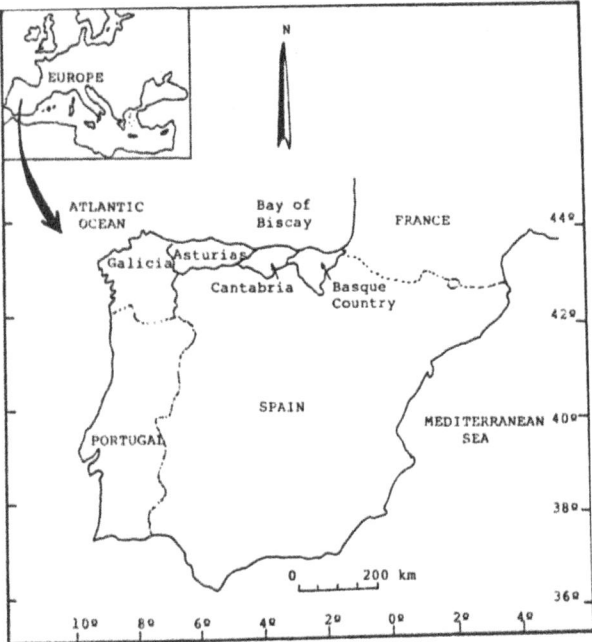

Fig. 1. Map of the Iberian Peninsula showing the location of the Spanish regions (Asturias, Basque Country, Cantabria and Galicia) involved in commercial harvesting of species of *Gelidium*.

Lately, increasing demands to better use and maintain this resource have resulted in an increased knowledge of the biology and ecology of the commercial species of *Gelidium*. This has increased the possibility of establishing rational programs for exploitation (Santelices, 1989).

In this work some of the biological aspects are analyzed and suggested guidelines for managing the resource are discussed.

Biology and ecology

The information used in this work for the formulation of a harvesting strategy for *Gelidium* covers five different aspects: growth (elongation), biomass production, reproduction and recruitment, agar yield, and regeneration capacity.

Growth

Growth, estimated as the increase in length of individual fronds, is strongly seasonal, with the highest rates of elongation occurring from March to October, and mainly during summer (June–September) (Seoane-Camba, 1966; Cendrero & Ramos, 1967; Juanes, 1983; Gorostiaga, 1990). The average growth per year of *G. sesquipedale* varies between *ca.* 7 and 9 cm in length. These values are higher than those registered from *G. latifolium* (Juanes, 1983).

Regardless of the branching pattern, these rates suggest slow rates of growth for these seaweeds; but Borja (unpubl. data) observed that individuals of *G. sesquipedale* remaining on the substratum after experimental harvesting (plucking), and seldom exceeded 18–20 cm, attained lengths of 35–40 cm within the following year. This suggests that growth of this species may not be as slow as previously considered.

Biomass and production

The annual cycle of biomass for these species (Anadón & Fernández, 1986; Juanes & Fernández, 1988; Gorostiaga, 1990) shows maximum values in summer (July–August) and minimum values in winter (March) (Fig. 2A). Mean biomass values recorded for different locales are summarized in Table 1; it must be stressed, however, that, at depths between 0 m and 5 m, average summer biomass values may be higher than 1 kg dry wt m^{-2} and exceeding 2 kg m^{-2} in some areas facing north and northwest (A. Borja, unpubl. data).

These areas are exposed with high water movement, and upwelling occurs off the west coast of the capes because of continuous easterly and northeasterly winds. Even though data are lacking, Borja (1988) suggested that high nutrient availability, resulting from the upwelling, favoured increased productivity of the *Gelidium* beds. In addition, high productivity has been reported during autumn (October–November) for both *G. latifolium* (Juanes & Fernández, 1988) and *G. sesquipedale* beds (A. Borja, unpubl. data), and during winter (January) in populations of *G. latifolium* (Juanes & Fernández, 1988).

The principal biomass losses occur during summer (August) and early winter (December) from intertidal populations of *G. latifolium* and mainly in late summer and autumn from *G. sesquipedale*

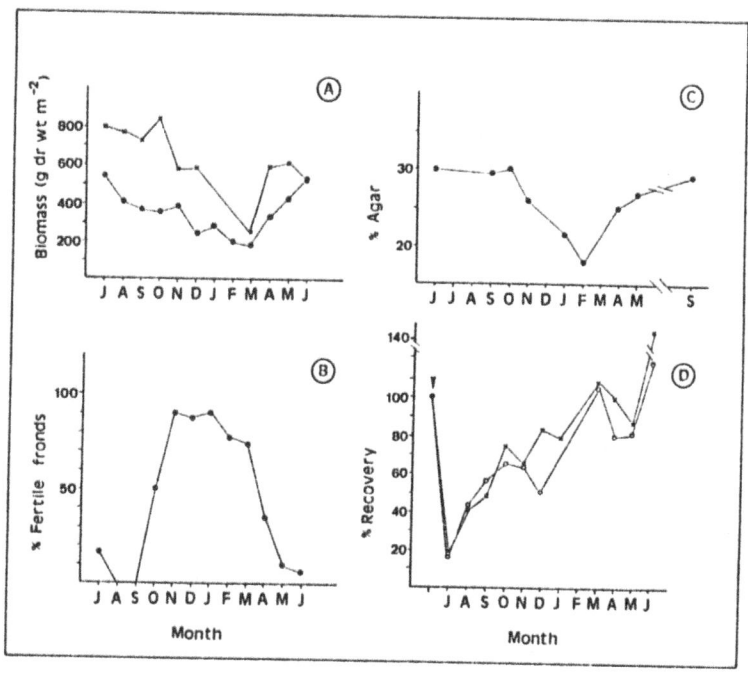

Fig. 2. A: Annual variation of biomass (g dry wt m⁻²) for *G. latifolium* (solid points), from an intertidal zone in Asturias (Juanes & Fernández, 1988), and *G. sesquipedale* (crosses) at 4-m depth in the Basque Country (A. Borja, unpubl.). B: Annual variation in percentage of fertile tetrasporophytes of *G. latifolium* from an intertidal bed in Asturias (Juanes, 1983). C: Annual variation in the agar yield of *G. sesquipedale* at 6-m depth from the Basque Country (García, 1988). D: Percentage biomass recovery of plucked populations at 4-m (circles) and 8-m (crosses) depths, relative to the biomass recorded monthly in control plots: arrow indicates harvest time (A. Borja, unpubl.).

beds (Gorostiaga, 1990). It is estimated that between 60% and 80% of the biomass produced in lost annually (Table 1).

The importance of cast seaweed, as the main source of collected biomass, has led to investigations on causes of natural detachment. Possible causes include: natural defoliation of the fronds (Seoane-Camba, 1966), the effects of grazers

Table 1. Summer and winter mean biomass (g dry wt m⁻²), net production (g dry wt m⁻² y⁻¹) and percentage of losses for different species of *Gelidium* on the Spanish coasts (A: average for the entire province; L: mean value corresponding to a limited area; I: intertidal, S: subtidal).

Specie	Area	Data	Mean biomass		Net production	% Losses	Author & Year
			Summer	Winter			
G. sesquipedale	Galicia	L,I				70	Seoane-Camba, 1969
	Asturias	L,I	355	1	509		Anadón & Fernández, 1986
	Guipúzcoa	A,S	592	180		73	Borja, 1987a
	Vizcaya	A,S	402				Borja, 1988
	Vizcaya	L,S				50–56	Gorostiaga, 1990
G. latifolium	Asturias	L,I	364	87	419		Anadón & Fernández, 1986
	Asturias	L,I	712	306	423		Anadón & Fernández, 1986
	Asturias	L,I	601	184	523		Anadón & Fernández, 1986
	Asturias	L,I	546–634	182–150	450–590	67–76	Juanes & Fernández, 1988
	Asturias	A,I				81	Fernández & Anadón, 1989

(Salinas *et al.*, 1976; Reguera *et al.*, 1978) and the incidence of storms (Juanes, 1983).

In considering effects of storms, Borja (1986, 1987a) reported that detachment is greater in shallower areas (81% between 0 and 5 m, and 64% between 10 and 15 m) and in those beds facing northwest; accordingly, the greatest quantities of cast seaweeds have been gathered after northwesterly gales (Borja, 1987b). Borja (1987a) has thus suggested that storms are an important factor in detachment.

In both species, the larger fronds show the greatest biomass increases, as well as the highest percentage of losses, (Juanes & Fernández, 1988; Gorostiaga, 1990). Tufts of the seaweed, which are most developed and bear greater epiphyte loads in summer, may offer more resistance to waves and thus increase the probability of detachment of either tufts or whole fronds from the rhizoids during autumn gales. In spring, although gales occur, detachment is not nearly as important, presumably because the fronds are less branched (A. Borja, unpubl. data).

Reproduction and recruitment

Theoretically, the life-history of *Gelidium* species corresponds to the '*Polysiphonia* type'. In the case of *G. sesquipedale*, this life-history is indirectly confirmed by the occurrences of fertile sporophytic and cystocarpic fronds, both producing viable spores (Alvarez *et al.*, 1978); in contrast, gametophytes of *G. latifolium* have not been found, whereas many fertile sporophytes were reported on a surface that had been devoid (scraped and burnt) of vegetation a year previously (Juanes, 1983). This may indicate either a possible modification in the theoretical life-history (production of mitotic tetraspores) or a capacity for survival of portions of rhizoidal filaments within the organic substratum that had been cleaned, and subsequent regeneration.

The highest proportion of fertile tetrasporophytic and cystocarpic fronds of *G. sesquipedale* (Gorostiaga, 1990) and tetrasporophytic ones of *G. latifolium* (Fig. 2B) are recorded in autumn and winter (October–April), although these may appear the year round (Juanes, 1983). In the case of *G. sesquipedale*, the sporophytic generation predominates, and is always greater than 9 : 1 (Gorostiaga, 1990).

There are no data for percentages of settling and survival of spores in the environment. Laboratory experiments on survival of sporelings obtained from tetraspores and carpospore of *G. sesquipedale* have fallen outside the autumnal-winter period (J. Juanes, unpubl. data).

The capacity for vegetative propagation is an important aspect in reproduction of *Gelidium*. The capacity for differentation of rhizoidal filaments from frond apical segments when in contact with the substratum is well-known. The formation of erect fronds from these filaments can contribute significantly to the recruitment of individuals of these species (Juanes, 1983; Gorostiaga, 1990). Gorostiaga (1990) indicated that there is a balance between recruitment and death of new individuals in *G. sesquipedale* populations. Both recruitment and death was estimated to be about 15%, resulting in a high degree of population stability. Production of erect fronds from the rhizoidal filaments appears to be the main mechanism for recruitment in *G. sesquipedale* (Gorostiaga, 1990); however, this must be experimentally assessed and compared to recruitment through spores.

Regenerative capacity

The capacity for regeneration of exploited fields of *Gelidium* depends on several factors, specifically on the harvesting method that is employed (*i.e.* methodology, intensity, time of the year, frequency) and on reproductive seasonality.

Different harvesting experiments were conducted on commercial populations of *Gelidium*. Although differences in the experimental procedures (*e.g.* month, area, intensity) make the comparison of results difficult, some general patterns can be noted. Where surfaces between 0.5 and 1 m² were cleared of biomass by scraping in summer, less than 10% of the biomass of *G. sesquipedale* populations reappeared a year later (Seoane-Camba, 1966); however, *G. latifo-*

lium (Juanes, 1983) and *Pterocladia capillacea* (Gmel.) Born. & Thur. (Seoane-Camba, 1966), recovered completely within the same time, regardless of harvesting period (winter or summer).

Similar plucking and cutting experiments started in summer (July and late September, respectively) showed that the biomass of *G. sesquipedale* reached its original level between one (Seoane-Camba, 1966) and two years (Gorostiaga, 1990) after being harvested. In the latter study, a more rapid recovery of biomass (% increase relative to original biomass) was observed for populations that had been cut 8 cm from the base of the thallus (153%), whereas no differences were recorded between intensively-plucked (115%) and the intensively-cut (2 cm from base: 110%) populations (Gorostiaga, 1990).

Cendreo & Ramos' (1967) plucking studies (during summer) on surfaces of 100 m² suggested that recovery to original biomass values in harvested populations of *G. sequipedale* took one year. Ongoing experimental harvesting in a 1-ha farm (A. Borja, unpubl.) indicates that the population plucked at the beginning of summer (July) recovers around March of the following year (*i.e.* 9 mo) to the level of biomass in the control surfaces , with similar results at depths of 4 and 8 m (Fig. 2D).

Agar content

Establier (1964) and García (1988) reported that the highest yields of agar from *G. sesquipedale* occur between May and July (Fig. 2C), whereas the greatest gel strengths (1600–1700 g cm^{-2}) are recorded from July to September.

Productivity and standing stock values

Until the 1980's, standing stock evaluations of *Gelidium* in Spain were limited to the studies of Seoane-Camba (1965) and Alvarez de Meneses (1972). More recently, applying Niell's (1983) methods of evaluation to larger areas, a survey of standing stocks of species of *Gelidium* has been partially completed in some areas on the north coast of Spain (Borja, 1987a, 1988; Fernández & Anadón, 1989; Gorostiaga, 1990) (Table 2).

Nowadays, biomass evaluations are being carried out in the two regions with the highest production of cast seaweeds: Asturias (E.M. Llera, *in verb.*) and Cantabria (Juanes & Gutiérrez, unpubl. data). With the incomplete information available at present, it is consequently very difficult to estimate standing stocks of the resources of *Gelidium* on the Spanish coasts.

The official quantitative harvest values (Anonymous, 1974–1987) range between 2000 and 6000 t dry-wt of cast seaweeds and between

Table 2. Standing stock values for different species of *Gelidium* on the Spanish coasts (A: data corresponding to the entire province; L: data from a limited area; I: intertidal; S: subtidal).

Area	Fresh weight (t)	Species	Data	Author & Year
Cádiz	44	*G. sesquipedale*	A, I	Seoane-Camba, 1965
Cádiz	7	*G. spinulosum*	A, I	Seoane-Camba, 1965
Cádiz	1	*G. attenuatum*	A, I	Seoane-Camba, 1965
Cantabria	2568	*G. sesquipedale*	L, S	Alvarez de Meneses, 1972
Guipúzcoa	14000	*G. sesquipedale*	A, S	Niell, 1983
Guipúzcoa	13943	*G. sesquipedale*	A, S	Borja, 1987a
Vizcaya	5068	*G. sesquipedale*	A, S	Borja, 1988
Asturias	1	*G. sesquipedale*	A, I	Fernández & Anadón, 1989
Asturias	253	*G. latifolium*	A, I	Fernández & Anadón, 1989
Vizcaya	400	*G. sesquipedale*	L, S	Gorostiaga, 1990
Cantabria	6290	*G. sesquipedale*	L, S	Juanes & Gutiérrez (unpubl. data)
Asturias	6074	*G. sesquipedale*	L, S	E.M. Llera (pers. comm.)

200 and 750 t of harvested seaweeds ('Programa de Orientacion Plurianual sobre Macroalgas', 1989. Unpubl.). It is estimated that only 18 to 35% of the detached seaweed is recovered (Borja, 1986, 1987a). The remaining biomass is lost along the inaccessible coastal areas, in non-exploited undersea wells or in deep water, indicated by harvesting seaweeds using trawls.

Methods and equipment used for harvesting

Gelidium resources are exploited by gathering cast seaweeds and direct harvesting of the beds (Fig. 3). In the former cases, diverse methods are employed, although the most common is manual dragging (Fig. 3A) of nets operated by one or several people; in rocky areas, cranes hoist the seaweeds, contained in baskets, up onto the shore. On the north coast, there are about 1200 gathering licenses, with an average daily harvest of 400 kg fresh-wt per person.

Dragging, using ships (Fig. 3B), is done exclusively in relatively shallow (<20 m) bays and inlets with sandy bottoms. The trawls consist of a metal rectangle 5 m long by 1 m high attached to a strong net. Sonar is used to detect masses of seaweeds, and about 60 ships harvest in the Basque country, each obtaining 1.5–2 t fresh-wt per day.

In order to exploit rocky bottoms with depressions where detached seaweeds collect (Fig. 3C), suction pumps are required. The ship is positioned over a depression and the suction hose is

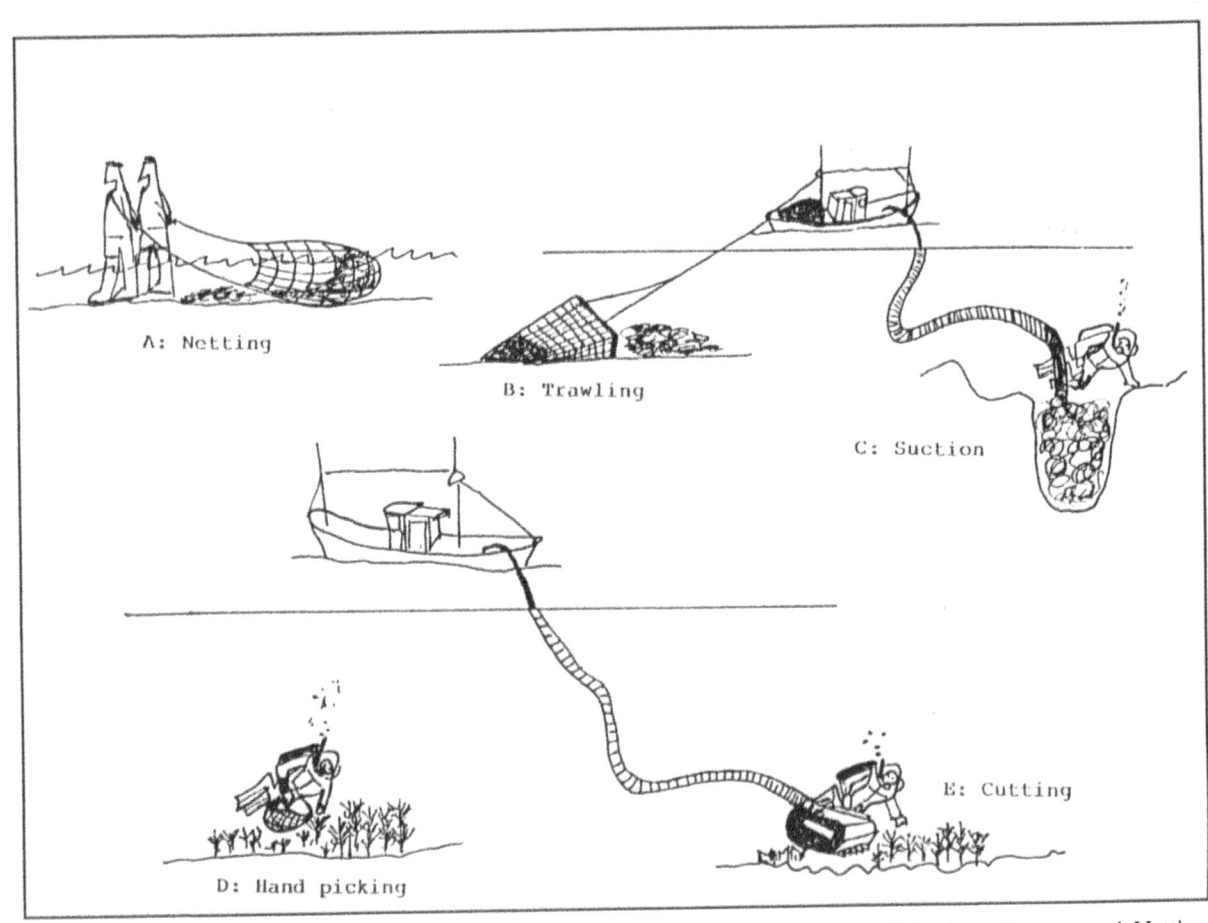

Fig. 3. Illustrations of harvesting methods used along the Spanish coasts for the exploitation of *Gelidium*. Cast seaweed: Netting (A), Trawling (B), Suction (C). Direct harvest: Hand plucking (D) Cutting (E).

manœuvred by divers. The number of ships employing this method is very limited, and each ship may harvest up to 15 t fresh-wt per day.

There are two types of direct harvesting: plucking and cutting. The former is more common and is used in the intertidal zone, or by divers in the subtidal zone (Fig. 3D). In the 1970's, 500 divers and about 100 ships were employed in plucking, with daily average harvests of 1 t fresh-wt per diver. The biomass gathered in this way represented, at that time, between 15 and 20% of the total harvested.

Cutting is a more complex and expensive task. Different devices have been developed which generally consist of blades operated electrically or hydraulically to increase the rate of harvesting. The cutting height is regulated and there is a suction system that draws the cut seaweeds up into the ship (Fig. 3E). Although these machines have not been used commercially, it is estimated that each can cut about 4 t fresh-wt of seaweeds per day.

Resource management

Traditionally, resource management in Spain has been concerned with controlling the direct exploitaion of *Gelidium*, through establishing quotas and setting seasons and areas to be harvested and in issuing licences for specific technologies (plucking, cutting). The lack of knowledge of the potential and dynamics of this resource has resulted in a conservative management program, with relatively small quotas being set and recent prohibition of harvesting by handpicking in most regions, until the effects of this method can be evaluated.

On the other hand, managers have promoted the design and development of industrial machines for cutting, some of which were licensed in 1989 (A. Vizcaino, *in verb.*), even without experimental assessment of the ecological impacts of commercial cutting.

Experiments on commercial plucking that are being currently developed (A. Borja, unpubl.) confirm the results of Gorostiaga (1990) on small

experimental surfaces (Fig. 4A). Commercial plucing results in changes in biomass (at 4 m depth from about 3200 to 550 g fresh-wt; at 8 m from about 2000 to 350 g fresh-wt), but not in density (A. Borja, unpubl. data). Reduction in biomass results from the harvesting of the longest individuals, in which most of the biomass is concentrated (Fig. 4B), while conserving a high percentage of small fronds (Fig. 4A) and maintaining the rhizoidal structure of the plants.

The few experiments carried out by Gorostiaga (1990) suggest that harvesting by cutting promotes multiple regrowth of the cut apices which results in increased production, while possible damage to the substratum is limited.

Discussion and conclusions

It is now possible to plan exploitation of *Gelidium* resources based on knowledge of the biology and ecology of these species. It will be necessary, however, to focus on and to promote further research to address questions that have been raised herein.

Assessment of standing stocks
Stock assessment of *Gelidium* along the north coast of Spain is an important objective, and evaluations will provide essential information on the potential of this resource. Assessment will provide a quantification of the resource of cast seaweed and the efficiency of the harvesting process. This information can be used to determine direct harvesting quotas for each area, according to the production rates, while precise techniques to estimate those rates will be required.

Harvesting cast seaweeds
As a high percentage of naturally-detached seaweed is not recovered by present harvesting protocols, Scoane-Camba's proposal (1966) should be followed, whereby the application of effective means of locating those seaweeds is employed. This may include for example, side-scan sonar (Borja, 1986; Salinas, 1986) or colour-sonar systems, already experimentally tested in *Gelidium* beds (Sanz & Rez, 1983). At the same

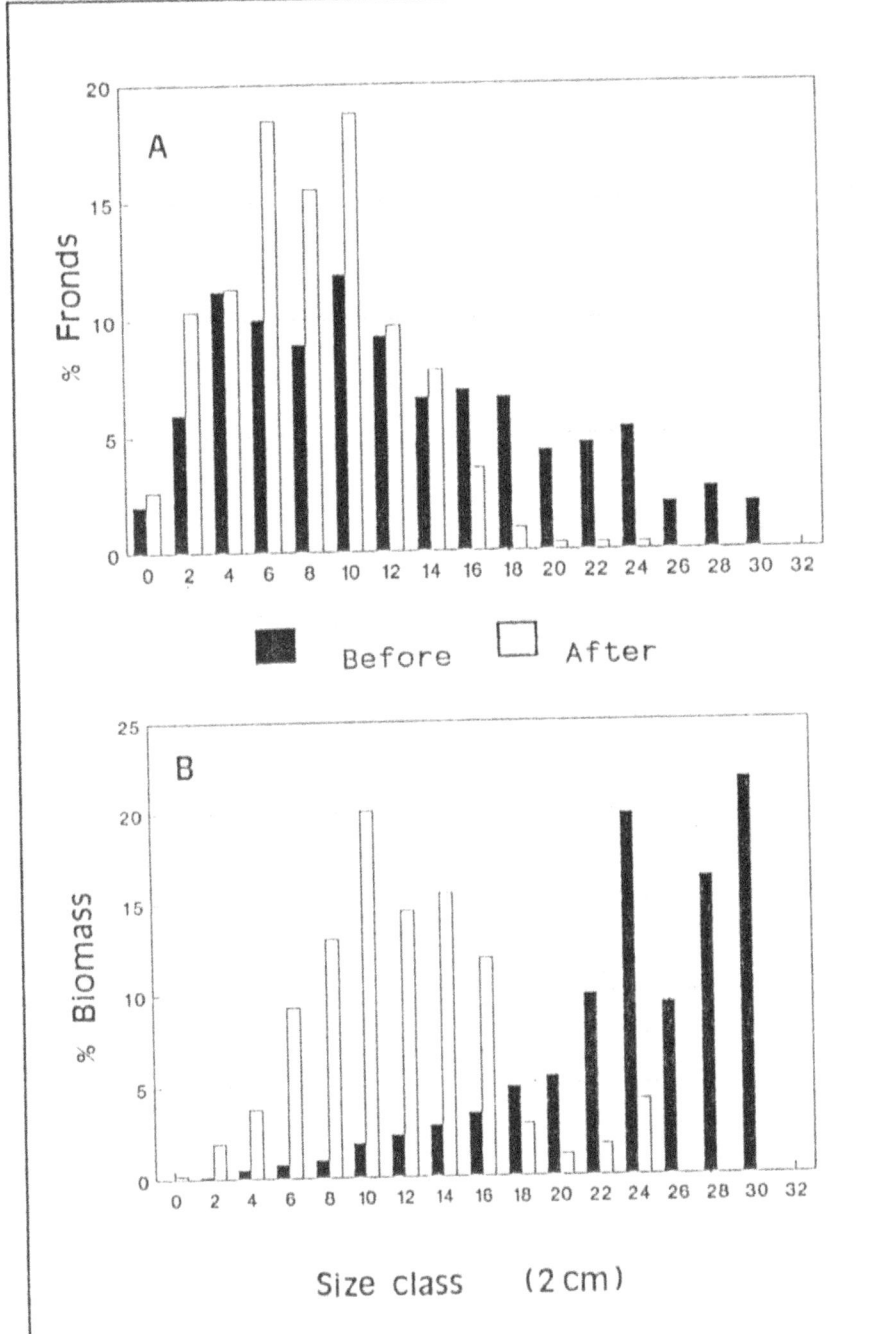

Fig. 4. Size distribution of fronds (%) (A) and biomass (dry wt) distribution per size class (2 cm) m^{-2} (B) of *G. sesquipedale* before (solid bars) and after (open bars) plucking experiments at 4-m depth, in the Basque Country (A. Borja, unpubl.).

time, equipment (such as suction pumps) which allows access to areas with masses of heretofore unexploited casts, should be carefully evaluated.

Direct exploitation

The rational regulation of the direct harvesting of *Gelidium* must take into account four basic aspects: methodology (how), season (when), intensity (how much), and frequency (how often).

As the propagation process in these species is not clearly understood (*i.e.* spores or rhizoidal propagation), it is imperative that this resource is protected from damage that may reduce recruitment capacities, until further information is available. The rhizoidal structure of the plants should be conserved (by prohibiting scraping technologies in harvesting), and during periods of maximum spore release (Septemer to April), harvesting should be prohibited.

Problems related to direct harvesting of this resource have been focussed on the convenience of the two different methods employed, plucking and cutting. In contrast to Salinas *et al.*, (1986) and Reguera *et al.* (1978), our data suggest that, because the biomass recorded in the exploited beds recovers within one year (Borja, unpubl. data), plucking of *G. sesquipedale* does not reduce the production capacity of the population. Moreover, large numbers of fronds exceeding 10 cm in length remain after harvesting and may become fertile during the reproductive season. Similarly, the recovery of biomass of cut populations of *G. sesquipedale*, together with possible increases in production (Gorostiaga, 1990), suggest greater possibilities for use of this method of harvesting. Nevertheless, for both technologies more experimental and observational data, regarding ecological impacts (*e.g.* changes in demographic structure, reproduction patterns, community composition) and commercial harvesting, are urgently required.

Intensive exploitation of the *Gelidium* beds in Spain appears to offer the best returns between the middle of June and the middle of August, when maximum biomass and the best yield of agar and gel strength are obtained. There are other reasons why harvesting should be limited to this period: a) natural biomass losses (Fig. 2A) begin in late summer, end of August–September (Juanes & Fernández, 1988; Gorostiaga, 1990); b) in September, a large percentage of *G. sesquipedale* fronds become fertile (Gorostiaga, 1990); c) populations harvested (plucked) during July show biomass increases until October, which

allows for partial recovery of the beds before the reproductive season; d) biomass of populations plucked during July (Fig. 2D) recovered one year later (Seoane-Camba, 1966; Cendrero & Ramos, 1967), whereas two years were required when harvesting (plucking or cutting) occurred in September (Gorostiaga, 1990).

Santelices (1989) has proposed that a minimum biomass, similar to the minimum biomass occurring during the annual cycle, should remain following a harvest. Commercial-plucking experiments (A. Borja, unpubl. data) demonstrated, however, that populations with lower post-harvest biomass (between one-half to one-third of the winter biomass) recovered to control levels in less than a year. Differences between recovery times of populations of *G. sesquipedale* plucked in July (Seoane-Camba, 1966; Cendrero & Ramos, 1967; this work) and in September (Gorostiaga, 1990) indicate the importance of the period at which harvesting is done. Differences were also found in the recovery of biomass following intensively (2 cm from the base) and moderately (8 cm from the base) cut populations (Gorostiaga, 1990) harvested in late summer; no data are available for cutting in early summer. Thus, time and intensity-interactive effects on biomass recovery of commercial-plucked and -cut populations of *Gelidium* remain to be evaluated.

Information is scarce on the effects of frequency of harvest. The experimental data of Gorostiaga (1990) based on small areas which were not replicated, and following annual, biannual and triannual harvests done in September suggest that biannual harvest results in the best returns for both plucked and cut beds, and that there was complete recovery of the demographic structure of the populations; nevertheless, as previously indicated, experiments on a larger scale, and at different times, are required. In the interim, the possibility of opening and closing zones for harvesting in alternate years can be considered. Such measures would have to be enforced through an effective control of harvesting effort and of harvest by management authorities.

54

Acknowledgements

This study was supported by a fellowship from Ministerio de Educación y Ciencia to J.A. Juanes and by funds from the Instituto de Investigación y Tecnología para la Oceanografía, Pesca y Alimentación (AZTI-SIO) del Gobierno Vasco. We thank J.M. Gorostiaga for kindly providing data from his Ph.D. Thesis. Our gratitude is also extended to E. M. Gonzalez and C. Cox for assistance in translation of the manuscript.

References

Alvarez de Meneses, A., 1972. Contribución al conocimiento de los campos de algas del Cantábrico. Bol. Inst. esp. Oceano. 154: 1–35.

Alvarez, A., M. Caloca, R. Gancedo, M. Mosquera & J. M. Salinas, 1978. Estudios sobre algas industrializables en España. Parte I. 'Contribución al conocimiento de la estructura microscópica de los filoides fértiles de *Gelidium sesquipedale* (Clem.) Born. et Thur. Bol. Inst. esp. Oceano. 250: 113–124.

Anadón, R. & C. Fernández, 1986. Comparación de tres comunidades de horizontes intermareales con abundancia de *G. latifolium* (Grev.) Born. et Thur. en la costa de Asturias (N de España). Inv. Pesq. 50: 353–366.

Anonymous, 1974–87. Anuario de Pesca Marítima. Ministerio de Agricultura, Pesca y Alimentación. Secretaría General Técnica. Madrid.

Borja, A., 1986. Evaluación de las arribazones de *Gelidium sesquipedale* producidas en la costa guipuzcoana. Inf. Dep. Agric. Pesca Gobierno Vasco, 14 pp.

Borja, A., 1987a. Cartografía, evaluación de la biomasa y arribazones del alga *Gelidium sesquipedale* (Clem.) Born. et Thur. en la costa quipuzcoana (N España). Inv. Pesq. 51: 199–224.

Borja, A., 1987b. El alga *Gelidium* en la costa vizcaína. Inf. técn. Dep. Agric. Pesca Gobierno Vasco 10, 57 pp.

Borja, A., 1988. Cartografía y evaluación de la biomasa del alga *Gelidium sesquipedale* (Clem.) Born. et Thur. 1876 en la costa vizcaína (N. de España). Inv. Pesq. 52: 85–107.

Cendrero, O. & F. Ramos, 1967. Trabajos sobre las algas del género *Gelidium* en la provincia de Santander. Publ. téc. J. Est. Pesca 6: 283–290.

Dixon, P. S. & L. M. Irvine, 1977. Seaweeds of the British Isles. Vol. I. Rhodophyta. Part 1: Introduction, Nemaliales, Gigartinales. British Mus. (Nat. Hist.) Publ. 781: 252 pp.

Establier, R., 1964. Variación estacional de la composición química, extracción y características del agar-agar de algunas algas (género *Gelidium*) de la costa sudatlántica española. Inv. Pesq. 26: 165–194.

Fernández, C. & R. Anadón, 1989. Cartografiado de biomasa de campos intermareales de *Gelidium sesquipedale* (Clem.) Born. et Thur. y *G. latifolium* (Grev.) Born. et Thur. en la costa de Asturias (N. de España). Inf. técn. Inv. Pesq. 149: 1–12.

García, I., 1988. Estudio de la variación estacional de la calidad y el rendimiento del agar obtenido del alga *Gelidium sesquipedale* (Clem.) Born. et Thur. en la costa quipuzcoana (N. de España). Inf. técn. Inv. Pesq. 146: 1–19.

Gorostiaga, J. M., 1990. Aspectos demográficos del alga *Gelidium sesquipedale* (Clem.) Born. et Thur.. Discusión sobre su adecuada gestión como recurso explotable. Tesis Doctoral Universidad del País Vasco. 313 pp.

Juanes, J. A., 1983. Contribución al conocimiento de la biología de *Gelidium latifolium* (Grev.) Born. et Thur. Tesis de Licenciatura Universidad de Oviedo. 101 pp.

Juanes, J. A. & C. Fernandez, 1988. Ciclo anual y producción de *Gelidium latifolium* (Grev.) Thur. et Born. (1876), en la región de cabo Peñas (Asturias, N. de España). Inv. Pesq. 52: 109–122.

Niell, F. X., 1983. Evaluación del tonelaje de algas productoras de agar en las costas de Euzkadi. Inf. Dept. Agric. Pesca Gobierno Vasco: 24 pp.

Reguera, B., J. M. Salinas & R. Gancedo, 1978. Biometría en *Gelidium sesquipedale* (Rhodophyta). Bol. Inst. esp. Oceano. 255: 101–138.

Salinas, J. M., 1986. Localización de campos industriales de *Gelidium sesquipedale*. Perspectivas de cultivo. Proc. J. acuicult. Com. Auton. Cantabria: 127–166.

Salinas, J. M., B. Reguera & R. Gancedo, 1976. Biometría de *Gelidium sesquipedale* (Rhodophyta) 1. Bol. Inst. esp. Oceano. 226: 1–70.

Santelices, B., 1989. Algas Marinas de Chile. Ed. Univ. Cat. Chile, Santiago. 400 pp.

Sanz, J. L. & J. Rey, 1983. Estudio de los campos de algas con sonar de barrido lateral. Bol. Inst. esp. Oceano., 1: 115–119.

Seoane-Camba, J., 1965. Estudios sobre las algas bentónicas en la costa sur de la Península Ibérica (litoral de Cádiz). Inv. Pesq. 29: 3–216.

Seoane-Camba, J., 1966. Algunos datos de interés en la recolección de *Gelidium sesquipedale*. Publ. técn. J. Est. Pesca 5: 437–455.

Seoane-Camba, J., 1969. Crecimiento, producción y desprendimiento de biomasa en *Gelidium sesquipedale* (Clem.) Thuret. Proc. int. seaweed Symp. 6: 365–374.

Hydrobiologia **221**: 55–66, 1991.
J. A. Juanes, B. Santelices & J. L. McLachlan (eds), International Workshop on Gelidium.
© 1991 *Kluwer Academic Publishers.*

Gelidium pristoides in South Africa

R.J. Anderson, R.H. Simons, N.G. Jarman & G.J. Levitt
Seaweed Unit, Sea Fisheries Research Institute, Private Bag X2, Roggebaai 8012, South Africa

Key words: Gelidium pristoides, commercial seaweeds, South Africa, agar, biomass, Rhodophyta

Abstract

Gelidium pristoides has been harvested commercially from the eastern Cape, South Africa, since 1951, with 40–80 t y^{-1} (dry wt) collected in recent years. This species has been intensively studied since 1983, and we briefly review knowledge of its biology in relation to harvesting. We describe a new study of intertidal epiphytic animals, showing that none is specific to *G. pristoides*, and that only 2.8% of these animals (numbers) inhabit this agarophyte, while the rest are found in other intertidal algal communities: harvesting is considered to have negligible effects on epifauna. Over the past 3 y, we have monitored, at two sites, the effects of the harvesting of *G. pristoides* on other benthic algae and animals. In only two of the seven main components analysed, did we find any difference between harvested and control plots. At one site only, the number of limpets and percentage cover of *Gelidium* was higher in harvested plots. These results show that harvesting has no significant biological effect. Regulations governing seaweed exploitation in South Africa were amended in 1988, to encourage local processing of products, and these changes are discussed in relation to the local *Gelidium* industry. Despite experimental results predicting a higher yield per unit effort if harvesting is limited to summer, harvesting continues throughout the year for practical reasons.

Introduction

Gelidium pristoides (Turn.) Kütz. is an endemic South African species of the littoral zone of rocky shores from Port Edward, near the border between Natal and Transkei, to Kommetjie on the west coast of the Cape of Good Hope (Fig. 1). It is often the dominant alga in the mid- to lower-eulittoral zone, particularly in the eastern Cape between Cape Seal and Transkei, where it has been harvested, together with three other *Gelidium* species, since at least 1951 (Isaac & Molteno, 1953). In the early 1980's public concern about the ecological effects of *Gelidium* harvesting led to a number of studies of *G. pristoides*, some of which are still in progress.

The aims of our paper are to:
1. Summarise the regulations governing seaweed harvesting in South Africa;
2. Review the biology of *Gelidium pristoides*, particularly in relation to commercial harvesting;
3. Describe previously unpublished studies on the effects of harvesting on epiphytic animals and associated benthic fauna and flora;
4. Discuss the management of commercial stocks of *G. pristoides* in relation to its biology and the practical problems experienced by harvesters.

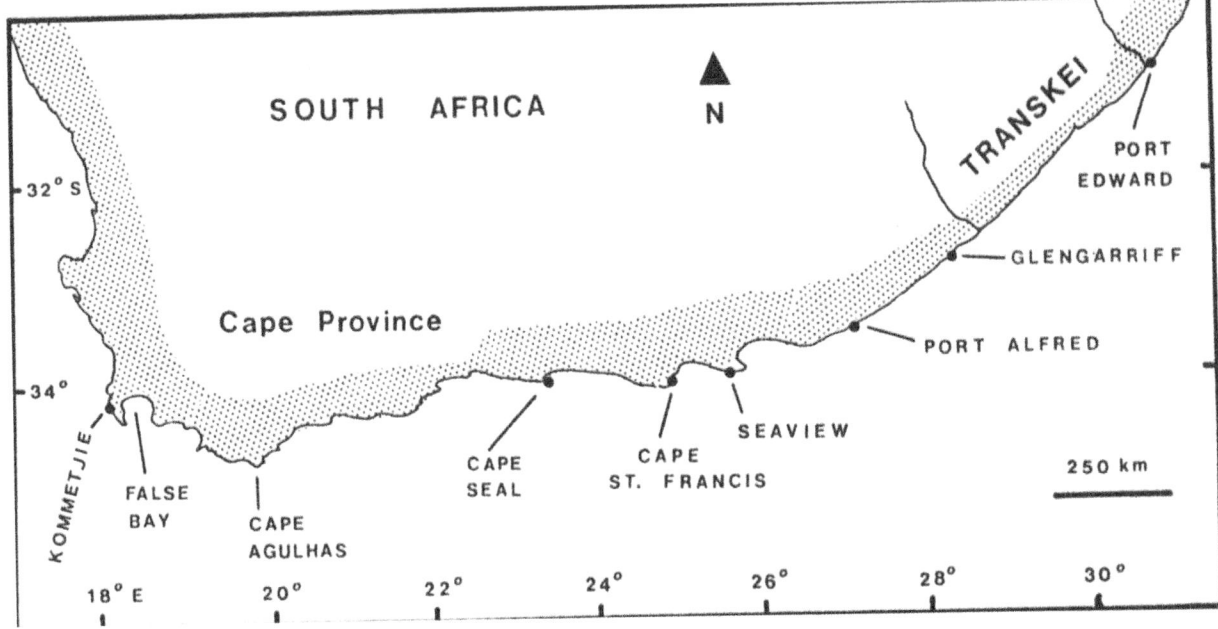

Fig. 1. Map showing southern part of the South African coast, with localities mentioned in text.

Regulations governing seaweed harvesting

The South African coast is divided into 17 Concession Areas for the purpose of seaweed exploitation, and these are administered according to the provisions of the Sea Fisheries Amendment Act of 1971, as revised in 1973 and in 1988. Concession Area 1 extends from the southern Transkei border to Cape Seal (Fig. 1), Concession Areas 2–4 lie between Cape Seal and Cape Agulhas, and Areas 5–9 extend from Cape Agulhas to 25 km north of Kommetjie. The regulations governing seaweed harvesting are discussed by Anderson *et al.* (1989), but the main points are that a permit for a concession area specifies the type of seaweed to be collected (*e.g.* beach-cast kelp) and is valid for 5 y. A fee of R 1 500 (about $570 US) per year for each concession, and a levy of R 4 (about $1.50 US) per t dry-wt of seaweed are payable. Permits are renewable indefinitely in cases where secondary processing (manufacture of a consumer product such as agar) is done in South Africa. The approval and issuance of a permit are subject to biological monitoring of stocks of the relevant seaweed,

either by Sea Fisheries staff or contracted researchers. Specific areas of the coast are closed to seaweed harvesting (*e.g.* national parks, military areas, and certain local authorities), but few such exclusions apply to the 135 km of rocky shore of Concession Area 1 where *Gelidium* is currently harvested.

Commercial yields

Data on total yields of *Gelidium* species are available from 1957 onwards (Fig. 2), and up until 1978 these data reflect harvests from both Transkei and South Africa. Transkei became politically independent in 1978, and subsequent data are for South Africa only, specifically for Concession Area 1, the part of the coast where almost all *Gelidium pristoides* harvesting has been done. There has been occasional harvesting of *G. pristoides* from Concession Area 2 and none from any of the other concession areas (3–9) in which it occurs.

The *Gelidium* yield data include *G. pristoides* as well as *G. pteridifolium* Norris, Hommersand et Fredericq, *G. amansii* (Lamour.) Lamour. and

G. *capense* (Gmel.) Silva. The last three species are collected from rock pools and the sublittoral fringe, and they also grow in the sublittoral zone. In Concession Area 1, about 70–80% of the harvest is comprised of *G. pristoides*, with the remainder made up of the other three species. In Transkei, negligible amounts of *G. pristoides* are collected, and the bulk of the 60–100 dry t harvested annually is made up of the three sublittoral-fringe species.

In South Africa, the *Gelidium* industry employs from 35–70 harvesters, who, with bonuses, earn about R 1 (about $0.40 US) per kg of seaweed, or an average of R 300–400 ($115–150 US) per month. Although this represents a minor contribution to the economy of the area, it is important in that the harvesters are drawn from impoverished rural communities, where poorly-paid farm labour is almost their only alternative (Carter, 1986). Also, the same concessionaires employ about 300 casual labourers to harvest *Gelidium* in the Transkei, where rural living standards are even lower.

We consider the large variations in annual yields (Fig. 2) the result of different harvesting efforts rather than from fluctuations in stocks; only since about 1980 have these yields become consistent. The concession-holders have now built up a team of experienced harvesters, who have become familiar with the collecting sites and have thus obtained yields that have increased in recent years. Choice of harvesting sites is based on experience and often on prior inspection. The

operators tend to visit certain favoured sites more frequently than others (up to 3 times per year) while other sites are visited once every few years.

The biology of *G. pristoides*

Morphology and reproduction

Gelidium pristoides grows in tufts composed of numerous individual juvenile and mature plants, sometimes of both reproductive phases (sporophytic and gametophytic), although usually of sporophytes (see below). Carter (1986) found that 10 tufts collected at Port Alfred were made up of 298 individual plants, each with one to several upright fronds. The vegetative 'juveniles' made up 15% of the sample and were all less than about 40 mm long; all plants less than 20 mm long were vegetative. Gametophytes ranged in length from 20–100 mm, and sporophytes from 20–120 mm. Mature plants consist of erect, flattened, irregularly-pinnately divided fronds arising from a system of branched, creeping axes. Fronds are about 2 mm wide and 50–100 mm high, although reaching 150 mm or longer when growing on vertical surfaces.

The vegetative and reproductive anatomy of this species has been described by Fan (1961) and Carter (1986). The genus *Gelidium* is unusual in that bispores are produced rather than tetraspores, which Carter (1985) has shown to be the result of meiosis in the sporangium. Carter (1985) suggested that if bispores are larger than tetraspores, they might be better able to survive and germinate in an intertidal habitat. Sporophytes predominate over gametophytes (male plus female plants) in natural populations, in the ratio of almost 3:1 (based on frequencies of fertile plants only) (Carter, 1985; Robertson, 1988), and Carter (1985) speculates on possible reasons for this imbalance. In culture, germination success-rate for carpospores is about thrice that for bispores, possibly as a result of the clumped-release pattern of carpospores, which may favour their germination (Carter, 1985). In culture, there is no difference in the growth rates of bispore and car-

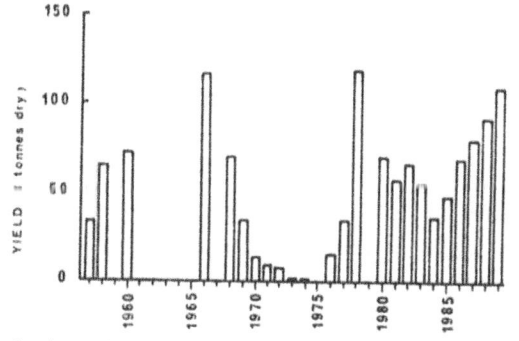

Fig. 2. Annual yield of *Gelidium* species including *G. pristoides*, from southern Africa.

58

pospore germlings and both grow best at temperatures between 15° and 23 °C (Carter, 1985). Robertson (1988) reported that the numbers of reproductive structures (tetrasporangial sori or cystocarps per weight of plant tissue) were about 30% higher in spring (September–November: 6500 tetrasporangial sori and 4800 cystocarps per g dry mass of tetrasporophytes and carposporophytes, respectively) than in late summer (January–March). From this, he deduced that recruitment must be highest in early summer; however, there is no information on seasonal spore output and, although Carter (1985) reported that, in culture, spores of both phases germinated throughout the year, he did not test their relative success on a seasonal basis. Furthermore, no seasonal demographic data are available, so that the relative importance of vegetative propagation versus recruitment from spores remains unknown.

Distribution

Gelidium pristoides is vertically-distributed on the shore from about 0.2 to 0.75 m above MLWS, with plants becoming progressively smaller with increasing elevation. Only *Porphyra capensis* Kütz. and a small species of *Gelidium* sometimes occur above *G. pristoides* on the shore. Carter and Anderson (in press) found that growth rates, recruitment and survival of germlings decreased with increasing elevation. All of these parameters were unaffected by exclusion of grazers, while adults transplanted to the sublittoral fringe were overgrown by encrusting coralline epiphytes. Exclusion of grazers from plots in the sublittoral fringe led to enhanced establishment of germlings, which were subsequently displaced by articulated corallines. Physical factors, therefore, control the upper distributional limits of *G. pristoides*, while its lower limits are controlled by biological factors (Carter & Anderson, in press).

The most important factor controlling the horizontal distribution of *G. pristoides*, within the vertical limits set by the above factors, appears to be grazing by limpets. These animals keep most

smooth rock surfaces clear of macroalgal growth, and largely restrict the tufts of this agarophyte to crevices, the shells of limpets and barnacles and coral-worm (*Pomatoleios krausii* Baird) and sandworm (*Gunnarea capensis* Schm.) tubes (Carter & Anderson, in press). Also, *G. pristoides* is more strongly attached to limpet and barnacle shells, which its rhizoids penetrate, than to rock, suggesting that resistance to wave dislodgement may also contribute to the disproportionate distribution of this alga on these substrata (Carter & Anderson, in press). Accidental removal of these animals (particularly limpets) occurs during harvesting, and has led to criticisms of the commercial harvesting methods (see below).

Growth rates

Measurements of linear elongation of individually tagged fronds, made over 25 mo at Port Alfred, showed a strongly seasonal growth-pattern in *Gelidium pristoides* (Carter & Anderson, 1986). A smoothed curve of their data (Fig. 3a) illustrates this; growth is lowest in late winter (July), rising in spring to a maximum in summer (January–February). Seasonal growth patterns in other *Gelidium* species are well-documented and have been correlated with factors such as photoperiod and temperature (Montalva & Santelices, 1981).

Regrowth of *G. pristoides* after experimental harvesting, to the same biomass as unharvested plants, took 3–4 mo in summer and 4–5 mo in winter, (Carter & Anderson, 1985). Robertson (1988) found that harvesting in late summer (February) resulted in the longest recovery period (5–8 mo), because growth slowed at the onset of winter. Tufts picked by hand (the normal commercial method) regenerated at the same rate as those clipped by hand with shears (Carter & Anderson, 1985; Carter & Simons, 1987; Robertson, 1988). Shearing was expected to result in faster regrowth because damage to the holdfast is avoided; however, plucking removes only part of the material in a tuft, and it appears that enough is left to protect the holdfast against desiccation (Carter & Anderson, 1985). As shear-

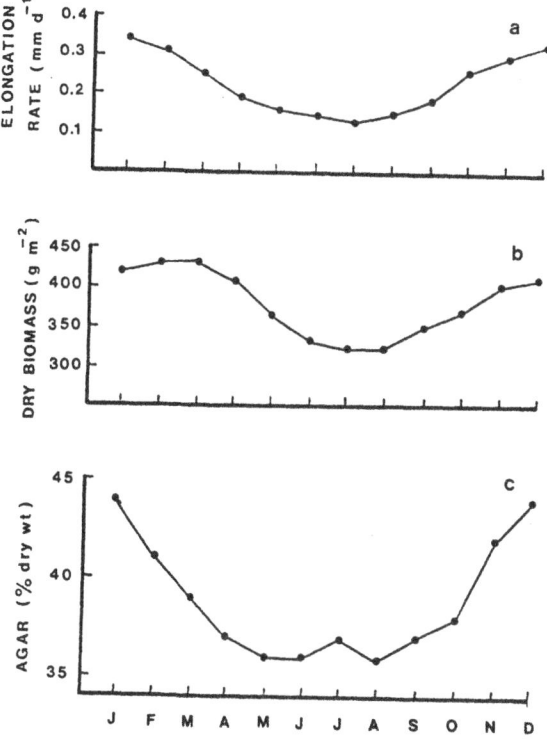

Fig. 3. Growth rate as frond elongation rate (Fig. 3a), biomass (Fig. 3b) and agar content (Fig. 3c) of *G. pristoides* at Port Alfred. Data are running means recalculated from Figs. 1, 3 and 5, respectively, of Carter & Anderson (1986).

ing is slower than hand-picking, it is not favoured by commercial harvesters. Carter and Simons (1987) studied the yields obtained from repeated harvesting of the same plants at various intervals (4, 6, 8, 10, 12 wk) over the summer season. They found that aggregate yields were highest with shorter intervals, whereas mean yield per harvest increased with longer intervals betwee harvests. In other words, although frequent re-harvesting stimulated overall production slightly, it was more efficient on a yield-per-unit-effort-basis to allow at least 8–12 wk recovery period between harvests. They point out that this study was based on re-peated harvesting of 50–80% of the biomass of the same individuals, whereas commercial harvesters choose only the larger plants, and pick only about 20–30% of the plants present at a site during one operation. In practice, therefore, exploitation of natural populations occurs at a

lower level than that employed in these experi-ments. At present levels of exploitation, it is, therefore, more efficient for the harvesters to spread their activities over a larger number of sites (as they do now) than to visit a few sites fre-quently.

Biomass

We have surveyed the total biomass of *Gelidium pristoides* in Concession Area 1 twice, using quad-rats (2 × 2 m) and belt transects (1 m wide). There is 135 km of rocky shore on the 450 km of coast, and we obtained estimates of 118 t dry wt in September 1983 and 126 t in December 1986. Many of the sites we visited differed with respect to shore topography, width of the *G. pristoides* zone and biomass of *G. pristoides* within this zone. As a result, the standing stock of this seaweed varied from 0.0–2.91 kg dry wt per run-ning m of shore, with a mean of 0.88 kg m⁻¹. Port Alfred, with a *G. pristoides* zone 1–3 m wide and a mean standing stock of 0.84 kg m⁻¹, was con-sidered representative of Concession Area 1.

The biomass of natural populations fluctuates seasonally (Fig. 3b), following a pattern similar to that of growth rate. Biomass is highest in late summer (January–March), declining to a mini-mum in late winter (July–August) (Carter & Anderson, 1986; Robertson, 1988). As a result, the total biomass on this shore probably approaches 135–140 t dry-wt in late summer.

Agar

Yield of agar per dry unit wt of *Gelidium pristoides* is highest in midsummer and lowest in midwinter (*e.g.* Fig. 3c). Carter and Anderson (1986) measured agar content monthly over 2 y, using mixed-phase samples, and obtained a (winter) minimum of 30% and a (summer) maximum of 48% agar. Their values are higher than those obtained by Onraet and Robertson (1987), who recorded between 23 and 39% agar. The latter authors suggest that these differences may reflect different localities, years or extraction methods.

We consider it likely that centrifugation of the boiled extract, as done by Carter and Anderson, (1986) yields more agar than filtration, as Onraet and Robertson (1987) used. Onraet and Robertson (1987) found no difference in agar content between sporophytes and gametophytes at any time of the year.

The quality of agar from sporophytes and gametophytes of *Gelidium pristoides* was analysed by Onraet and Robertson (1987) on a monthly basis for 1 y. Gel strength varied from 400–800 g cm^{-2}, which is high and, as the single most important measure of gel quality (Tsuchiya & Hong, 1966), imparts an exceptional value to this product. The contention of Onraet and Robertson (1987) that overall gel quality (gel strength, 3,6-anhydrogalactose level, gel rigidity, and cohesiveness) is highest in late summer (February–March) is not fully reflected in their figures. When the data from their figures for gametophytes and sporophytes are averaged and re-plotted as running means, gel strength is shown to be highest in midwinter, 3,6-anhydrogalactose in late winter, and gel rigidity and cohesiveness highest in late summer-early winter. Overall, these gel-quality data do not indicate seasonal trends sufficient to warrant a change in the timing of harvesting.

Ecological effects of harvesting

In this section, we describe studies on the effects of *Gelidium* harvesting on other plants and animals in the intertidal zone. Because these studies have not been published previously, they are described in detail.

Removal of limpets and barnacles

One of the criticisms leveled against commercial *Gelidium* harvesters is that unacceptable numbers of limpets are removed along with the epizoic seaweed. Commercial harvesters are instructed to tap limpet shells before picking *Gelidium*, an action that causes the limpets to cling tightly to the rock and reduces the chance of their accidental removal. Nevertheless, official inspections show that some, invariably small, limpet shells and barnacle shells are present in harvests. We, therefore, determined what proportion of these animals are removed by simulated commercial harvesting.

All *Gelidium pristoides* tufts (each consisting of numerous plants) were harvested by hand-picking from 27 quadrats (each either 2 or 4 m^2) at 5 sites in the eastern Cape. For each quadrat, we recorded the substratum to which each tuft was attached, the weights of any animal removed and the number of limpets without *Gelidium* growing on their shells. Limpet shells were not tapped, so that we obtained the highest possible rate of inadvertent limpet removal.

Gelidium pristoides was found on various substrata, at all of the 5 sites. Overall percentage frequencies (numbers of tufts) on the substrata were: limpet 29.4, rock 26.8, barnacle 14.3, sandworm and coral-worm tubes 29.5%. A total of 2252 tufts of *Gelidium pristoides* were picked in the 27 quadrats. Of these, 661 were borne on limpets, and a further 107 limpets bore no *Gelidium* (limpets = 768). Fifty-five limpets (almost all *Patella oculus* Born.) were removed with the harvested seaweed, and 51 of these (92%) were smaller than 7 mm in length. Larger limpels (30–50 mm long) were common and far less prone to removal than small limpets. In these quadrats, 2% of the barnacles were accidentally removed, and 75% of these were dead, comprised only of hollow shells.

Although 7% of the limpets were removed in this 'worst-case' harvesting, 92% of these were so small that they would be reproductively inactive (Branch, 1974). Also, the tapping of the shells practiced by the commercial harvesters, should considerably reduce accidental removals. We consider the ecological impact of accidental removal of barnacles and limpets to be negligible.

Epiphytic animals

These are animals which spend all or part of their lives living on seaweeds. In the eastern Cape, these animals are eaten by certain reef fish and birds (Beckley & McLachlan, 1980). Beckley (1982), in a study of the intertidal epiphytic animals ('algal epifauna') of St. Croix Island, Algoa Bay, expressed concern that commercial harvesting of *Gelidium pristoides* might adversely affect this group of animals. She was particularly concerned that juveniles of the brown mussel *Perna perna* L., which appear to use *G. pristoides* at St. Croix as a nursery site, were either being removed in large numbers or deprived of a nursery. We, therefore, assessed whether the fauna which is removed during picking differs from that left in the remnant *Gelidium*, and how important *G. pristoides* is as a habitat for intertidal epiphytic fauna.

The study site was a 250-m length of rocky shore at Glengarriff, eastern Cape, which had been protected from commercial harvesting for 2 y as part of another study. On three occasions (July 1984, January 1985, March 1986), 40 large individuals tufts of *G. pristoides*, growing on rock, limpets and barnacles, were haphazardly sampled as follows: a tuft was hand picked, and the material quickly placed in a labelled plastic bag. The remnant of the tuft was then carefully scraped off the substratum, allowing as few animals as possible to escape, and placed in a separate bag. The samples were preserved in 5% formalin in seawater. In the laboratory, the seaweed was rinsed repeatedly with fresh water, over 0.2 mm mesh plankton netting, to wash out the animals. In practice, this retrieved the macrofauna (greater than 0.5 mm) and some of the meiofauna (0.5 mm–0.1 mm). The animals were sorted into taxa and counted under a dissecting microscope. It was not practical to try and identify most of the taxa, and animals were simply grouped into major taxa such as classes. The seaweed in each sample was oven-dried, weighed, and the results were expressed as number of animals per dry wt of seaweed.

To assess the proportional distribution of animals in *G. pristoides* and the other littoral algal communities, a survey was carried out at Glengarriff in March 1986, a month when the biomass of *G. pristoides* is high (Carter & Anderson, 1986). The areas occupied by the major communities in a 250 m-long strip of shore were determined from hand-drawn maps and 6 line transects laid perpendicular to the shore. Faunal samples were obtained either from ten 0.1 × 0.1 m quadrats (*Hypnea spicifera*/articulated coralline community) or 20-26.5-mm diameter cylindrical cores taken with a sharpened steel pipe (algal turf communities). Data on *G. pristoides* fauna were obtained by combining counts from the March 1986 harvest and remnant samples.

Animals were washed out of the algae over 0.2-mm mesh. Although most taxa could not be named, we checked whether any appeared specific to *G. pristoides*. *Perna perna* was analysed separately from other bivalves in the samples, because of its possible importance (Beckley, 1979, 1982). Animals were counted and total numbers of the major taxa calculated for each algal community per running metre of shore. The percentages of each of these groups of fauna occurring in each community are shown in Table 1.

The most numerous fauna were isopods, amphipods and gastropods. Other important groups (included with the above in 'Total Fauna') were harpacticoid copepods (*e.g.*, *Orthopsyllus* sp.), ostracods and nematodes. *Perna perna* occurred in only 28% of the tufts sampled: 1 g of *Gelidium pristoides* contained a mean of 0.2 *Perna*, all less than 5 mm in length and usually found in the remnant; there were too few *Perna* for statistical comparisons with the harvested portion.

In general, there were no differences between the numbers of epiphytic fauna in harvested and remnant portions of *Gelidium pristoides* tufts, except in March 1986 when there were more gastropods and total fauna in the remnant portion (Fig. 4). The same taxa of animals were found in the harvested and remnant portions, in general, although more nematodes tended to occur in the remnant when there was accumulated sand among the holdfasts.

The transect survey revealed 5 distinctive algal communities. In wave-exposed areas (*e.g.* rocky points), a mixed *Hypnea spicifera* (Suhr) Harv. and articulated coralline community occupied the lowest zone on the shore. Above this was a patchy

Table 1. Percentages of main groups of epiphytic animals found in algal communities at Glengarriff.

Algal community	Isopods	Amphipods	Gastropods	Perna	Total animals
G. pristoides	2.8	1.7	1.9	0.8	2.8
Hypnea/articulated coralline	90.8	84.1	26.4	43.1	50.8
Sandy coralline turf	5.8	10.3	66.1	21.7	35.6
Mixed algal turf	0.3	1.9	3.6	3.9	6.1
Gelidium turf	0.3	1.9	1.9	30.6	4.7

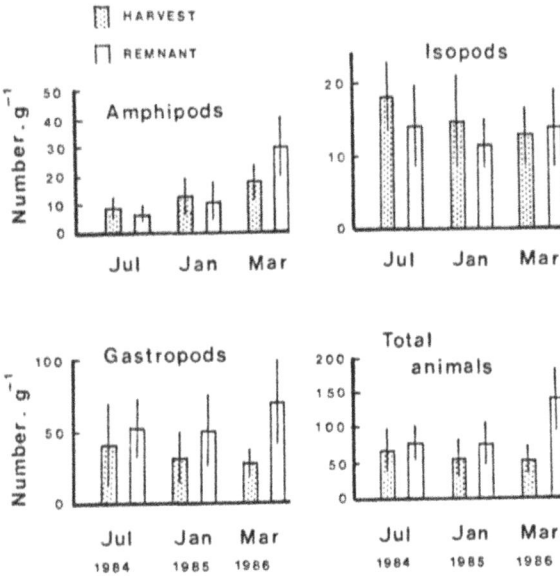

Fig. 4. Epifauna (numbers g^{-1} dry *G. pristoides*) in harvested and remnant portions of *G. pristoides* plants at Glengarriff. Vertical bars show 95% confidence limits of means.

sandy turf made up mostly of mixed articulated corallines, and extending down into the sublittoral zone. Above this was a zone dominated by limpets (usually *Patella oculus*), with *G. pristoides* growing on the limpet shells, providing almost the only algal cover. A narrow *G. pristoides* zone occurred above this. The upper eulittoral zone of both sheltered and exposed areas supported patches of a very short, carpet-like turf of an unidentified *Gelidium* which Carter (1986) considered to be a *G. pristoides* ecotype.

Most of the epiphytic fauna occurred in the *Hypnea*/articulated-coralline and sandy-coralline turf communities (Table 1). Only 2.8% of the total fauna was found in *G. pristoides* tufts. As commercial harvesters remove only about 25% of the *G. pristoides* standing crop, only about 0.7% of the total epiphytic fauna would be removed at each harvest.

The usually similar numbers of epiphytic fauna in harvested and remnant portions of the *G. pristoides* plants may partly be because hand-picking of tufts removes some holdfast material, along with the frond material which makes up the bulk of the harvest; however, in general, the major groups (amphipods, isopods, gastropods and total epifauna) appear to be distributed throughout the plants. An exception to this pattern is the mussel *Perna perna*, which was usually found in the remnant portion of our samples, and according to Beckley (1982) lives among the hold-fasts of *G. pristoides*. The concentration of gastropods and total fauna in the remnant portion in

algal turf/limpet community, comprising small areas of bare rock, each with one or two limpets, surrounded by a low algal turf, made up largely of *Caulacanthus ustulatus* (Turn.) Kütz., *Ceramium* spp., *Ulva* sp., *Polysiphonia* spp. and articulated corallines. *G. pristoides* usually grew on the limpet shells. A zone dominated by *G. pristoides* occurred above this, extending up to the balanoid zone. In more sheltered areas (*e.g.* the sides of gullies), the lowest zone was occupied by a short,

March 1986 may have been caused by downward migration of the animals to escape the high temperatures we noted during low tide on that particular day, although this is speculative. Certain taxa have the ability for vertical migration on the *G. pristoides* fronds, as, for example, the harpacticoid copepod *Porcellidium* does on fronds of *Gigartina radula* (Esper). J. Ag. in the western Cape (Gibbons, 1989).

High numbers of amphipods and isopods were recorded in *G. pristoides* tufts from exposed shores at St. Croix Island, eastern Cape, by Beckley (1982), and Gibbons (1988) found gastropods to be an important component of the epiphytic fauna of this species in False Bay, western Cape. The far greater numbers of copepods, ostracods and particularly nematodes which they recorded probably resulted from their use of a 63 μm mesh sieve, compared with the 200 μm mesh that we used. This explanation cannot account for Beckley's (1982) record of higher numbers of *Perna perna* per g dry *Gelidium* at St. Croix Island (10 on the exposed shore compared to our mean of 0.2 at Glengarriff). This probably reflects different numbers of this bivalve at the respective study sites, as the Glengarriff shore is certainly exposed.

Our results clearly indicate that only 2.8% of epiphytic animals (in terms of numbers) in the intertidal occur in *G. pristoides*, and at harvesting, only a fraction of these are removed. We predict that the same conditions apply at St. Croix where, by extrapolation from data in Beckley and McLachlan (1980), we calculate that 2.4% of the total epiphytic fauna occurs in *G. pristoides*. In our study, the lower eulittoral algal communities supported the greatest numbers of these animals, which may be a permanent distribution-pattern, or in the case of highly-mobile taxa, may reflect movement down the shore with the receding tide, as shown in a species of the amphipod *Stenothoe* (Wieser, 1952); however, this does not affect the issue here, as harvesting takes place at low springtides, when *G. pristoides* is emersed.

Effects of harvesting on benthic organisms

Gelidium pristoides usually dominates the flora in the lower- to mid-eulittoral zone of eastern Cape shores, with a number of other algae and various benthic animals, including limpets and large herbivorous gastropods such as *Oxystele* found here. The control of littoral seaweed communities by herbivores is well-documented (Hawkins & Hartnoll, 1983), and limpets are important in determining the vertical distribution pattern of *G. pristoides* (Carter & Anderson, in press); however, the effects of periodically reducing the biomass and cover of *G. pristoides* (*i.e.* harvesting) on benthic organisms which form part of this community were not known. We tested the hypothesis that harvesting 3 to 4 times per y (maximum commercial harvesting intensity) would cause gradual changes in the proportions of the different species which make up this community.

Sites at Seaview and Port Alfred were chosen as typical of eastern Cape shores. At each site the corners of 8 plots were marked by stainless-steel bolts driven into holes drilled in the rock. The plots were placed subjectively in parts of the *G. pristoides*-dominated community that were as homogeneous as possible and at the same vertical level. *Gelidium pristoides* was harvested from alternate plots, those in between serving as controls. Parameters recorded were the percentage cover of algae and the numbers of animals. On each occasion, cover of *G. pristoides* was estimated before harvesting in the appropriate plots. Sampling was done about every 3 mo.

At Seaview, the plots were 1 × 1 m, but at Port Alfred, where the *G. pristoides* zone was narrower, 0.5 × 0.5-m plots were used. At both sites, sampling accuracy was improved by dividing each plot into quarters for cover estimations, and later calculating the mean. Results for each taxon were subjected to two-way analysis of variance to test the null hypotheses that there were no differences between harvested and control plots and between samples (*i.e.* temporal differences). Comparisons of means were made using the Student-Newman-Keuls (SKN) test. Percentage cover data were arcsine transformed for analysis. We did not test for differences between sites. For each taxon, we compared mean percentage cover (algae) or numbers (animals) in harvested and control plots, over the 3-y period, but only data for *G. pristoides* cover (Fig. 5) are shown. Most of the 10 algal species, besides *G. pristoides*, were present in small amounts (0–2% cover) so these data were combined as 'total other algae' for statistical analysis; similarly, of the 18 animal species recorded, three were combined for analysis (limpets) and 11 ignored as too rare or highly variable in numbers.

The percentage cover of *G. pristoides* was significantly higher in harvested than in control plots at Port Alfred, although there was no significant difference at Seaview (Table 2). The SNK test showed no difference at either site between *G. pristoides* cover at the beginning (March 1987) and end (January 1990) of the experiment (Fig. 5). Cover was significantly lower than usual at Port Alfred during June and September 1989, and significantly higher than usual at Seaview in June 1988.

Algae other than *G. pristoides* ('other algae') showed no temporal changes and were unaffected by harvesting of *G. pristoides*. Unidentified encrusting corallines were the largest component of this group, with a mean of about 8% cover at both sites. Other taxa included *Ralfsia verrucosa* (Aresch.) J. Ag. (3% cover), *Caulacanthus ustulatus*, *Ulva* sp. and various small filamentous red algae, which, together with small articulated corallines, formed patchy turfs.

The 'limpet' group comprised *Patella oculus* and *P. longicosta* Lam. At Port Alfred, there were significantly more limpets in the harvested plots throughout the course of the experiment, and at Seaview numbers were similar in harvested and control plots (Table 2). *Oxystele variegata* Anton., *Patella granularis* Linn., *Siphonaria* spp. and

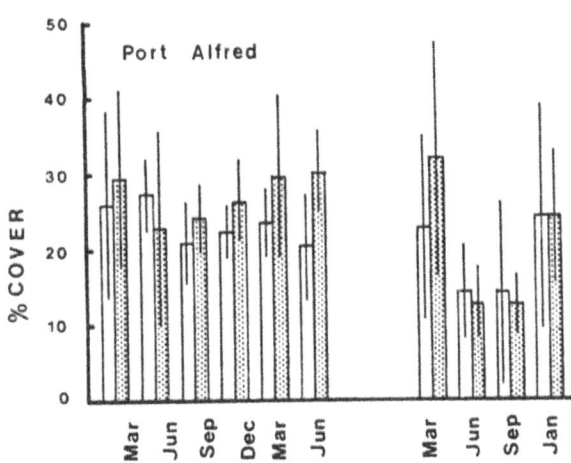

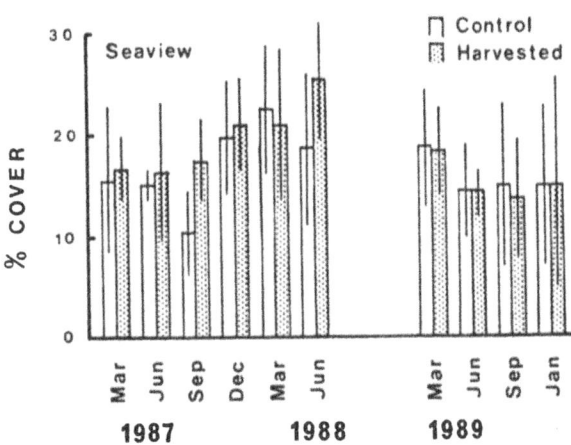

Fig. 5. Percentage cover of *G. pristoides* in harvested and unharvested plots at Port Alfred and Seaview. Vertical bars show 95% confidence limits of means.

Table 2. Mean algal cover and mean number of animals in harvested and control plots. Significant difference between harvested and control values: # – at 5%, ## – at 1% confidence level.

	Port Alfred		Seaview	
	Control	Harvested	Control	Harvested
% Cover				
G. pristoides	21	24 #	18	18
Other algae	11	10	20	22
Number m^{-2}				
Limpets	12.8	22.4 ##	25.1	23.0
Patella granularis	9.2	9.6	6.2	4.8
Oxystele variegata	8.0	11.2	3.6	6.1
Siphonaria spp.	18.4	13.2	7.5	7.6
Tetraclita serrata	465.6	729.2	54.1	56.8

Tetraclita serrata Darwin were unaffected by harvesting.

We recorded small and highly variable numbers of chitons, whelks and small starfish, but did not attempt to analyse these data statistically. The small barnacle *Chthamalus dentatus* Krauss was present in large numbers (>1000 m^{-2}) at both sites at the start of the study, and these numbers showed a linear decline with time, and no difference between harvested and control plots.

We conclude from these results that harvesting of *G. pristoides* at this intensity has negligible effects on other benthic organisms in the lower- to mid-eulittoral community. The harvested *G. pristoides* always recovered to control levels within 3 mo, and none of the associated organisms decreased in numbers or cover over this 3-y period; furthermore, our experimental harvesting was more thorough than that practiced commercially, because we removed material from every tuft in each harvested plot every 3 mo. The commercial pickers choose only the larger tufts, and visit any one site between once and thrice yearly.

It is perhaps not surprising that temporary reductions in the cover of *G. pristoides* did not affect the numbers of *Patella* spp. These animals have reduced tooth-rows and feed by scraping small algae off the rock (Creese & Underwood, 1982). We never observed them to feed directly on adult *Gelidium* plants; similarly, we never observed *Siphonaria* spp. to feed on *G. pristoides*, rather they fed on low algal turfs which have a plain upper surface, perhaps allowing them easier mobility. Many intertidal animals, particularly *Oxystele*, were observed to shelter under algal fronds and in crevices during low tide; however, harvesting of *G. pristoides* did not reduce their numbers, possibly because not all of this seaweed is removed and there are other refugia available (*e.g.* crevices, other seaweeds).

General discussion

The research that has been done so far in South Africa provides information both on the effects of commercial harvesting, as currently practiced, and on how the commercial harvesters could increase their yield, or improve their harvesting efficiency. It is clear from the results of these studies that neither epiphytic animals nor benthic seaweeds and animals are significantly affected by harvesting of *Gelidium pristoides* at current levels of commercial activity. Sites harvested several times per year for 3 decades have strikingly similar intertidal communities (and *G. pristoides* biomass) compared with sites which have seldom or never been harvested. Perhaps the largest gap in our knowledge of this species is the extent to which a *G. pristoides* population is maintained by recruitment from spores and vegetative regrowth from old plants. Established tufts may provide ideal refugia for sporeling growth, borne out by the occurrence of juveniles together with the adults. Harvesting, in this case, simply opens up the 'canopy', allowing juveniles to grow rapidly. However, if the spring peak in fertility of adult plants (Robertson, 1988) results in a seasonal peak in the establishment of new sporelings, then clearly harvesting during peak fertility should be avoided.

There are several reasons why intensive harvesting in late summer (December–February) appears to offer the best returns per unit-effort. The peak fertility period is over, and biomass and agar content are high; however, such a concentration of effort is simply not practical for the commercial operators, whose team of labourers want employment throughout the year; furthermore, the eastern Cape coast is a popular holiday area between mid-December and mid-January, rendering some of the most productive sites temporarily inaccessible. Commercial harvesting is therefore likely to continue as in recent years. In view of the high value of *G. pristoides* (about R 7500 of $2700 US per dry t), it is not clear why it has not been collected to any extent from areas south of Concession Area 1. There are no data on the biomass of this species from such areas, although indications are that it is abundant at some localities that might be worth harvesting. We recommend that biomass studies of accessible areas be undertaken.

At this stage, mariculture of *G. pristoides* is impractical, and probably the only way of greatly increasing its biomass would be by large-scale removal of limpets from certain areas, allowing *G. pristoides* to colonise rock surfaces currently kept bare by limpets grazing off sporelings; however, the ecological effects of such actions are unknown, and they would certainly arouse adverse public reaction. The large subtidal *Gelidium* species, *G. pteridifolium* and *G. capense*, may well have potential for mariculture.

As biologists our main concern is to protect local resources from over-exploitation, while encouraging their better local utilization. It is thus disappointing that after several decades *G. pristoides* is still not processed in South Africa, rather exported after drying, sorting and baling. The local industry claims that demand for high quality agar in southern Africa is too small to justify the expense of establishing the necessary plant, and that the high value of the *G. pristoides* harvest, and the usefulness of its agar for blending, make it more worthwhile to export the dried material than to process it locally.

Acknowledgements

We gratefully acknowledge the support of the Director, Sea Fisheries Research Institute, and thank Dr. J.J. Bolton for helpful comments on this manuscript.

References

Anderson, R. J., R. H. Simons & N. G. Jarman, 1989. Commercial seaweeds in southern Africa: A review of utilization and research. S. afr. J. mar. Sci. 8: 277–299.

Beckley, L. E., 1979. Primary settlement of *Perna perna* (L.) on littoral seaweeds on St. Croix Island. S. afr. J. Zool. 14: 171.

Beckley, L. E., 1982. Studies on the littoral seaweed fauna of St. Croix Island. 3. *Gelidium pristoides* (Rhodophyta) and its epifauna. S. afr. J. Zool. 17: 3–10.

Beckley, L. E. & A. McLachlan, 1980. Studies on the littoral seaweed epifauna of St. Croix Island. 2. Composition and summer standing stock. S. afr. J. Zool. 15: 170–176.

Branch, G. M., 1974. The ecology of *Patella* Linnaeus from the Cape Peninsula, South Africa. 2. Reproductive cycles. Trans. r. Soc. S. Afr. 41: 111–160.

Carter, A. R., 1985. Reproductive morphology and phenology, and culture studies of *Gelidium pristoides*

(Rhodophyta) from Port Alfred in South Africa. Bot. mar. 28: 303–311.

Carter, A. R., 1986. Studies on the biology of the economic marine red alga *Gelidium pristoides* (Turner) Kuetzing (Gelidiales: Rhodophyta). Ph. D. Thesis, Rhodes University. 188 pp.

Carter, A. R. & R. J. Anderson, 1985. Regrowth after experimental harvesting of the agarophyte *Gelidium pristoides* (Gelidiales: Rhodophyta) in the eastern Cape Province. S. afr. J. mar. Sci. 3: 111–118.

Carter, A. R. & R. J. Anderson, 1986. Seasonal growth and agar contents in *Gelidium pristoides* (Gelidiales, Rhodophyta) from Port Alfred, South Africa. Bot. mar. 29: 117–123.

Carter, A. R. & R. J. Anderson, 1991. Biological and physical factors controlling the spatial distribution of the intertidal alga *Gelidium pristoides* in the eastern Cape, South Africa. J. mar. biol. Ass. U.K. 71, in press.

Carter, A. R. & R. H. Simons, 1987. Regrowth and production capacity of *Gelidium pristoides* (Gelidiales, Rhodophyta) under various harvesting regimes at Port Alfred, South Africa. Bot. mar. 30: 227–231.

Creese, R. G. & A. J. Underwood, 1982. Analysis of the inter- and intraspecific competition among intertidal limpets with different methods of feeding. Oecologia 53: 337–346.

Fan, K. C., 1961. Morphological studies of the Gelidiales. Univ. Calif. Publ. Bot. 32: 315–368.

Gibbons, M. J., 1988. The impact of wave exposure on the meiofauna of *Gelidium pristoides* (Turner) Kuetzing (Gelidiales: Rhodophyta). Estuar. coast. shelf Sci. 27: 581–593.

Gibbons, M. J., 1989. Tidal migration of *Porcellidium* (Copepoda: Harpacticoida) on fronds on the rocky shore alga *Gigartina radula* (Esper) J. Agardh (Gigartinales: Rhodophyta). S. afr. J. mar. Sci. 8: 3–7.

Hawkins, S. J. & R. G. Hartnoll, 1983. Grazing of intertidal marine algae by marine invertebrates. Oceanogr. mar. biol. Ann. Rev. 21: 195–282.

Isaac, W. E. & C. J. Molteno, 1953. Seaweed resources of South Africa. J. s. afr. Bot. 19: 85–92.

Montalva, S. & B. Santelices, 1981. Interspecific interference among species of *Gelidium* from central Chile. J. exp. mar. Biol. Ecol. 53: 77–88.

Onraet, A. C. & B. L. Robertson, 1987. Seasonal variation in yield and properties of agar from sporophytic and gametophytic phases of *Onikusa pristoides* (Turner) Akatsuka (Gelidiaceae, Rhodophyta). Bot. mar. 30: 491–495.

Robertson, B. L., 1988. Management of *Gelidium pristoides* resources. In Branch G. M. & L. Y. Shackleton (eds), Research needs in the Transkei and Ciskei coastal zone. Rep. S. Afr. natn. scient. Progms 155: 16–18.

Tsuchiya, Y. & K. Hong, 1966. Agarose and agaropectin in *Gelidium* and *Gracilaria* agar. In Young E. G. & J. L. McLachlan (eds), Proceedings of the Fifth International Seaweed Symposium, Halifax, August 1965. Pergamon Press, Oxford, pp. 315–321.

Wieser, W., 1952. Investigations on the microfauna inhabiting seaweeds on rocky coasts. 4. Studies on the vertical distribution of the fauna inhabiting seaweeds below the Plymouth Laboratory. J. mar. biol. Ass. U.K. 31: 145–174.

Hydrobiologia **221**: 67–75, 1991.
J. A. Juanes, B. Santelices & J. L. McLachlan (eds), International Workshop on Gelidium.
© 1991 *Kluwer Academic Publishers.*

Field studies and growth experiments on *Gelidium latifolium* from Asturias (northern Spain)

José M. Rico
Departamento de Biología de Organismos y Sistemas, Laboratorio de Ecología, Universidad de Oviedo, Oviedo, Spain

Key words: abiotic factors, cultivation, *Gelidium latifolium*, physiology, seasonality

Abstract

Seasonal cycles of environmental factors (temperature, day-length, nutrient concentration) and changes in *Gelidium latifolium* biomass, percentage reproduction and size are given, and non-parametric correlation is used to quantify possible relationships. The results are compared with growth experiments, testing effects of total light dosage, agitation, temperature and Photon Flux Density (PFD). Results of total light dosage × agitation growth experiment show that maximum growth is obtained when plants are cultured at a long photoperiod (16:8 L/D) with agitation. Results of temperature × PFD experiment show that maximum growth is obtained at PFD values higher than $50 \, \mu E \, m^{-2} \, s^{-1}$ at temperatures between 20–25 °C. Possible applications of field studies and culture experiments in management of wild resources and industrial cultivation are proposed.

Introduction

Gelidium resources are used worldwide as a source of agar. Spain is a major producer of agar from two *Gelidium* species: *G. sesquipedale* (Clem.) Born. & Thur. and *G. latifolium* (Grev.) Born. & Thur. The former is the most abundant species, while the latter is restricted, as an exploitable resource, to the central coast of Asturias (northern Spain). Work on *G. latifolium* from Asturias has included studies on abundance and productivity (Fernández & Anadón, 1989; Juanes & Fernández, 1988; Anadón & Fernández, 1986).

Several studies have recently been done on *Gelidium* species (Fredriksen & Rueness, 1989; Rueness & Fredriksen, 1989; Macler & West, 1987; Carter & Anderson, 1986; Carter, 1985; Correa *et al.*, 1985; D'Antonio & Gibor, 1985;

Rao & Kaliaperumal, 1983; Oliger & Santelices, 1981; Maihr & Rao, 1978). These investigations deal with aspects of laboratory cultivation, life-histories in culture and ecological characterization of species using culture data. These studies point to the importance of attempting to complete successfully the life-history of *Gelidium* in culture (Macler & West, 1987) as a means to developing large-scale cultivation systems.

Management procedures must include, as an initial step, biological data (*e.g.* biomass and reproductive cycles) from field studies (Santelices, 1989). Recent use of multivariate and correlation analysis of seaweed field studies (Kautsky & van der Maarel, 1990; Murray & Horn, 1989; Oliger & Santelices, 1981) facilitates these types of studies. Combinations of field work and studies of phenology in culture have been made on some rhodophycean species (Breeman & Guiry, 1989; Guiry et al., 1987; Maggs & Guiry, 1987).

This paper describes a field study on a *Gelidium latifolium* population and its correlation with environmental factors; comparisons between field studies and results of culture experiments are made. Results from growth responses in culture are given as a first step for further studies on laboratory and large-scale cultivation.

Materials and methods

Gelidium latifolium is a rhodophyte present along the European Atlantic coast, from southern Norway (Rueness & Fredriksen, 1989) to the Mediterranean (Cecere & Perrone, 1987). On the central coast of Asturias (northern Spain) this species forms a dense intertidal belt between +0.4 and +0.8 m above Lowest Astronomical Tide (Fernández & Niell, 1981), with summer biomass values up to 0.7 kg m^{-2} (Fernández & Anadón, 1989).

Samples for field studies were collected monthly during low-water spring-tides between January 1988 and October 1990 at Aramar (42° 37′ N, 5° 46′ W). The collecting site is a rocky shore facing NE, located on the eastern face of Cape Peñas (Fig. 1a, b). This area is partially protected from the most severe storms, which occur from October to April. Climatic conditions in this area are fairly predictable. Tides are semi-diurnal, with low-water spring-tides occurring between 0800 and 1100 h GMT.

At each sampling date, the water temperature was taken, and samples for nutrient (NO_3^- + NO_2^-) analysis were collected. Fronds for growth experiments were transported under refrigeration. Nutrient samples were stored frozen in the laboratory (< -20 °C) until analyzed using a Technicon II autoanalyzer. For each sampling period, 150–200 fronds were measured and 50–100 were collected to determine frequencies of reproduction. The term 'frond' refers to an individual thallus arising from one rhizoid. Rhizoids were obtained from 50 × 50 cm quadrats randomly placed in the study area.

In the laboratory, fronds for culture experiments were washed thoroughly with steam-sterilized seawater. Apical sections from healthy, non-reproductive fronds were excised and placed in 500 mL flasks filled with half-strength PES (Guiry & Cunningham, 1984; Maggs, 1983; McLachlan, 1973) prior to use.

Day-length values at the sea surface have been calculated for Aramar, using the equations from Kirk (1983). Biomass values have been calculated as means based on monthly values from Fernández and Anadón (1989), Arrontes (1987) and Anadón and Fernández (1986). Percentages of reproductive structures were calculated as the number of fronds with tetrasporangia in the total number of fronds per sample.

To examine the effect of total light dosage and agitation on growth, the following experimental design was used: 10, 5–6 mm apical segments were placed in each of 16 experimental units (4 photoperiods × 2 agitation levels × 2 flasks per treatment). Photoperiods used were 8:16, 8:7.5:1:7.5 (NB: night break), 12:12 and 16:8 light/dark. Irradiance (55 μE m^{-2} s^{-1}) was provided by cool-white 36 W fluorescent tubes. The use of two flasks per treatment permitted testing of the effect of the culture vessel. The experiment was done in a controlled temperature chamber (20 °C during 'day' conditions, 18 °C during 'night' conditions). Agitation was provided by continuous aeration. The medium was changed weekly and the experiment ran for 10 wks. At the end of the experiment, apical segments were weighed individually, and maximum length and number of branches or branch initials per segment were recorded. Data were analyzed using ANOVA. Logarithmic transformation satisfied the homogeneity of variance assumptions (Cochran's test, $p > 0.05$; Winer, 1971). Statistical analyses were performed using BMDP and STATGRAPHICS statistical packages.

A cross-gradient growth-table (Yarish et al., 1979) was used to test the combined effect of PFD and temperature. It provided 7 temperatures (10–30 °C) and 4 PFD values (50–300 μE m^{-2} s^{-1}). Photoperiod was 16:8 L/D. At each combination of PFD and temperature, 5-apical portions were inoculated in 25 ml polystyrene Petri dishes filled with full-strength PES (Fredriksen & Rueness, 1989). A total of 28 dishes were used. No replication was considered. The medium was changed every 2 d and the experiment ran for 30 d. Measurements and analysis were done as above.

Results

Figure 1 shows the seasonal cycle for temperature and day-length at the surface, and Fig. 2 the $NO_3^- + NO_2^-$ concentrations at Cape Peñas.

Maximum day-length, highest temperature and PFD values, and lowest nutrient concentrations

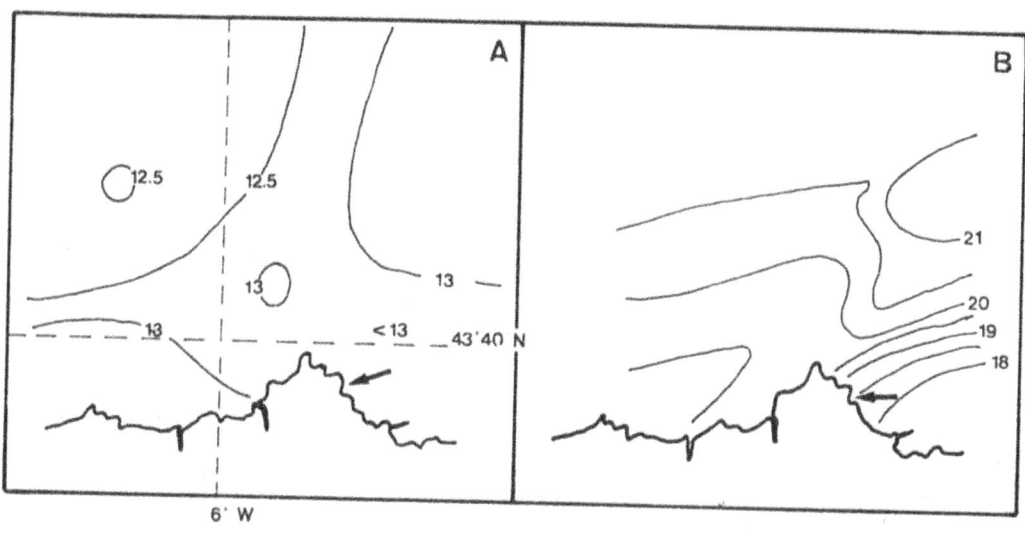

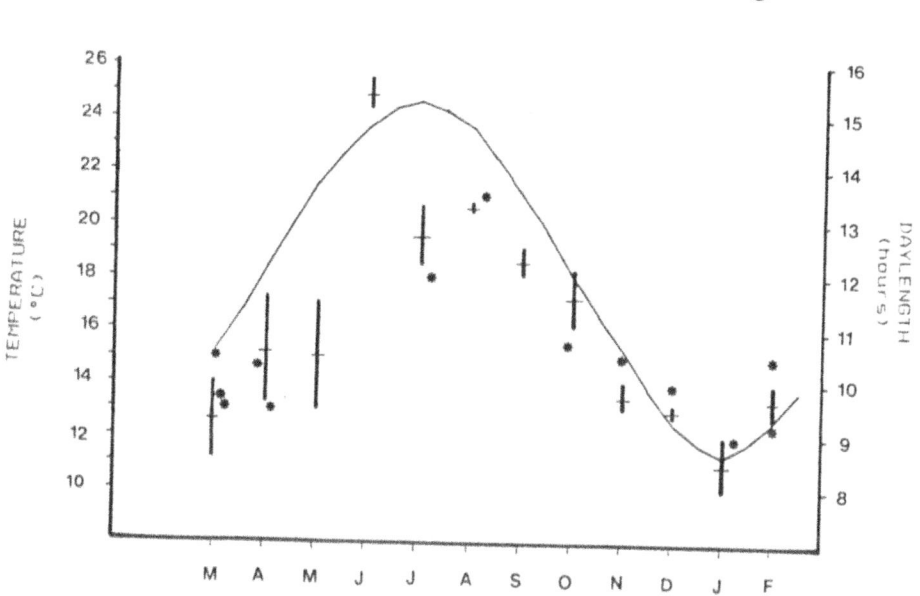

Fig. 1. Values of surface seawater temperature in March (A) and September (B). Arrows show location of study. (C) Monthly values of seawater temperature (*), temperature in the *G. latifolium* belt (bars; Arrontes, 1987) and day-length ———.

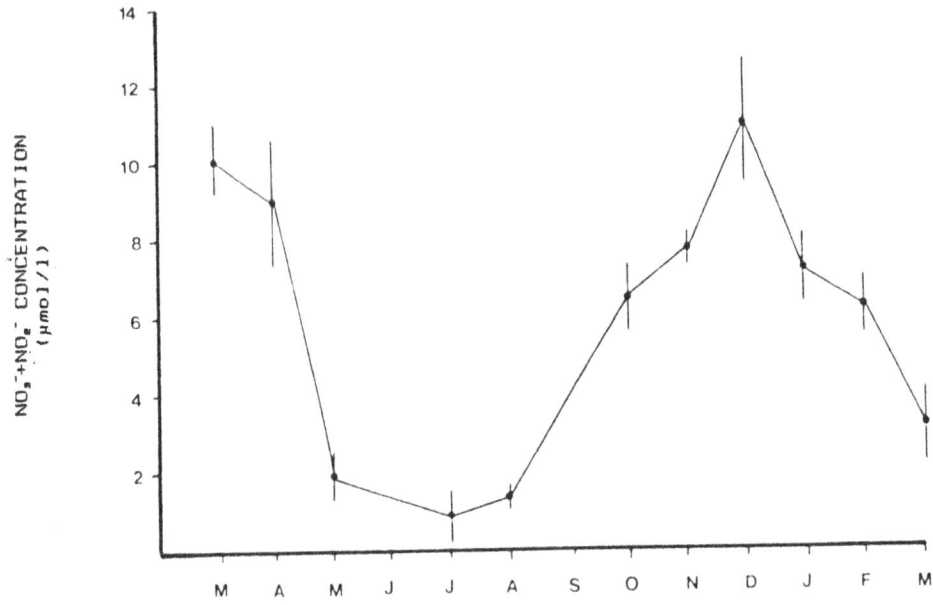

Fig. 2. Monthly mean values ± S.D. of nitrate + nitrite concentrations (µmoles/L) at Aramar. *n* = 6.

occur in summer. During winter, nutrient concentrations are maximal while light conditions are low. Changes in maximum and minimum values in abiotic factors are correlated with maximum values in *Gelidium* biomass and percentage reproduction (Table 1). Growth (as increase in biomass) and reproduction (as percentage of reproductive fronds) also show a seasonal pattern (Fig. 3a). No seasonality appears in monthly variations in frond size (Fig. 3b).

Using values from Figs. 1–3, Spearman rank correlation analysis was performed as an example. Only paired data from October 1989 to September 1990 have been included in the analysis, except for biomass values, which were included as mean monthly values from Fernández and Anadón (1989), Arrontes (1987) and Anadón and Fernández (1986). Results are shown in Table 2. Both biomass and percentage reproduction show strong correlation with abiotic factors. Highest values for correlation coefficients are for biomass and surface day-length and for percentage reproduction and seawater temperature. There is also a strong inverse correlation between biomass and percentage reproduction.

Laboratory experimental data are consistent

Table 1. Maximum (Max.) and minimum (Min.) monthly values for abiotic and *Gelidium latifolium* variables. month (M).
a: data from oceanographic stations
b: measured on the horizon during low tides
c: mean for the previous 15 days
1: this work; 2: Botas *et al.* 1989; 3: Arrontes, 1987; 4: Anadón & Fernández, 1986; 5: Fernández & Anadón, 1989; 6: Juànes, 1983.

Variable	Max.	(M)	Min.	(M)	Source
Temp. (°)	21	Aug.	12	Jan.	1
	20	Sep.	11.2	Jan.	2a
	24.8	Jun.	11	Jan.	3b
Day length (h) c	15.04	Jun.	8.87	Jan.	1
NO_3^- (µM)	10.13	Mar.	0.86	Jul.	1
	5.98	Oct.	0.04	Jun.	2a
PO_4^{3-} (µM)	1.74	Oct.	0.08	Jul.	1
	1.20	Oct.	0.05	Jul.	2a
Mean size (mm)	90	Jul.	62	Jan.	1
% reproductive	95	Jan.	8	Sep.	1
	100	Feb.	0	Sep.	6
Biomass (g/m²)	712.3	Jul.	306.6	Feb.	4
	822.5	Jul.	144.4	Jan.	3
	791	Jul.	144	Jan.	5

with field data. Growth measured as increases in total biomass (Fig. 4), final frond weight (Fig. 5a) and final frond length (Fig. 5b) was highest when fronds were cultured at the longest day-length.

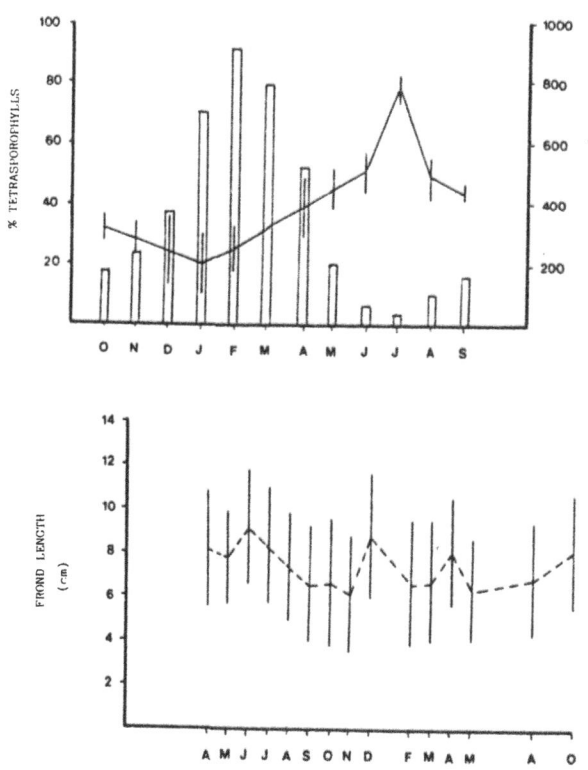

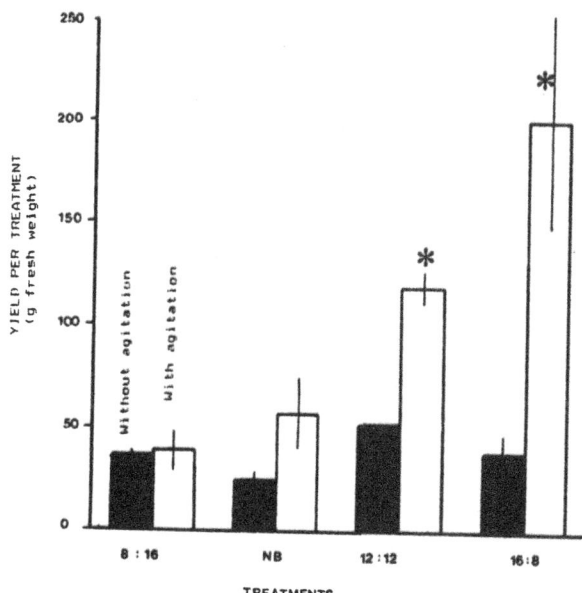

Fig. 4. Yield (g fresh weight per treatment) of apical segments of *G. latifolium* cultured under different photoperiods. *means significant (*p* < 0.01, Student-Neumann-Keuls test) differences in mean yield of two replicates.

Fig. 3. Monthly values of biomass from Fernández & Anadón (1989), Arrontes (1987) and Fernández & Anadón (1986); and percentage reproduction of *G. latifolium* (white bars) sampled between October 1989–September 1990 (A). B: Mean ± S.D. length of *G. latifolium* fronds collected at Aramar.

Table 2. Matrix of Spearman rank correlation coefficients between variables described in Figs 2–6. ST: seawater temperature; N: NO_3^- + NO_2^- concentration; dL: daylength; % rep.: percentage reproduction. (Degree of significance in parenthesis).

	ST	N	dL	Biomass	% rep.
ST	1				
N	− 0.7280	1			
	(0.0158)				
dL	0.6962	− 0.8601	1		
	(0.0210)	(0.0043)			
Biomass	0.7774	− 0.8951	0.9650	1	
	(0.0099)	(0.0030)	(0.0014)		
% rep.	− 0.7986	0.6643	− 0.7063	− 0.8252	1
	(0.0081)	(0.0276)	(0.0192)	(0.0062)	

Table 3. ANOVA results for transformed variables Ln (final weight) (A) and Ln (final length) (B).
a: flask and error values pooled as *p* value for flasks >0.25 (Underwood, 1983).
ns: *p* > 0.05
*: *p* > 0.05
**: *p* > 0.01

Source	df	F-statistic	
Photoperiod	3	11.04***	A
Agitation	1	38.43***	
(*P* × A)	3	4.55*	
Flask			
(nested (*P* × A)	8	2.93*	
Error	126		
Photoperiod	3	4.24*	B
Agitation	1	4.73*	
(*P* × A)	3	8.25***	
Pooled flask			
+ Error (a)	134		

These values increased when the alga was cultured in medium with constant agitation. Table 3

shows the results of ANOVA. Note also (Fig. 5b) the effect of night break without agitation in final frond length, even when total light dosage was higher in the night-break treatment. There were no

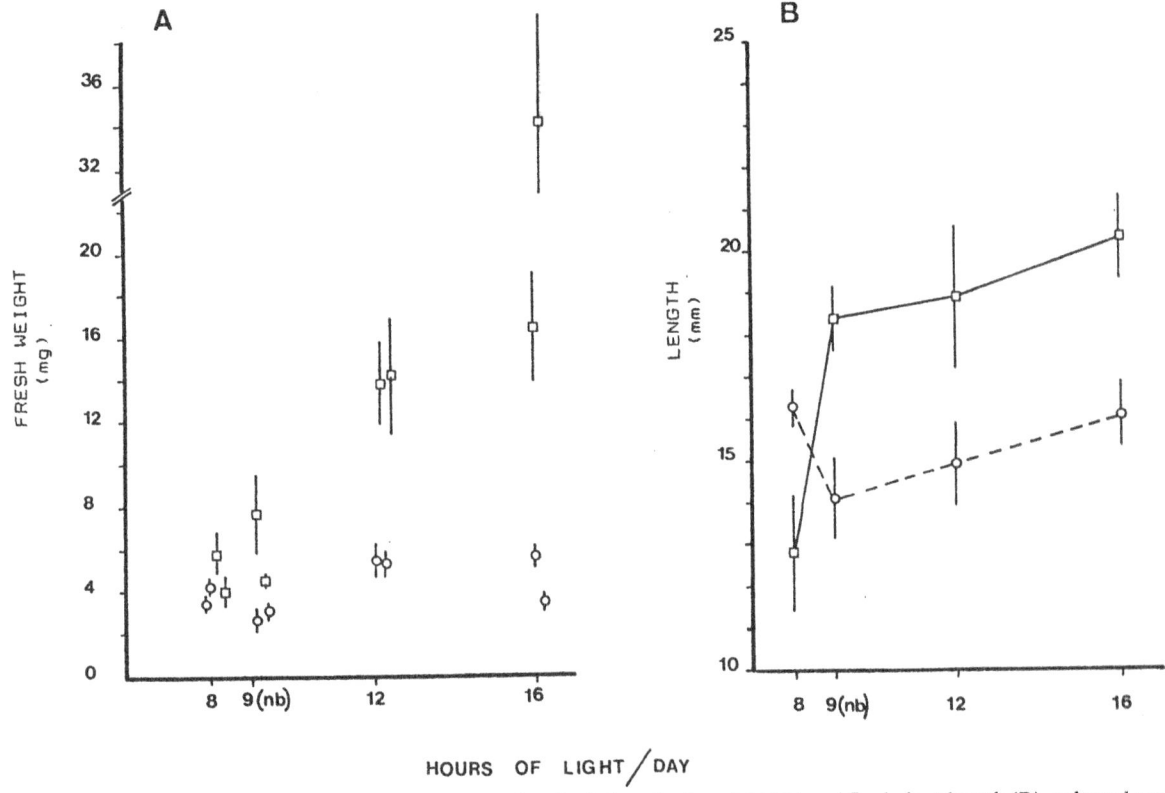

Fig. 5. Growth of *G. latifolium* apical portions, measured as final plant fresh weight (A) and final plant length (B), cultured under different photoperiod conditions. Values are mean ± S.E. (*n* = 10). (□): with agitation; (○): without agitation.

significant effects related to the number of branches, branch initials or reproductive structures in this experiment.

Results from temperature × PFD experiments show that growth responses to temperature depended on PFD. No variation appears in growth response of plants cultured at the lowest PFD (Fig. 6a), while plants cultured at higher PFD values show a broad optimum between

Table 4. ANOVA results for temperature and PFD effects on total biomass per treatment. No interaction is shown in the absence of replication.

*: $p < 0.05$
**: $p < 0.01$

Source	df	F-statistic
Temperature	6	5.44**
PFD	3	3.68*
Error	18	

20–25 °C (Fig. 6b). ANOVA results for this experiments are shown in Table 4.

Discussion

Relationships between environmental factors and seaweed abundance and phenology have been discussed elsewhere (Kautsky & Van der Maarel, 1990; Brophy & Murray, 1989; Egan *et al.*, 1989; Guiry *et al.*, 1987; Maggs & Guiry, 1987). *Gelidium latifolium* abundance and reproductive patterns can be related to environmental factors. Maximum biomass values were coincident with summer seawater temperature, longer day-lengths and lowest nutrient concentrations; on the other hand, maximum reproduction occurs in winter, when seawater temperatures are lowest and day-lengths are shortest. The same has been described for other intertidal *Gelidium* species (Montalva &

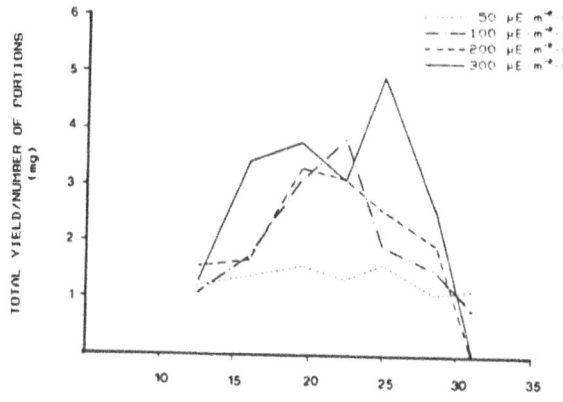

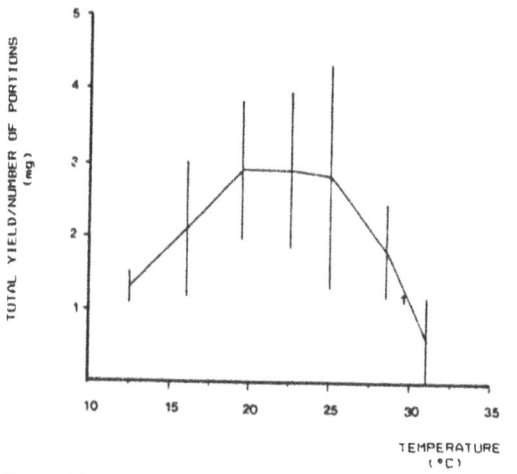

Fig. 6. Fresh weight per dish/number of apical portions per dish of *G. latifolium* fronds cultured under different temperature and PFD conditions (A). B: Mean ± S.D. fresh weight per apical portion of fronds cultured under different temperatures, after pooling results for different PFD conditions.

Santelices, 1981). 'Anticipative'-type of reproductive response (*sensu* Kain, 1989) is contrasted to 'responder'-type of growth response. No correlations have been found between frond size and environmental factors in this or in other studies in the same area (Juanes & Fernández, 1988). Other factors, abiotic (*e.g.* frequency of storms) and biotic (*e.g.* grazing pressure), could result in monthly variations in plant size.

Statistical-rank analysis served to quantify predicted relationships (direct or indirect), although, in this case, it was only used as *a priori* approximation. However, it must be remembered that this kind of analysis yields not causal but

correlative relationships (Murray & Horn, 1989). Single and multiple correlation procedures have been recently used in studies of seaweed seasonality (Kautsky & Van der Maarel, 1990; Murray & Horn, 1989).

Differences in response to environmental factors between biomass and percentage reproduction could result from differences in the nature of control processes. Growth of perennial species is usually a direct response to factors fulfilling primary physiological requirements, while reproduction is the result of multiple interactive complex processes activated by 'trigger' signals, such as short day-length or low temperature (Lüning & tom Dieck, 1989). The existence of this restricted 'reproductive window' is frequent in rhodophytes (Breeman & Guiry, 1989; Maggs & Guiry, 1987).

Data from Fernández and Anadón (1989), Arrontes (1987) and Anadón and Fernández (1986) show that maximum standing stock values of *Gelidium latifolium* occur in late summer, even when nutrient levels are $<2 \mu M\ NO_3^-$. Akatsuka (1986) cites various examples of effects of previous winter temperatures on summer biomass of Japanese Gelidiales. Maximum presence of tetrasporangia is favoured by low seawater temperatures ($<13\ °C$) and short day-lengths (<10 h). Presence of reproductive structures, however, does not necessarily indicate existence of viable spores or gametes (Maggs & Guiry, 1987). Studies on the viability of spores so produced should be made. Also, the effect of microclimatic local conditions caused by turbidity, tidal cycles, wave action and other factors (Breeman & Guiry, 1989) should be included in further field studies.

Enhanced growth of plants cultured under long photoperiods have been cited for *G. pusillum* (Maihr & Rao, 1978) and two Chilean species of *Gelidium* (Correa *et al.*, 1985). Results from *G. latifolium* seem similar, although combined effects of temperature and total light dosage (h light/day × PFD) have been proposed as having more influence (Oliger & Santelices, 1981). Results from my study indicate that maximum summer biomass could result from optimum seawater temperature, light dosage and day-length for growth, as shown by culture experiments. These results are consistent with others for *Gelidium* species of temperate areas (Fredriksen & Rueness, 1989; Oliger & Santelices, 1981). Studies of the effects of these abiotic factors on reproductive responses are needed for completion of the life-history in culture (Macler & West, 1987).

Culture experiments on *Gelidium* species show that environmental factors have to be considered when undertaking laboratory cultivation (Macler & West, 1987; D'Antonio & Gibor, 1985; Maihr & Rao, 1978). Recent reviews (Santelices, 1988; Akatsuka, 1986) note the scarce data available on successful attempts of *Gelidium* cultivation. Mariculture of *G. sesquipedale* has been described recently (Seoane-Camba, 1989). No examples of cultivation procedures for *G. latifolium* are found. Large-scale procedures, succesfully applied to other rhodophyte species (Lignell *et al.*, 1987), should be tried. However, laboratory culture and pilot trials are needed before industrial cultivation is attempted. Field studies could improve quantity and quality of yield and could be used as systems to test effects of variations of abiotic factors on resource responses. Santelices (1989) gives a biologically-based model for seaweed resource management. Field studies are clearly a necessary step prior to development of harvesting programmes.

Acknowledgements

This work is taken from a PhD. Thesis on *Gelidium latifolium* and supported by P.F.P.I. (Spanish Ministry of Science) and included in a project on 'Temática asturiana', Universidad de Oviedo. Comments of Drs J. Rueness, S. Fredriksen and Dr C. Fernández were very invaluable in preparing and designing the experiments. Corrections of Dr Arrontes and L. Gutiérrez on an early version of the manuscript and comments and criticism from two anonymous reviewers are acknowledged.

References

Anadón, R. & C. Fernández, 1986. Comparación de tres comunidades intermareales con abundancia de *Gelidium latifolium* (Grev.) Born. et Thur. en la costa de Asturias. Inv. Pesq. 50: 353–366.

Arrontes, J. M., 1987. Estrategias adaptativas de isópodos en la zona intermareal. PhD. Thesis. Universidad de Oviedo, Oviedo, Spain.

Akatsuka, I., 1986. Japanese Gelidiales (Rhodophyta), especially *Gelidium*. Oceanogr. Mar. Biol. annu. Rev., 24: 171–263.

Botas, A., E. Fernández, A. Bode & R. Anadón y E. Anadón, 1989. Datos básicos de las campañas 'COCACE'. I.- Hidrografía, nutrientes, fitoplancton y seston en el Cantábrico central. Rev. Biol. Univ. Oviedo (supl.) Vol. 1.

Breeman, A. M. & M. D. Guiry, 1989. Tidal influences on the photoperiodic induction of tetrasporogenesis in *Bonnemaisonia hamifera* (Rhodophyta). Mar. Biol. 102: 5–14.

Brophy, T. C. & S. N. Murray, 1989. Field and culture studies of a population of *Endarachne binghamiae* (Phaeophyta) from southern California. J. Phycol. 25: 6–15.

Carter, A. R., 1985. Reproductive morphology and phenology, and culture studies of *Gelidium pristoides* (Rhodophyta) from Port Alfred in South Africa. Bot. mar. 28: 303–311.

Carter, A. R. & R. J. Anderson, 1986. Seasonal growth and agar contents in *Gelidium pristoides* (Gelidiales, Rhodophyta) from Port Alfred in South Africa. Bot. mar. 29: 117–123.

Cecere, E. & C. Perrone, 1987/88. First contribution to the knowledge of macrobenthic flora of the Amendolara seamount (Ionian sea). Oebalia, 14: 43–68.

Correa, J., M. Avila & B. Santelices, 1985. Effects of some environmental factors on growth of sporelings in two species of *Gelidium* (Rhodophyta). Aquaculture, 44: 221–227.

D'Antonio, C. M. & A. Gibor, 1985. A note of some influences of photon flux density on the morphology of germlings of *Gelidium robustum* (Gelidiales, Rhodophyta) in culture. Bot. Mar. 28: 313–316.

Egan, B., A. Vlasto & C. Yarish, 1989. Seasonal acclimation to temperature and light in *Laminaria longicruris* de la Pyl. (Phaeophyta). J. exp. mar. Biol. Ecol. 129: 1–16.

Fernández, C. & R. Anadón, 1989. Cartografiado de biomasa de campos intermareales de *Gelidium sesquipedale* (Clem.) Born. et Thur. y *G. latifolium* (Grev.) Born. et Thur. en la costa de Asturias (N. de España). Inf. Técn. Inv. Pesq. 149.

Fernández, C. & F. X. Niell, 1981. Distribución espacial del fitobentos en los horizontes inferiores del sistema intermareal rocoso de cabo Peñas (Asturias). Inv. Pesq., 46: 121–141.

Fredriksen, S. & J. Rueness, 1989. Culture studies of *Gelidium latifolium* (Grev.) Born. et Thur. (Rhodophyta) from Norway. Growth and nitrogen storage in response to varying photon flux density, temperature and nitrogen availability. Bot. mar. 32: 539–546.

Guiry, M. D. & E. Cunningham, 1984. Photoperiodic and temperature responses in the reproduction of northeastern *Gigartina acicularis* (Rhodophyta: Gigartinales). Phycologia 23: 357–367.

Guiry, M. D., W. R. Kee & D. J. Garbary, 1987. Morphology, temperature and photoperiodic responses in *Audouinella botryocarpa* (Harvey) Woelkerling (Acrochaetiaceae, Rhodophyta) from Ireland. Gior. Bot. Ital., 121: 229–246.

Juanes, J., 1983. Contribución al conocimiento de la biología de *Gelidium latifolium* (Grev.) Thuret et Born. Memoria de Licenciatura. Universidad de Oviedo, Oviedo, Spain.

Juanes, J. & C. Fernández, 1988. Ciclo anual y producción de *Gelidium latifolium* (Grev.) Thur. et Born. (1876), en la región de cabo Peñas (Asturias, N de España). Inv. Pesq. 52: 109–122.

Kain, J. M., 1989. The seasons in the subtidal. Br. phycol. J., 24: 203–215.

Kautsky, H. & E. Van der Maarel, 1990. Multivariate approaches to the variation in phytobenthic communities and environmental vectors in the Baltic Sea. Mar. Ecol. Prog. Ser. 60: 169–184.

Kirk, J. T. O., 1983. Light and Photosynthesis in Aquatic Ecosystems. Cambridge University Press, London. 323 pp.

Lignell, Å., P. Ekman & M. Pedersen, 1987. Cultivation technique for marine seaweeds allowing controlled and optimized conditions in the laboratory and on a pilot scale. Bot. mar. 30: 417–424.

Lüning, K. & I. tom Dieck, 1989. Environmental triggers in algal seasonality. Bot. Mar. 32: 389–397.

McLachlan, J., 1973. Growth media-marine. In: Handbook of Phycological Methods- Culture Methods and Growth Measurements. J. R. Stein (ed.), Cambridge University Press, London, pp. 26–51.

Macler, B. A. & J. A. West, 1987. Life history and physiology of the red alga, *Gelidium coulteri*, in unialgal culture. Aquaculture, 61: 281–293.

Maihr, O. P. & P. S. Rao, 1978. Culture studies on *Gelidium pusillum* (Stackh.) Le Jolis. Bot. Mar. 21: 169–174.

Maggs, C. A., 1983. A phenological study of the epiflora of two maerl beds in Galway Bay. PhD. Thesis, Galway College, Galway, Ireland.

Maggs, C. A. & M. D. Guiry, 1987. Environmental control of macroalgal phenology. In: Plant Life in Aquatic and Amphibious Habitats. R. M. M. Crawford (ed.), Blackwell Scientific Publications, London. pp. 359–373.

Montalva, S. & B. Santelices, 1981. Interspecific interference among species of *Gelidium* from central Chile. J. exp. mar. Biol. Ecol. 53: 77–88.

Murray, S. N. & M. H. Horn, 1989. Seasonal dynamics of macrophyte populations from an eastern North Pacific rocky-intertidal habitat. Bot. mar., 32: 457–473.

Oliger, P. & B. Santelices, 1981. Physiological ecology studies on chilean Gelidiales. J. exp. mar. Biol. Ecol. 53: 65–76.

Rao, M. U. & N. Kaliaperumal, 1983. Effects of environmental factors on the liberation of spores for some red algae of Visakhapatnam coast. J. exp. mar. Biol. Ecol. 70: 45–53.

Rueness, J. & S. Fredriksen, 1989. Field and culture studies of *Gelidium latifolium* (Grev.) Born. & Thur. (Rhodophyta) from Norway. Sarsia 74: 177–185.

Santelices, B., 1989. Algas Marinas de Chile. Distribución Ecología, Utilización y Diversidad. Ediciones Universidad Católica de Chile. Santiago de Chile. 393 pp.

Santelices, B., 1988. Synopsis of biological data of the seaweed genera *Gelidium* and *Pterocladia* (Rhodophyta). FAO Fish. Synop., 145: 55 pp.

Seoane-Camba, J. A., 1989. On the possibility of culturing *Gelidium sesquipedale* by vegetative propagation. Proceedings of the Second Workshop of COST 48 subgroup 1. J. M. Kain, J. W. Andrews and B. J. McGregor (eds), Brussels.

Underwood, A. J., 1981. Techniques of analysis of variance in experimental marine biology and ecology. Ecol. Monogr. 59: 433–463.

Winer, B. J., 1971. Statistical Principles in Experimental Design. McGraw-Hill, New York. 907 pp. (Second edition).

Yarish, C., K. W. Lee & P. Edwards, 1979. An improved apparatus for the culture of algae under varying regimes of temperature and light intensity. Bot. mar. 22: 395–397.

Hydrobiologia **221**: 77–82, 1991.
J. A. Juanes, B. Santelices & J. L. McLachlan (eds), International Workshop on Gelidium.
© 1991 *Kluwer Academic Publishers.*

Photosynthesis of *Gelidium sesquipedale*: effects of temperature and light on pigment concentration, C/N ratio and cell-wall polysaccharides

Maria Torres, F. Xavier Niell & Patricia Algarra
Department of Ecology, Faculty of Sciences, University of Malaga Campus de Teatinos, 29071 Malaga, Spain

Key words: Gelidium sesquipedale, photoadaptation, pigments, cell-wall polysaccharides, C : N ratio

Abstract

Gelidium sesquipedale, the most important raw material for the extraction of Spanish agar, was studied ecophysiologically. Plant material came from a stressed intertidal system of high temperatures and high photon fluences. The plants were photoresistant and showed saturation of photosynthesis near a PFD of 300 μmol m^{-2} s^{-1} and a compensation point under 100 μmol m^{-2} s^{-1}. Temperature and irradiance have a synergistic effect on photosynthesis. Pigment concentration decreased with irradiance except in the case of carotenes. Carbon and nitrogen were maximal at maximal photosynthetic rates, and the C : N ratio increased with increasing irradiances together with increases in cell-wall polysaccharides.

Introduction

Gelidium sequipedale (Clem.) Born. et Thur., the most important raw material used for the extraction of Spanish agar, grows on rocky shores along the Spanish and Portuguese Atlantic coasts. Seoane-Camba (1965) within our population located on the Strait of Gibraltar. Previous studies carried out on this population (Establier, 1964; Seoane-Camba, 1965) did not deal with the ecophysiological aspects presented in this paper, and no previous literature is available concerning the behaviour of this species.

The aim of our work was to acquire basic knowledge to assist in future development of extensive cultivation of local populations of *Gelidium sesquipedale* which are adapted to the extreme conditions of irradiance and temperature on the southern coasts of Spain. Photosynthesis is used as the main descriptor of the responses of *G. sesquipedale*. The adaptive changes are related to total cell-wall polysaccharides and to phycobilin content in the thalli. Nitrogen can be removed from phycobilins in cases of nitrogen starvation (Frederiksen & Rueness, 1989). Plants living in the transitional region within the Atlantic Ocean and Mediterranean Sea are under severe stress (Niell *et al.*, 1989). In this region, the specimens of *Gelidium sesquipedale* are found only in limestone crevices of the low intertidal zone; the tidal regime has a low amplitude (1 m). The splash of the waves is very important in preventing desiccation of the algae, which otherwise would be exposed to air temperatures of 30° to 40 °C in summer.

Nutrient limitation in seawater results in restriction in growth of *G. sesquipedale*. Nitrate concentrations were maximal in February, August and November, while throughout the rest of the year concentrations were very low. Phosphate concentrations were maximal as nitrogen concentrations were maximal, and often phosphate fell below 1 μg-at l^{-1}. Mean Photon Fluence Density measured in the crevices in which *G. sesquipedale* plants live was 140 $\pm$ 68 μmol m^{-2} s^{-1}.

In this peculiar environment, the flora is impoverished in species related to both Atlantic and Mediterranean floras, and plants of all species are stunted. In the populations of *G. sesquipedale* no fronds of more than 0.15 m in length were ever found.

Materials and methods

Plants collected on the coast were transported to the laboratory and cleaned of epiphytes, rinsed with filtered seawater and placed in double-walled incubation chambers of 0.118 l capacity. A constant temperature (12° or 22 °C) was maintained by circulation of water within the interwall space (Selecta cooler Model 398).

To study irradiance-photosynthetic relationships, different irradiances were obtained by a set of neutral-density filters prepared as smoked slides. The light source was a slide projector with a 150 W halogenic lamp; irradiances were tested carefully with in the incubation chamber; outside the values are greater. Photosynthetic measurements were made under a range of PFD's, between darkness (to estimate respiration) and 1900 μmol m^{-2} s^{-1} to determine maximal photosynthetic rates (P_m), saturation irradiance (I_k) and α: P_m/I_k, where α is an estimation of photosynthetic efficiently measured from the exponential slope of the P : I curve.

Oxygen evolution, measured with a YSI 5750 electrode, was determined after 20 min incubation (Aranda *et al.*, 1984), using the method of Kanwisher (1966).

Pigments were extracted into acetone and their concentrations determined spectrophotometrically by the method of Talling and Driver (1963) for chlorophyll *a*, and the equations of Strickland and Parsons (1968) were used to estimate carotenoid concentrations. To estimate concentrations of water-soluble phycobiliproteins and to avoid interferences by acidification in the extracts

used to estimate lipo-soluble pigments, the extraction was done in buffer, using sodium phosphate at pH 6.5 (Algarra & Niell, 1987). Parallel with samples from the same thallus, calculations were made using the trichromatic expressions of Rosenberg (1981). Phycobiliprotein concentrations are expressed either on a weight to weight basis (mg g^{-1}) or as the number of chromophores (Kursar & Alberte, 1983). The equivalence of chromophores to total molecules was based on O'Carra and O'Heocha (1976).

To study C : N ratios, samples (3 to 4 g) of fronds were ground in a mortar and dried at 60 °C; 3.5 $\times$ 10^{-3} g of the homogenate was used for the C : N analysis (Perkin-Elmer 240 C Elemental Analyzer); the remaining material was used for agar extraction.

A subsample of 0.5–1 g (dry wt) of plant material was used for extraction of cell-wall polysaccharides, assumed to be agar. The method described by Santos and Doty (1983) was used, although plant materials were ground and rehydrated in a proportion of 1 : 10 for 8 to 12 h. A similar modification was carried out by Establier (1964) for plants of the same area.

Results

Maximal photosynthetic rates were three times greater at 22 °C than at 12 °C (Fig. 1). *Gelidium sesquipedale* withstood high irradiances without photoinhibition at the low temperature, while PFD's exceeding 250 μmol m^{-2} s^{-1} resulted in a decrease in photosynthesis at 22 °C.

The compensation point occurred at 40 μmol m^{-2} s^{-1} in plants incubated at 12 °C, and at 80 μmol m^{-2} s^{-1} for plants incubated at 22 °C. This difference resulted from increased respiration at the higher temperature.

Photosynthesis based on chlorophyll *a* showed a pattern similar to that based on dry weight. The synergistic effect of irradiance and temperature is also evident (Tukey test; α = 0.01; Poole, 1974) in this plot.

Pigment concentrations were greater at 22 °C than at 12 °C. All pigment concentrations were

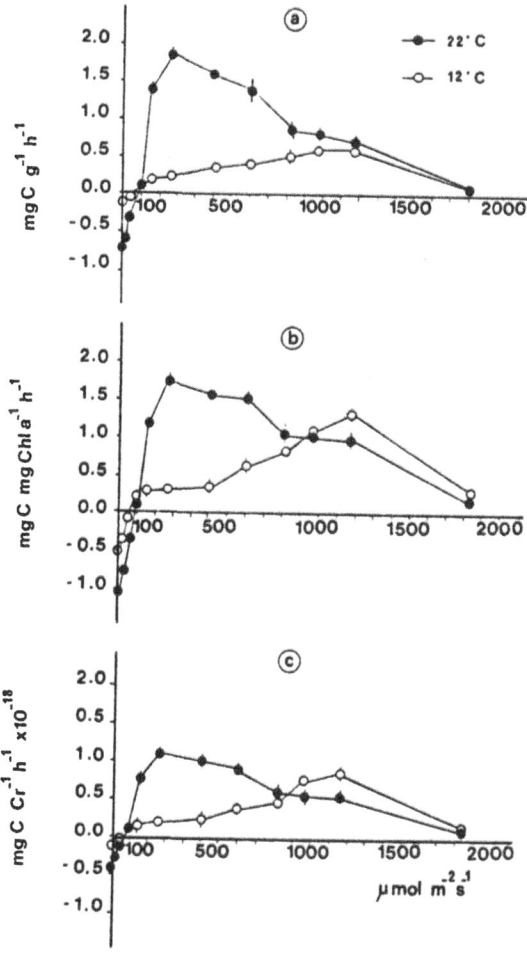

Fig. 1. Photosynthetic rates (Ps) of *Gelidium sesquipedale* as functions of PFD (I) at two temperatures, 22 °C (solid circles) and 12 °C (open circles). Photosynthetic values are expressed in terms of dry weight (a), chlorophyll *a* content (b) and chromophores (c). PFD is μmol m^{-2} s^{-1}.

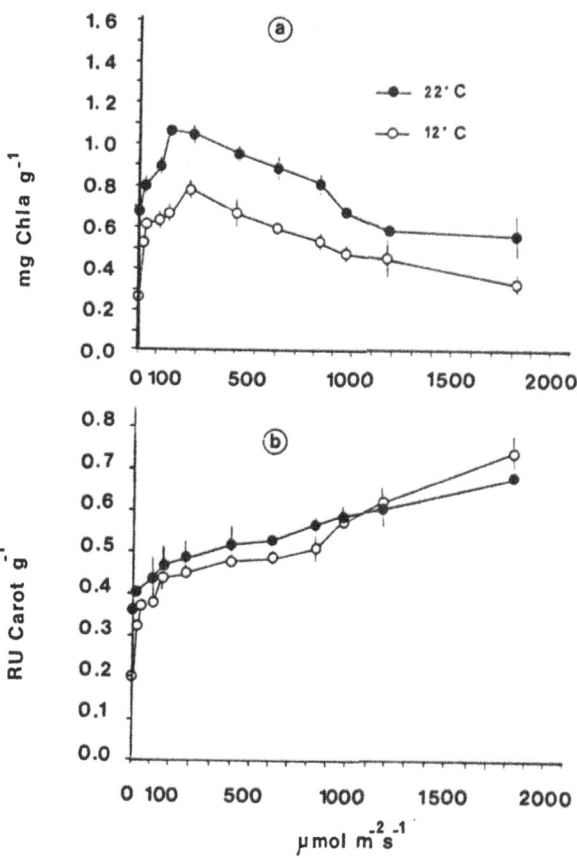

Fig. 2. Concentrations of chlorophyll *a* (a) and carotenoids (b) as a function of temperature and PFD (μmol m^{-2} s^{-1}).

less at the high irradiance, except carotenes (Fig. 2, 3) (Anova; $\alpha = 0.01$; Sokal & Rohlf, 1969).

No significant differences (X^2; $\alpha = 0.01$) were detected in C:N ratios with temperature (Fig. 4). The carbon concentration was maximal at 280 μmol m^{-2} s^{-1}. Nitrogen concentrations declined slowly with increasing irradiances, and differences between nitrogen concentrations with temperature were not significant (Anova; $\alpha = 0.01$; Sokal & Rohlf, 1969).

The quantity of polysaccharides increased with increasing irradiance (Fig. 5), and the increment

was greater at 22 °C than at 12 °C. Below the compensation point, the percentage of polysaccharides in *G. sesquipedale* was very low; when the plants were incubated at or above 100 μmol m^{-2} s^{-1} the concentration was about 20%, and thereafter increased with increasing temperature.

Discussion

As a working hypothesis, we have assumed that physiological strains of *Gelidium sesquipedale* occur in relation to latitude. Although *G. sesquipedale* did not exhibit clear photoinhibition, it appears to be less tolerant of high irradiances than other intertidal species in the same area; therefore, the photosensitivity of several species of

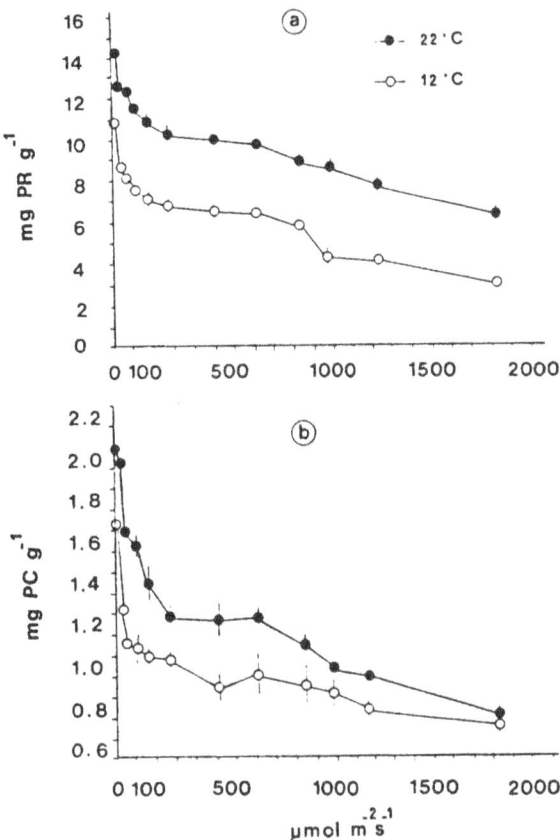

Fig. 3. Concentrations of phycoerythrin (a) and phycocyanin (b) as a function of temperature and PFD (μmol m^{-3} s^{-1}).

Gelidium (Chapman, 1966; Ogata & Matsui, 1965) is also applicable to *G. sesquipedale*.

Saturation irradiances for *Gelidium amansii* Lamouroux were found to be lower than those obtained for *G. sesquipedale* at 22 °C while similar to those obtained at 12 °C. Other species occurring on the same coast as *G. amansii*, showed saturation PFD's at 378 μmol m^{-2} s^{-1} (Ogata & Matsui, 1965). In the south of Spain, *G. sesquipedale* showed variable values, depending upon temperature, suggesting different seasonal responses. Compared with other species, no definitive statements can be made regarding the photosensitivity of *G. sesquipedale*.

It is generally known that the compensation point also changes with temperature, because of lineal increases in respiration as temperature

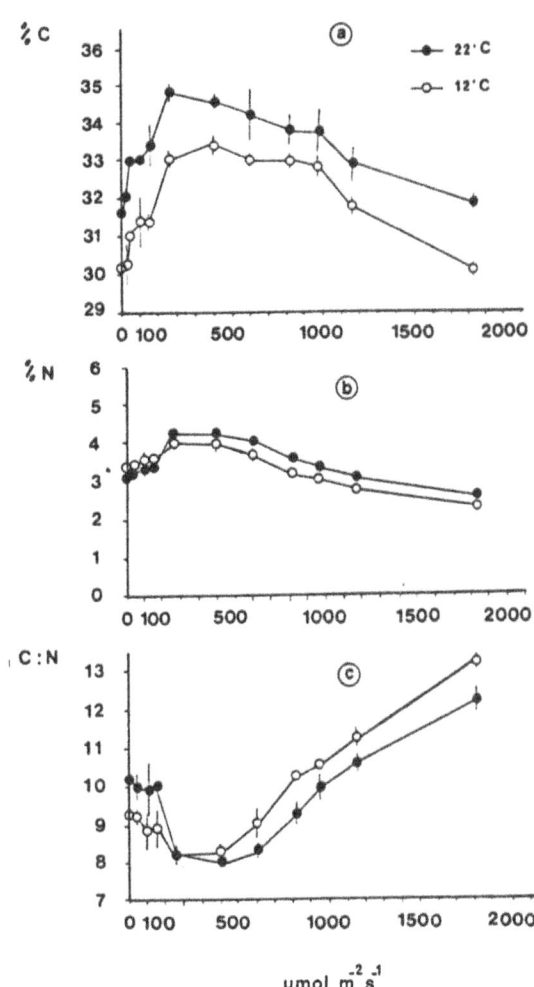

Fig. 4. Carbon (a) and nitrogen (b) concentrations (% on dry matter) in *Gelidium sesquipedale* and C:N ratios (c) as a function of temperature and PFD (μmol m^{-2} s^{-1}).

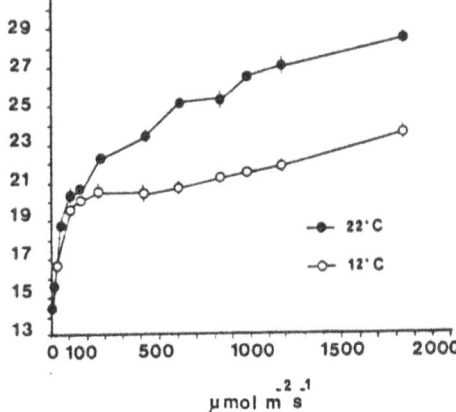

Fig. 5. Changes in cell-wall polysaccharides (% on dry matter) as a function of temperature and PFD's (μmol m^{-3} s^{-1}).

increases. Values obtained for different species in the south of Spain seem to be related to their intertidal positions, while species occurring at similar tidal levels as *G. sequipedale* have compensation points below 100 μmol m^{-2} s^{-1}.

Our results have indicated a synergistic effect between light and temperature on photosynthetic performance, and future studies of photosynthesis in this species must take this interaction into account.

Curves expressing rates of photosynthesis related to chlorophyll *a* were well correlated with those expressing the same rates on a dry-weight basis, but their shape is smoothed. The meaning of both functions is not the same, given that chlorophyll *a* was more concentrated at the higher temperature and was not a conservative variable during the experiment. The most reliable estimation of photosynthetic efficiency is, thus, the relation between this parameter and the number of chromophores of all pigments.

C:N ratios increased at photoinhibition intensities and declined at irradiances at which photosynthesis was higher. This is consistent with previous findings in other species (Niell, 1976). It is interesting that some exceptions have been noted (Jimenez & Niell, in press); no significant differences were related to temperature.

More studies on the quality of the carbon found in *Gelidium* must be undertaken in order to determine if increases of polysaccharides in the cell wall were agar or other products. Macler (1986) suggested that agar is not a storage of carbon as a terminal metabolite, but our results suggest the contrary.

Acknowledgements

This research was supported by project n° 12710 funded by the National Government of Andalucia and is a part of the Ph D dissertation of Maria Torres.

References

Algarra, P. & F. X. Niell, 1987. Structural adaptations to light reception in two morphotypes of *Corallina elongata* Ellis et Soland. Mar. Ecol. 8: 253–261.

Aranda, J., F. X. Niell & J. A. Fernandez, 1984. Production of *Asparagopsis armata* (Harvey) in a thermally-stressed intertidal system of low tidal amplitude. J. exp. mar. Biol. Ecol. 84: 285–295.

Chapman, V. J., 1966. The physiological ecology of some New Zealand seaweeds. Proc. Int. Seaweed Symp. 5: 29–54.

Establier, R., 1964. Variación estacional de la composicion química, extracción y caracteristicas del agar-agar de algunas algas (género *Gelidium*) de la costa sudatlántica española. Inv. Pesq. 26: 165–194.

Frederiksen, S. & J. Rueness, 1989. Culture studies of *Gelidium latifolium* (Grev.) Born. et Thur. (Rhodophyta) from Norway. Growth and nitrogen storage in response to varying photon flux density, temperature and nitrogen availability. Bot. mar. 32: 539–546.

Jimenez, C. & F. X. Niell. Influence of temperature and salinity on carbon and nitrogen content in *Dunaliella viridis* Teodoresco under nitrogen sufficiency. Biomass (in press).

Kanwisher, 1966. Photosynthesis and respiration in some seaweeds. In: Barnes H. (ed), Some Contemporary Studies in Marine Science. George Allen + Unwin Ltd., London, pp. 407–420.

Kursar, T. A. & R. S. Alberte, 1983. Photosynthetic unit organization in a red alga. Relationships between light harvesting pigments and reaction centers. Plant Physiol. 72: 409–414.

Macler, B. A., 1986. Regulation of carbon flow by nitrogen and light in the red algae *Gelidium coulteri*. Plant Physiol. 82: 136–141.

Niell, F. X., 1976. C:N ratio in some marine macrophytes and its possible ecological significance. Bot. mar. 19: 347–350.

Niell, F. X., M. Espejo, J. A. Fernandez & P. Algarra, 1989. Performances and use of energy in intertidal systems related to random disturbance and thermal stress. Scient. mar., 53: 293–299.

O'Carra, P., O'Heocha, 1976. Algal biliproteins and phycobilins. In: Goodwin, T. W. (ed). Chemistry and Biochemistry of Plant Pigments. Academic Press. London. 270 pp.

Ogata, E. & T. Matsui, 1965. Photosynthesis in several marine plants of Japan as affected by salinity, drying and pH with attention to their growth habitats. Bot. mar. 8: 199–217.

Poole, R. W., 1974. An Introduction to Quantitative Ecology. McGraw-Hill Kogakusha. Ltd. (eds) Tokyo. 518 pp.

Rosenberg, G., 1981. Ecological Growth Strategies in the Seaweeds *Gracilaria foliiera* (Rhodophyceae) and *Ulva* sp. (Chlorophyceae) PhD. Thesis, Yale University, New Haven, CT. 151 pp.

82

Santos, G. A. & M. S. Doty, 1983. Agar from some Hawaiian red algae. Aquat. Bot. 16: 385–389.

Seoane-Camba, J., 1965. Estudios sobre las algas bentonicas en la costa sur de la Península Ibérica (litoral de Cadíz). Inv. Pesq. 29: 3–216.

Sokal, R. R. & F. J. Rohlf, 1969. Biometry. The Principles and Practice of Statistics in Biological Research. W. H. Freeman & Co., San Francisco. 776 pp.

Strickland, J. D. H. & T. R. Parsons, 1968. A Practical Handbook of Seawater Analysis. Fish. Res. Bd Can. Bulletin no. 167, Ottawa.

Talling, J. F. & D. Driver, 1963. Some problems in the estimation of chlorophyll in phytoplankton. Proceedings of Conference of Primary Productivities Measurement. Marine & Freshwater. Honolulu. Hawaii USA. Atomic Energy. 7633: 142–166.

Hydrobiologia **221**: 83–90, 1991.
J. A. Juanes, B. Santelices & J. L. McLachlan (eds), International Workshop on Gelidium.
© *1991 Kluwer Academic Publishers.*

Physiological basis for the cultivation of the Gelidiaceae

Bruce A. Macler & John R. Zupan*
*U.S. Environmental Protection Agency, Region 9, 75 Hawthorne street, W-6-1, San Francisco, CA 94105, USA; *University of California, Department of Plant Biology, Berkeley, CA 94720, USA*

Key words: Gelidium, mass culture, photosynthesis, plant physiology, *Pterocladia*

Abstract

An understanding of the physiological factors important to growth and agar production of the Gelidiales would be useful for successful mariculture of these commercially valuable plants. Several environmental factors, including light, nitrogen, carbon, temperature and water motion, have been shown to have potential significance for growth rates, reproduction and carbon partitioning in defining optimal conditions for cultivation. Limiting and optimal growth conditions, where known, are presented, and evaluation of data reported in the literature is addressed.

Introduction

The Gelidiaceae have been exploited for centuries for their agar-type polysaccharides, which are used in a variety of industrial, cosmetic and food applications. Until recently, the harvest of natural populations of species of this family has provided source material for the extraction of agar. Typical of natural populations, the quality and quantity of the desired species have varied markedly over time and between locations. As well, an increasing demand for agar has led to overharvesting of the more desirable and available plant populations, and this had led to commercial problems of inconsistent availability and agar quality. Markets for highly-characterized agars, developing among biotechnology companies, point to the use of modern scientific methods to isolate or engineer algal strains that produce pure agars and then to grow these strains in monoculture. Attempts were made throughout the 1970's and 1980's, to develop intensive land-based mariculture systems for the domestication of agarophytes (Silverthorne, 1977; Hansen, 1980). This work utilized generally applicable results from algal physiology and algal mariculture (Mathieson, 1980; Hansen *et al.*, 1981) to supplement the existing Gelidiaceae research, which has primarily been on the genera *Gelidium* and *Pterocladia*.

Basic research on the physiology of agarophyte species is necessary to develop the process of their domestication; the benefits are explicit. Preliminary identification of promising strains in the laboratory, without resorting to production-scale activity, will result in considerable cost-savings. Environmental factors critical for growth in culture can be identified in the laboratory and means for controlling them can be included in the design of the mass-culture system. Once suitable strains have been selected, laboratory physiological studies can help us to understand the results from outdoor mass-culture, and perhaps predict responses where many factors vary simultaneously. Our own experiences have shown that if reported research has been performed carefully and with sufficient controls, the data may be useful for describing broad physiological cultivation factors and limits; however, many studies reported in the literature have proved to be incomplete or poorly-controlled. In addition, it has frequently been difficult to utilize even the best-quality data directly, as the situations in large-scale mariculture systems often do not allow the level of control possible in laboratory experiments; finally, no one species has been studied extensively, and the physiological differences between studied and utilized species may be large.

While considerable work on red algal physiology in culture has been reported, particularly for the carrageenophyte *Chondrus crispus* Stackh. (Neish *et al.*, 1977; Simpson *et al.*, 1978; Simpson & Shacklock, 1979), and for the agarophyte genus, *Gracilaria* (Edelstein, 1977; Lapointe & Ryther, 1979; Bird *et al.*, 1981), the physiological requirements of gelidiacean species in culture have not been comprehensively studied. Bird (1976) determined nitrogen-uptake rates in *Gelidium nudifrons* Gardner. Mair and Rao (1978) examined daylength and media effects on growth rate in *G. pusillum* (Stackh.) Le Jolis. Light, temperature and water movement were examined in *Pterocladia caerulescens* (Kütz.) Santelices, *P. capillacea* (Gmel.) Bornet & Thuret and *Gelidiella acerosa* (Forsskål) Feldmann & Hamel (Santelices, 1978) and in *Gelidium filicinum* Bory, *G. lingulatum* Kützing and *G. spinulosum* (Ag.) J. Agardh (Santelices *et al.*, 1981). Hansen (1980) examined photosynthetic O_2 evolution as a function of temperature and light. Environmental effects on sporeling development were reported by Correa *et al.* (1985). Macler (1986) reported on the effects of nitrogen and light on photosynthesis and agar biosynthesis in *G. coulteri* Harvey in laboratory studies. Macler and West (1987) described the life history of *G. coulteri* and its physiology with respect to light, temperature, pH and nitrogen. The effects of salinity on photosynthesis, nitrogen assimilation and carbon flow in *G. coulteri* were reported by Macler (1988). Frederiksen and Rueness (1989) reported on light, temperature and nitrogen effects on *G. latifolium* (Grev.) Bornet & Thuret. Much of this work has been reviewed by Santelices (1988).

While the overall interest in gelidiacean mariculture generally is to produce agar, our view is that maximizing biomass production will maximize agar production, even though, under certain conditions, percent agar on a dry-weight basis may not be maximal. These conditions will be addressed in this paper; however, the overall focus will be on growth and biomass production. Five general factors have proved important in defining algal mariculture conditions: light, temperature, water motion, carbon supply (pH) and nitrogen. In the following sections, we assess the results of physiological studies that are relevant to the mass-culture of species of the Gelidiaceae. Species differences and variability within a species are addressed. We consider interactive effects of these factors, both from the perspective of how they compromise research results and of what we can learn from the interactions; finally, we will consider the relationship between laboratory studies and mass-culture.

Light

Light is the most obvious variable to consider when working with plants. Light is important as the source of energy for photosynthesis, thus affecting growth; it also affects reproduction and morphology. Light studies have included work on daylength, quantity (photon flux density (PFD), illumination, intensity, *etc.*) and light quality (spectral characteristics).

Mair and Rao (1978) studied the effects of daylength on *Gelidium pusillum*. They grew plants under five daylengths: 8 : 16, 16 : 8, 20 : 4 and 24 : 0 L : D, at 2000–2500 lux. The reported work was apparently from a single experiment with one flask per condition and insufficient mixing; as such, these results are not particularly applicable to mariculture. Studies on *G. coulteri* (Macler & Matsumoto, unpubl. data) have shown no differences in growth rate at different daylengths when compensated for absolute number of photons. Macler and West (1987) reported experiments with *G. coulteri* that showed increasing growth rates and percent-agar yield with increasing continuous PFD up to 150–250 μmol m^{-2} s^{-1} in 16 l aerated tank-culture. Growth inhibition and bleaching were seen at PFD's $> 400 \, \mu$mol m^{-2} s^{-1}. Plants were free-floating in 20 l carboys and maintained at about 1 g fresh wt (f.wt.) l^{-1}, below the maximal density of 5 g l^{-1}, to eliminate density-induced inhibition. Light affected morphology as well, with PFD's $< 50 \, \mu$mol m^{-2} s^{-1} producing linear plants and PFD's $> 150 \, \mu$mol m^{-2} s^{-1} stimulating branching. The published data, as well as our unpublished work, suggest an optimum time-weighted average PFD well below full sunlight.

The relationship between light and photosynthesis is a fundamental aspect of the physiological basis for cultivation. Typical mass-culture systems use tanks or troughs with some type of mixing-system to allow for high plant-loading densities. Owing to shading in this optically-dense environment, individual plants will circulate from near-full sunlight at the surface to near-total darkness at deeper depths. The amount of time spent at any particular PFD will be a function of the mixing characteristics of the mass-culture system; consequently, understanding of whether a cultivar is capable of photoacclimation (and, if so, does it acclimate to the lowest light it experiences, the highest light or some average PFD?) is critical. The range of PFD's, over which a species is capable of acclimation, as well as the timecourse of acclimation, are also of interest.

The photosynthesis: irradiance (PI) relationship has been described for field plants of *Pterocladia capillacea* collected over a depth-gradient (Coutinho & Yoneshigue, 1988). Plants

from 0-m depth had higher rates of biomass-specific, light-saturated photosynthesis than plants from the subtidal region. Subtidal plants, however, exhibited higher photosynthetic rates in the light-limited portion of the PI curve, such that the initial slopes of the PI curves were steeper for subtidal plants than for intertidal plants. Based on the convergence of PI curves for deep-water plants transplanted to the surface, and those of shallow-water plants, *P. capillacea* was suggested to have the potential to acclimate to different light levels. The reciprocal transplantation was not performed; therefore, it is not possible to conclude whether or not the shallow-water plants are genetically adapted to a high-light environment. The ability of this species to increase its photosynthetic capacity at low light (apparently with the concomitant cost of lowered photosynthetic capacity at high light), however, indicates its rate of biomass production can potentially become optimal in a high-density culture.

Photoacclimation has also been described in a cultivated strain of *P. capillacea* (Table 1). Photosynthesis was a saturating function of growth-irradiance that was satisfactorily described by the hyperbolic tangent (Zupan *et al.*, unpubl. data). Differences in photoacclimation can be described by comparison of curve parameters. The asymptotic maximal photosynthetic rate, P_{max}, was a saturating function of growth-irradiance. Dark respiration increased with increasing growth-irradiance and did not appear to saturate. The initial slope, α, did not differ among treatments. By the subsequent comparison of PI curves of mass-cultured plants with these laboratory results, the effects of stocking density and mixing rate of photosynthetic capacity were determined.

Culture density (mass/area) is important in commercial algal mariculture. System-sizing to produce commercial amounts of *Gelidium* and the capital costs associated with this are major components of an economically viable project. Optimization of continuous yield is a function of density versus biomass production rate, where production rates of the Gelidiaceae are generally observed to increase with density, exhibit a broad maximum, and decline at higher densities. To the

Table 1. The effect of growth irradiance on parameters of the hyperbolic tangent.

Growth irradiance[a]	P_{max}[b]	Respiration[b]	α[c]
29	0.24	−0.017	0.008
53	0.48	−0.051	0.009
128	0.46	−0.053	0.01
271	0.62	−0.078	0.009
467	0.59	−0.141	0.011

[a] μmol m^{-2} s^{-1};
[b] mmol O$_2$ g dry wt^{-1} hr^{-1};
[c] mmol O$_2$ m^2 s g dry wt^{-1} hr^{-1} μmol^{-1}

extent that growth rate is limited by nutrients or water motion, the maximum will be at lower density than it would be under unlimited or saturating conditions; however, an upper boundary appears to exist, where increasing density yields decreasing production rates even under saturating conditions. Such cultures are probably so dense that individual plants are light limited for significant amounts of time. Proper data can be used to construct curves of growth rate versus density and production versus density, from which the optimal culture density can be determined; for example, while for *G. coulteri*, *G. floridanum* Taylor and *P. capillacea*, exponential growth rates of 10–12% day^{-1} could be maintained in the laboratory over months for stocking densities maintained at 1–2 g f.wt. l^{-1}, and cultures continued to grow at low rates up to 15 g f.wt. l^{-1}; optimal continuous production of biomass was found at near 5 g f.wt. l^{-1} (Macler, unpubl. data).

Temperature

Temperature is a variable that offers some opportunities for maximizing growth in mass culture of the Gelidiaceae. Ocean temperatures are seldom optimal. Depending on location, temperature may vary markedly, from near freezing in winter to lethal levels in late summer, or remain constant at some level, which may be suboptimal. Controlled-temperature mariculture systems, on the other hand, can be maintained at temperatures supporting maximal growth. For the Gelidiaceae, there has been a consistency in results from laboratory experiments examining the effects of temperature on growth rate and overall plant physiology. Growth increases with increasing temperature for all studied *Gelidium* species up to a broad maximum. For *G. coulteri*, growth-rates increase from 5° to 20 °C and remain maximal from 20° to 27 °C (Macler & West, 1987). Growth rates of *G. latifolium* increase from 10° to 14–15 °C and are maximal from 14° to 28° (Rueness & Tanager, 1984; Frederiksen & Rueness, 1989); when temperatures exceed 27–28°, growth is inhibited, and

bleaching and death occur at about 34°. For *G. coulteri*, temperatures greater than 30° result in decreased photosynthetic O_2 evolution and increased respiration above the compensation point (Macler, unpubl. data). At the biochemical level, C and N assimilation are inhibited and low molecular-weight C compounds decrease in concentration at elevated temperatures. Agar, a non-metabolizable polymeric carbohydrate in these plants, increases as total-percent dry weight only as other cellular materials are lost.

It is possible that maintaining the optimal temperature for growth rate for some species is important only at stocking densities too low to be useful for mass culture. It has been seen in *G. latifolium* (Frederiksen & Rueness, 1989) and in *P. capillacea* (Zupan, unpubl. data) that growth rate: irradiance curves determined over a range of temperatures converged at low PFD; thus, under the conditions of light limitation typically seen for high-density cultures, growth rate may be independent of temperature for temperatures below inhibition.

Carbon source (pH)

Carbon nutrition is an important factor that can easily be controlled in culture by pH-regulated additions of carbon. The variables that must be defined are the chemical form (bicarbonate or carbon dioxide) in which the carbon is added and the pH setpoint at which it is added. These practical considerations for mass culture have their basis in the photosynthetic physiology of the species to be cultured. Photosynthetic activity in unbuffered seawater causes the pH to rise. This could be the result of the plants taking up CO_2 from the water and the concomitant dehydration of HCO_3^- to CO_2 and OH^-. Alternatively, the plants may take up HCO_3^-, requiring cells to release OH^- to maintain electrical neutrality and pH balance. Cook *et al.* (1988) have presented compelling evidence that *G. crinale* (Turn.) Lamouroux and *P. capillacea* primarily take up HCO_3^- from the medium as carbon source. They suggest that the bicarbonate is taken up by an active transport mechanism. These results are important in culture-system design. At pH 7–8, the addition of HCO_3^- will raise the pH, whereas CO_2 addition will lower the pH; consequently, CO_2 should be used to adjust the pH. Above pH 7.8, less than 5% of the carbon in seawater is present as CO_2. As pH is lowered, the equilibrium favors CO_2. If the cultured strain primarily utilizes HCO_3^-, then the setpoint for CO_2 addition can be set at, or possibly above, the pH of seawater (about 8). We found, however, that pH > 8.5 is inhibitory to *G. coulteri*; also, carbonate, which is not assimilable by plants, becomes the dominant chemical species at higher pH values.

Nitrogen

Nitrogen nutrition is critical to red algal mass culture both for its role in growth and in the regulation of metabolism and reproduction. While detailed work has only been reported for *G. coulteri* (Macler, 1986; 1988; Macler & West, 1987), our unpublished work on *G. floridanum* and *Pterocladia tenuis* Okamura is consistent with those results. N is required for amino acid, nucleic acid and protein biosynthesis; N-deficiency clearly limits growth. The phycobiliproteins have been indicated as reservoirs for excess N (Bird *et al.*, 1982; Frederiksen & Rueness, 1989), as N is depleted from a *G. coulteri* culture medium and limitation begins, plants first lose phycobiliproteins. As growth becomes N-limited, and photosynthesis continues, a metabolic shift occurs from amino acid and protein synthesis to synthesis of the non-nitrogenous polysaccharides, agar and floridian starch. With continued N-starvation, chlorophyll is lost, photosynthesis declines and growth rate decreases. Eventually, plants catabolize their reserve materials and die.

As a result of these studies and the obvious bleaching and decreased growth rates observed in large cultures limited for N, N-supplementation of culture media is commonly practiced; yet the optimal N-concentration remains unknown. While N as nitrate or ammonium can be used by species of

the Gelidiaceae, both these ions are toxic or inhibitory at relatively low concentrations, particularly if those concentrations are maintained over time. Most culture practices add N at intervals, such that the absolute N concentration falls over time and inhibition may not be seen unless the initial N concentration is very high or intervals are frequent. There appears to be species variability in tolerance to N. *P. capillacea* growth rates were not inhibited by sodium nitrate maintained up to 1.4 mM, although growth rate was saturated at 100 μM; ammonium was toxic to *P. capillacea* at 0.07 mM (Zupan, unpubl. data). For *G. coulteri* cultures, concentrations greater than 600 μM nitrate or above 100 μM ammonium caused bleaching and decreased photosynthesis and growth rate (Macler, unpubl. data); in addition, higher N concentrations frequently enhance growth of microalgal and bacterial contaminates of culture systems. Optimal N concentrations for mass culture are dependent on overall culture conditions and typically range from 10–100 μM nitrate.

There is also interest in the regulation of agar biosynthesis by N. Depending on the species, the presence or absence of N in the medium can alter percent agar on a dry-weight basis by 40–50%. Plants limited for N will route a significantly-larger fraction of photosynthetically-fixed C into storage and cell-wall carbon than plants grown with luxury levels (100–300 μM) of N (Macler, 1986). Total photosynthesis, growth and biomass may be sufficiently greater in N-sufficient plants, so that absolute yields of agar over time may be higher; in addition, agar quality in terms of gel strength is lower in plants limited for N (Zupan, unpubl. data).

Water motion and exposure to air

It has frequently been reported that water motion is essential to allow high growth rates and culture densities in algae (Wheeler, 1980; Bidwell *et al.*, 1985; Guerin & Bird, 1987). Wheeler's analysis (1980; *in litt.*) indicates that nutrient limitation, especially inorganic carbon, is not necessarily the result of low concentrations in the environment, but rather because of passive transport resistance through water into the plant. Diffusion across the surface boundary layer is the limiting process, especially in still water and as the concentrations of nutrients local to the plant surface become lower: therefore, water motion is necessary to minimize the effective boundary layer thickness, largely through increasing turbulence, leading to higher local nutrient concentrations and increased growth rates. In a report of tank-culture studies, Guerin & Bird (1987) gave evidence that, for *Gracilaria* sp., water movement in the light was more effective than that in darkness and that increasing aeration periods in the light led to increased yields. They did not observe aeration effects on aguar quality or quantity. Besides increasing nutrient fluxes from water to plant, water motion has been frequently observed to decrease epiphytic contamination and damage from grazers.

Attempts have been made to grow some species of the Gelidiaceae, and other algae exposed to the air, with various forms of misting or short periods of submergence to prevent desiccation. In part, this results from interest in avoiding the high capital-costs of seawater tank-culture systems; there has also been effort to minimize epiphyte and grazer problems. As many of the Gelidiaceae are intertidal species, their exposure to air has not been considered a problem. We examined photosynthesis of *G. coulteri* in air at 100% relative humidity in sealed flasks (Macler & Zupan, 1985), and found that photosynthesis decreased markedly after 30–60 min at both ambient and elevated CO_2 levels. Diffusion of CO_2 into the plants may be limited under these conditions or carbonic anhydrase activity may be lacking. Typical products of photorespiration, including glycolate, serine and glycine, increased substantially during this period. This would be expected as the ratio of O_2 to CO_2 in the air is relatively high compared with that in seawater, and there could be a considerable metabolic liability for cultivation under air-exposed conditions.

Physiological effects on reproduction

One question for those considering mariculture of the Gelidiaceae is whether to propagate plant material vegetatively or via spores. Advantages of spore propagation include the potential ease of generating many plants in small volumes and the ability to genetically alter plant material. Advantages of vegetative propagation include the ease of working with a simpler cultivation system and agar quality consistency is frequently found in cloned mature vegetative plants. Several physiological factors can influence this decision: for some species of the Gelidiaceae, it has proved impossible to find reproductive plants or induce reproduction in culture, so that spore production is impossible; for other species, environmental conditions to induce reproduction may need to be different from those maximizing growth or yield. For *G. coulteri*, reproduction could be induced either by starving plants of nutrients until growth is limited, then relieving the limitation with nutrient addition, or by growing plants first under low-light conditions (50 μmol m^{-2} s^{-1}, then raising light levels to 150 μmol m^{-2} s^{-1} (Macler & West, 1987). Plants remained vegetative at PFD < 100 μmol m^{-2} s^{-1} or at temperatures <20 °C. Culture conditions were different for spore germination and development. Successful germination occurred only at PFD < 100 μmol m^{-2} s^{-1}; above this PFD, spores bleached and died. Spores required a surface to adhere to in order to germinate – spores kept in suspension by aeration failed to germinate. It was also found that sporelings would reach reproductive maturity at 1–2 mm in length if cultured at 150 μmol m^{-2} s^{-1} and 26 °C, a size much smaller than would be desirable for mass culture. Thus, light or temperature could be used to control reproduction in *G. coulteri*.

Experimental design

Appropriate design of experimental apparatus and methodology is essential for generating a critical test of a physiological hypothesis. This is especially true if the experiment involves a growth response or requires growing plants under experimental treatments prior to measuring some physiological response. Unfortunately, some published data on various aspects of the physiology of the Gelidiales are of little value because the experimental esign introduced too many artifacts. Among the most common errors in experimental design are: inappropriate use of field-collected plants; lack of sufficient water motion; lack of control for carbon nutrition; growth of plants at PFD's that are below saturation levels for growth; insufficient replication and lack of or improper experimental controls. The most common error in analysis is using an inappropriate indicator of growth.

If an experiment is designed to test the effect of some environmental factor on growth, then only that factor should be tested at limiting levels; for example, Wheeler (1980) demonstrated that water motion is necessary to reduce the effects of boundary layers on carbon assimilation. If the effect of, say, PFD on growth is to be examined in the laboratory, plants must be grown in apparatus with turbulent mixing. In still culture, reduction of carbon or nutrient diffusion to the surface of the plant may limit growth to the extent that the effect of PFD may be quantitatively in error or completely obscured; likewise, inadequate water motion or PFD may produce artifactual results in a nutrient experiment. The use of field material in physiological experiments cannot be condoned in most instances. Plant-to-plant variabilities within a species may yield statistical variations as great or greater than those between species (unpubl. data); in addition, the nutrition and light requirements and stages in the life history of the plant are often unknown. Adaptation to defined laboratory conditions may take days to weeks (Macler, 1988). Unless experiments are long enough to preclude the possibility that results include significant periods of acclimation to laboratory conditions, cloned, cultured plants from defined growth conditions should be used.

In experiments designed to determine the effect of some environmental factor on growth, the only acceptable dependent variable is growth rate, because environmental factors affect the rate at which the biochemical reactions and physiological processes (whose sum is growth) proceed. Looking at final plant size is inadequate for two reasons: first, the amount of material at the end of an experiment depends to some extent on how much material is present at the beginning; second, insignificant differences in growth rate may result in large differences in final yield if the experiment is of sufficient duration. If the ultimate goal of an experiment is to provide information for the purpose of mass culture, then the production of biomass is the variable of interest. Growth rate should be calculated from change in biomass, not some linear dimension. For the reasons discussed above, including possible morphological variation from experimental treatments (Macler & West, 1987), measuring final plant length is unacceptable. The rate of change in length might be used only if all main axes and branches are included, if biomass measurements are inaccurate for some reason, and, most importantly, if all aspects of plant geometry (e.g., radius and cross-sectional shape and area) are factored in, such that the relationship of length to biomass is well characterized.

References

Bidwell, R. G. S., J. McLachlan & N. D. H. Lloyd, 1985. Tank cultivation of Irish moss, Chondrus crispus Stackh. Bot. mar. 28: 87–97.

Bird, K. T., 1976. Simultaneous assimilation of ammonium and nitrate by Gelidium nudifrons (Gelidiales: Rhodophyta). J. Phycol. 12: 238–241.

Bird, K. T., M. D. Hanisak & J. Ryther, 1981. Chemical quality and production of agars extracted from Gracilaria tikvahiae grown in different nitrogen enrichment conditions. Bot. mar. 24: 441–444.

Bird, K. T., C. Habig & T. DeBusk, 1982. Nitrogen allocation and storage patterns in Gracilaria tikvahiae (Rhodophyta). J. Phycol. 18: 344–348.

Cook, C. M., T. Lanaras & K. A. Roubelakis-Angelakis, 1988. Bicarbonate transport and alkalization of the medium by four species of Rhodophyta. J. exp. Bot. 39: 1185–1198.

Correa, J., M. Avila & B. Santelices, 1985. Effects of some environmental factors on growth of sporelings in two species of Gelidium (Rhodophyta). Aquaculture 44: 221–227.

Coutinho, R. & Y. Yoneshigue, 1988. Diurnal variations in photosynthesis *vs.* irradiance curves from 'sun' and 'shade' plants of *Pterocladia capillacea* (Gmelin) Bornet et Thuret (Gelidiaciaceae: Rhodophyta) from Cabo Frio, Rio de Janeiro, Brazil. J. exp. mar. Biol. Ecol. 118: 217–228.

Edelstein, T., 1977. Studies on *Gracilaria* sp: experiments on inocula incubated under greenhouse conditions. J. exp. mar. Biol. Ecol. 30: 249–259.

Frederiksen, S. & J. Rueness, 1989. Culture studies of *Gelidium latifolium* (Grev.) Born. et Thur. (Rhodophyta) from Norway. Growth and nitrogen storage in response to varying photon flux density, temperature and nitrogen availability. Bot. mar. 32: 539–546.

Guerin, J. M. & K. T. Bird, 1987. Effects of aeration period on the productivity and agar quality of *Gracilaria* sp. Aquaculture 64: 105–110.

Hansen, J. E., 1980. Physiological considerations in the mariculture of red algae. In Abbott I. A., M. S. Foster & L. F. Eklund (eds), Pacific Seaweed Aquaculture. California Sea Grant College Program, pp. 80–91.

Hansen, J. E., J. E. Packard & W. T. Doyle, 1981. Mariculture of red seaweeds. Report #T-CSGCP-002, California Sea Grant College Program.

Lapointe, B. E. & J. H. Ryther, 1979. The effects of nitrogen and seawater flow rate on the growth and biochemical composition of *Gracilaria foliifera* var. *angustissima* in mass outdoor cultures. Bot. mar. 22: 529–537.

Macler, B. A., 1986. Regulation of carbon flow by nitrogen and light in the red alga, *Gelidium coulteri*. Plant Physiol. 82: 136–141.

Macler, B. A., 1988. Salinity effects on photosynthesis, carbon allocation and nitrogen assimilation in the red alga, *Gelidium coulteri*. Plant Physiol. 88: 690–694.

Macler, B. A. & J. A. West, 1987. Life history and physiology of the red alga, *Gelidium coulteri*, in unialgal culture. Aquaculture 61: 281–293.

Macler, B. A. & J. R. Zupan, 1985. Photosynthetic carbon fixation under emerged and submerged conditions for two red algae. 1st Blinks Conference, Abstracts. Pacific Grove, California, p. 8.

Mairh, O. P. & P. S. Rao, 1978. Culture studies on *Gelidium pusillum* (Stackh.) Le Jois. Bot. mar. 21: 169–174.

Mathieson, A. C., 1980. Seaweed cultivation: a review. In Sindermann C. J. (ed), Proc. Sixth U.S. Japan Meeting on Aquaculture, Santa Barbara, California. U.S. Dept. Commerce Rep. NMFS Circ. 442, pp 25–66.

Neish, A. C., P. F. Shacklock, C. H. Fox & F. J. Simpson, 1977. The cultivation of *Chondrus crispus*. Factors affecting growth under greenhouse conditions. Can. J. Bot. 55: 2263–2271.

Rueness, J. & T. Tanager, 1984. Growth in culture of four red algae from Norway with potential for mariculture. Hydrobiologia 116/117: 303–307.

Santelices, B., 1978. Multiple interaction of factors in the distribution of some Hawaiian Gelidiales (Rhodophyta). Pac. Sci. 32: 119–147.

Santelices, B., 1988. Synopsis of biological data on the seaweed genera *Gelidium* and *Pterocladia* (Rhodophyta). FAO Fish. Synop. 145: 55 p.

Santelices, B., P. Oliger & S. Montalve, 1981. Production ecology of Chilean Gelidiales. Proc. int. Seaweed Symp. 10: 351–356.

Silverthorne, W., 1977. Optimal production from a seaweed resource. Bot. mar. 20: 75–98.

Simpson, F. J., A. C. Neish, P. F. Shacklock & D. R. Robson, 1978. The cultivation of *Chondrus crispus*. Effect of pH on growth and production of carrageenan. Bot. mar. 21: 229–235.

Simpson, F. J. & P. F. Shacklock, 1979. The cultivation of *Chondrus crispus*. Effect of temperature on growth and carrageenan production. Bot. mar. 22: 295–298.

Wheeler, W. N., 1980. Effect of boundary layer transport on the fixation of carbon by the giant kelp *Macrocystis pyrifera*. Mar. Biol. 56: 103–110.

Hydrobiologia **221**: 91–106, 1991.
J. A. Juanes, B. Santelices & J. L. McLachlan (eds), International Workshop on Gelidium.
© 1991 *Kluwer Academic Publishers.*

Gelidium cultivation in the sea

R.A. Melo[1], B.W.W. Harger & M. Neushul
*Department of Biological Sciences and Marine Science Institute, University of California, Santa Barbara,
CA 93106, USA;* [1]*present address: Departamento de Biologia Vegetal, Faculdade de Ciências da
Universidade de Lisboa, Campo Grande, Edifício C2 – Piso 4, 1700 Lisboa, Portugal*

Key words: agarophyte, cultivation, *Gelidium*, growth, hydrodynamics, mariculture

Abstract

Gelidium fronds were grown in the sea under a variety of experimental conditions: on rigid, damped and tensioned test farms of various designs, in calmer and more turbulent habitats, at various depths, with and without commercial fertilizer supply. Initially, the effectiveness of a given cultivation strategy was based on the survival and growth of the fronds, here termed 'bio-assay' mariculture. Ambient seawater temperature, nutrient availability, hydrodynamics and other environmental parameters were measured periodically. In-the-sea irrigation of test plants with commercial fertilizers was apparently effective, at least with some farm designs, and when ambient nutrient levels were low. Under optimal conditions, achieved through experimental manipulation of farm design, specific growth rates of over 2% per day were recorded. However, considerable variation in growth rates and in plant performances was observed. It was not always possible to correlate these variations with design modifications or other experimental parameters. In view of these findings, we have reviewed our initial 'bio-assay' approach, namely the assumptions about the design and operation of farm structures and their interactions with the water and the fronds. Methods were developed to quantify these interactions. We advocate a quantitative, 'hydro-dynamic' approach in developing an effective cultivation strategy for gelidioid algae and are optimistic about progressing from test-to-commercial scale farms in the near future.

Introduction

The only successful attempts to grow species of *Gelidium* in the sea have been on test farms. In order to move from test to commercial farms many problems must yet be solved. Suitable farming sites must be identified, farm structures installed, seed stock obtained, outplanted and cultivated and the crop harvested. We are, nevertheless, optimistic about the possibility of advancing from *Gelidium* test farms to commercial ones in the near future. The macroalgal crop plants that have been commercially farmed may serve as examples of successful farming strategies.

These are, for example, *Porphyra* (Oohusa, 1984), *Laminaria* (Tseng, 1987), *Eucheuma* (Doty, 1986) and *Gracilaria* (Santelices & Doty, 1989). Matsumoto (1959), a pioneer in the farming of nori, *Porphyra tenera* Kjellm., recognized the importance of carefully measuring plant responses to hydrodynamic conditions, nutrients, light and temperature. He then used these measurements to define optimum-growth conditions and searched for locations in the sea where these conditions existed. Test and ultimately commercial farms were installed at these locations. We are hopeful that a similar transition can be made with gelidioid algae.

Several excellent reviews of gelidioid algae have recently been published (Akatsuka, 1986; Santelices, 1988). The approaches to *Gelidium* cultivation have either been in tanks or in the sea (Santelices, 1988). Small-scale strategies for enclosed or tank cultivation of free-floating *Gelidium* and related genera have been reported from California (Hansen, 1980), Chile (Santelices, 1976), India (Mairh & Sreenivasa Rao, 1978) and Norway (Fredriksen & Rueness, 1989). Attempts to grow *Gelidium* in the sea have been made in Japan (Akatsuka, 1986) and more recently reported from Chile (Santelices, 1987), and China (Fei & Huang, this volume). The methods used for open sea cultivation have been: (i) to increase natural *Gelidium* beds either by extending the substrate (Akatsuka, 1986) or by removing competing seaweeds and grazers (Santelices, 1987), (ii) to suspend plants or fronds from submerged ropes (Akatsuka, 1986), lines and baskets (Friedlander & Lipkin, 1982), or rafts (Fei & Huang, this volume) and (iii) net cultures in intertidal gullies and rapids (Santelices, 1987). None of these experimental schemes has been expanded into larger-scale commercial cultivation, although some of the growth rates reported are high enough, in our opinion, to justify attempts to cultivate *Gelidium*: 5–7% d^{-1} (Akatsuka, 1986); 8% d^{-1} (Friedlander & Zelikovitch, 1984, with *Pterocladia capillacea* (Gmel.) Born. et Thur.); 1.23–4.58% d^{-1} (Fei & Huang, this volume).

The test farming discussed here was undertaken coincidentally with attempts to farm the giant kelp, *Macrocystis pyrifera* (L.) C. Ag. (Neushul & Harger, 1987). Other investigations that were particularly relevant to this one, include a study of the hydrodynamic interaction between moving water and the sea palm, *Eisenia arborea* Aresch., (Charters *et al.*, 1968) where motion diagrams made in natural habitats were first introduced. A preliminary report of work on domestication and cultivation of Californian macroalgae was presented earlier (Neushul, 1981). The objective of this work was to develop the effective planting and cultivation techniques necessary to commercially farm marine algae used in the industrial production of agar. Efforts were made to take advantage of advances in the commercial cultivation of other algae and to apply procedures used by agriculturalists, including the following: documenting environmental conditions; designing, building and measuring performance of farm structures; collecting, propagating, growing, culling and analyzing plants; and performing field-test experiments.

Materials and methods

The *Gelidium* experimental farming discussed here began with test plantings made at three sites near the campus of the University of California Santa Barbara (UCSB), over a 3-year period (1979–1981). Work on agarophyte domestication and marine-farm engineering has continued in the area to the present. Work began in the sheltered Goleta Bay and another more exposed site at Campus Point, both adjacent to the UCSB campus. A third experimental farm site was constructed adjacent to Ellwood Pier, about 5 km from the UCSB campus. Work was carried out from small outboard-motor boats launched from piers at Goleta and Ellwood and from a larger diesel fishing boat used for farm installation and maintenance.

Weekly measurements of seawater temperature, nutrient concentration, transparency and current speed were made at each test farm site. Divers also measured horizontal visibility (using a Secchi disk extended horizontally) and current speed with a small Eckman-type current meter. Water collection took place at the surface, bottom and at 4–5 m from the bottom. Water samples were taken with a model 1120-C45 Van-Dorn water sampler and frozen for later analysis using standard methods (Strickland & Parsons, 1972). These data are not presented but are available from the authors (MN) upon request.

Seedstock collection and farm seedling. The first test plantings included the giant kelp *Macrocystis pyrifera*, *Sargassum muticum* (Yendo) Fensh. and the agarophyte *Gracilaria sjoestedtii* Kyl. in addition to *Gelidium* species. The *Gelidium* fronds grown on the test farms were taken, by divers, from natural populations of *Gelidium robustum* (Gardn.) Hollenb. et Abb., at Ellwood Pier, Ellwood Beach, Naples Reef, Goleta Point and Hendry's Beach – all near the UCSB campus – and from near Coches Prietos on the southern side of Santa Cruz Island, Santa Barbara Co. Divers selected *Gelidium* fronds, cutting them and placing them in nylon mesh bags for transport to a seawater-supplied greenhouse (Charters & Neushul, 1979). Only those fronds that were robust, well-pigmented and free from epiphytes were selected for cultivation.

Single *Gelidium* fronds were inserted into short segments of polypropylene rope by twisting it open Each rope segment was heat-sealed at one end, the other end was slipped through a hole in a 3-inch diameter PVC pipe ring, 3 cm wide (Fig. 1) and melted flat so that it would not pull out of the ring. The rings were split opposite the point of rope insertion, so that they could be

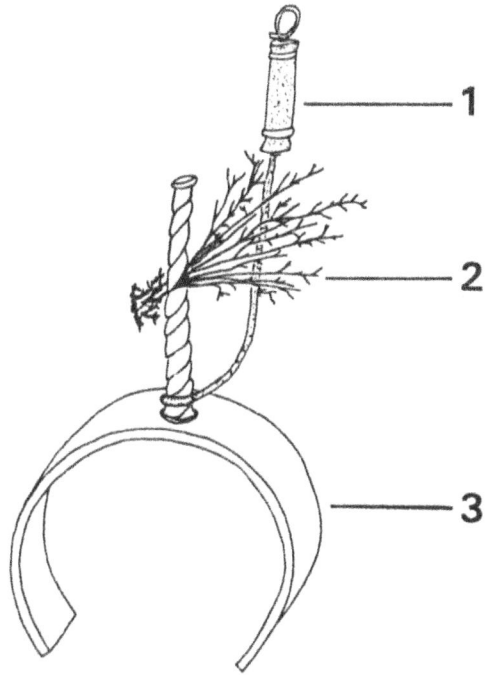

Fig. 1. Single *Gelidium* fronds were planted on test farms in the sea by inserting them in a short segment of rope attached to a split ring of PVC pipe, called a collar. Frond numbers and substrate identification codes were written on the collars (1 – small fertilizer tube, 2 – *Gelidium* frond, 3 – split PVC collar).

spread and fitted around planting substrate made of the same diameter PVC pipes (Fig. 1). Frond numbers and substrate identification codes were written on the collars. A number of these frond-and-ring sets (collars) could be placed on short pipe segments with enlarged ends to prevent the collars from slipping off. Sets of fronds marked in this way were placed in tubs and transported with minimal exposure to the air from the greenhouse tanks to be planted on test farms in the sea. Similarly, after a grow-out period in the test farms, divers collected fronds for transport to the laboratory, where growth measurements were performed.

Growth data collection and analysis. Morphological characteristics of seed *Gelidium* fronds were recorded photographically. A Pesola spring scale was used to weigh each PVC collar alone and the collar and frond together before and after outplanting and growout in the sea. If epiphytes were present after a growout in the sea, these were removed after the initial weighing and the frond was re-weighed. Arithmetic means were used to indicate the average growth rate obtained. Some of the means were back-transformed from the data used in the analysis of variance (ANOVA) calculations. Since arithmetic and back-transformed means of the same data can differ considerably, ANOVA calculations were used in comparing treatments.

As growth in *Gelidium* is non-linear, the specific growth rate (GR) (Brinkhuis, 1985) was calculated:

$$GR = 100 \times \frac{\ln W_t - \ln W_o}{T_t - T_o},$$

where W_o is the starting weight, W_t the ending weight and $T_t - T_o$ the number of cultivation days. The growth units are given as percent per day ($\% \ d^{-1}$).

Individual *Gelidium* fronds were harvested by hand, using scissors, weighed and after some were cut at the base, some at mid-frond and some near the tip, the fronds were reweighed and then replanted. Growth was again measured after a subsequent growout period in the sea. A second experimental harvesting method involved the use of a Naruse seaweed harvester, donated by the Naruse Harvest Co., Japan. This machine, mounted on a boat, would mechanically harvest *Gelidium* fronds attached to rope lines, or to nets stretched over PVC panels in location over the farms.

Experimental farm designs. Several kinds of test farms identified by their basic shape as 'H', 'I' and 'Z' (Fig. 2a, b, c) were installed and tested. All the experimental farms were designed to allow for depth adjustment. The first in-the-sea farming experiments used 'H' farms. Their design was continuously improved and streamlined and in their final form they were constructed by securing a 6-m horizontal pipe – the planting substrate – between two 3-m long vertical spar-buoys

anchored to the bottom (Fig. 2a). Both spars and substrates were made with 5-cm diameter PVC pipe. The horizontal planting substrates were seeded with numbered PVC collars with single *Gelidium* fronds (Fig. 1), placed in line 30 cm apart and at a defined depth, usually 2 m from the bottom. The simple 'H' farm could also be shortened during stormy winter periods. Nutrient irrigation devices could be added to the farms, either in the form of wave-actuated pumps or fertilizer tubes (Fig. 2a, d) wrapped around the horizontal planting substrates. Even in their improved versions, 'H' farms still required the use of divers for planting, harvesting and maintenance.

The simplest design, the 'I' farm (Fig. 2), was a single spar buoy constructed with a sealed

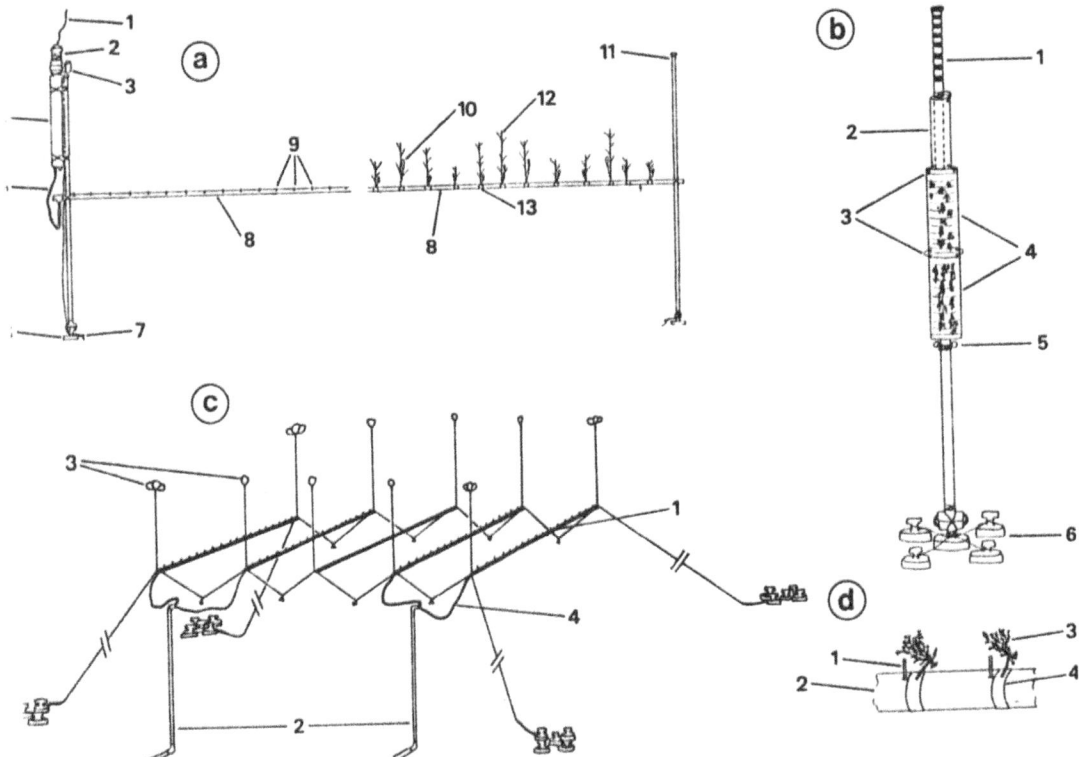

Fig. 2. Test farming was done by attaching *Gelidium* fronds and other plants to farm structures identified as 'H', 'I' and 'Z' farms, corresponding to their general shapes: (a) 'H' farm with wave pump (1 – line to surface float, 2 – wave pump canister float, 4 – fertilizer canister, 5 – hose, 6 – anchor, 7 – screw-in anchor, 8 – planting substrate, 9 – emitters) and with tube fertilizers (10 – small fertilizer tube, 11 – vertical spar buoy, 12 – *Gelidium* frond, 13 – collar); (b) 'I' farm (1 – 10 m spar buoy, 2 – 1.5 × 7 cm upper stop that is boat accessible, 3 – connecting lines, 4 – 1.5 m hancore sections with fronds on ropes, 5 – bottom stop, 6 – cement anchors); (c) 'Z' farm (1 – planting substrate with emitters, 2 – nutrient irrigation pipe line from pier, 3 – surface floats, 4 – hose); d) detail of planting substrate (1 – emitter, 2 – planting substrate, 3 – *Gelidium* frond, 4 – collar).

T-fitting at the base of the spar and a pipe cap at the top to preserve buoyancy. Over this spar slid either a series of double-walled pipe segments roped together, or a long sleeve-like piece of corrugated pipe. The farms were planted by wrapping ropes seeded with *Gelidium* fronds around the corrugated pipe. On a monthly basis, resin-coated fertilizer pellets (Osmocote, Sierra Chemical Co.), were placed inside the double-walled sleeves. The dissolved fertilizer was allowed to escape through a series of small pin-holes drilled in the corrugated pipe. Planting and maintenance could be performed by workers in a boat rather than by divers. Also, fronds grown on the sleeves could be retrieved from the boat without having to dive, simply by pulling the sleeves to the surface with the attachment rope.

The 'Z' configuration farms consisted of a rope-ladder-like array of horizontal pipe rungs supported 3 m below the surface by surface floats (Fig. 2c). The lateral rope lines were weighted, to give a 'Z'-like configuration when seen from the side. This float and weight counter-balance system stretched and retracted when a wave passed, much like an accordion. The horizontal pipes were planted with numbered PVC collars holding single *Gelidium* fronds (Fig. 1), as in the 'H' farms. A farm could be pulled to the surface for maintenance from a small boat and the horizontal pipes could be used for nutrient irrigation (Fig. 2c, d). When the side ropes were attached eccentrically and wrapped around the horizontal pipes, the 'accordion' action rotated the pipes and fronds attached to them around the axis of the pipe.

Other designs tested included float-supported horizontal long lines (Fig. 3a, b) from which hung planting substrates made of a PVC pipe framework over which were stretched ropes or nets planted with fronds (Fig. 3c). These long-line farms were designed after (i) the commercial kelp farms built by the Nichimo Co. in Japan (Fig. 3a) and (ii) the raft kelp farms used in China (Fig. 3b). The long-line farms were built and installed at Ellwood Pier so that the panels hung at different angles to the prevailing swell: perpendicular, parallel and oblique.

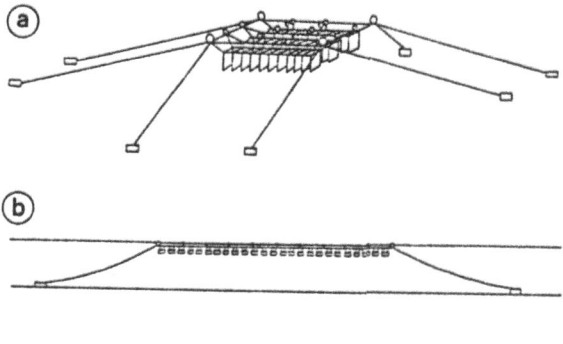
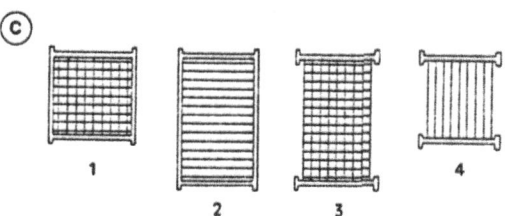

Fig. 3. Farm designs tested included float-supported horizontal long lines from which hung planting substrates made of a PVC pipe framework over which were stretched ropes or nets planted with fronds: (a) Japanese long line farm, 10 m × 20 m, capable of handling 33 panels, taut-moored by lines 4 × the depth to 8 slab anchors; (b) Chinese long line farm, 55 m long single line with capacity for 25 panels, moored by lines 4 × the depth to 2 slab anchors; (c) long line farm planting panels made of PVC pipe (1 – rigid 2 m × 2 m panel with 3-strand nylon net, 2 – rigid 2 m × 3 m panel with polypropylene or nylon line, 3 – collapsible 2 m × 3 m panel with net, 4 – collapsible 2 m × 2 m panel with line.

Nutrient irrigation. The simplest system of nutrient irrigation was to take fronds from the farms and soak them weekly in a nutrient-rich medium in a container on a boat. By measuring the nutrient concentration in the medium before and after dipping the fronds, it was possible to determine the amount of nutrients taken up by the algae. Similar experiments were run in the laboratory to compare nutrient uptake under light and darkness and to study the effect the *a priori* nutrient state of the fronds (starved or not) had on further nutrient uptake (data not shown). Other attempts to fertilize test farms involved filling fabric tubes or porous rubber pipe with slow-release fertilizer and tying these next to the single *Gelidium* fronds inserted on the PVC collars (Fig. 1). In all trials involving the use of slow-release Osmocote, the resin-coated pellets were weighed both before placing them in the sea and afterwards to measure the amount of nutrient that had been leached from them. The Osmocote was made with a 'marine mix' without potassium having 10% nitrate and 14% ammonia by weight. A sewage-based fertilizer (Milorganite, Milwaukee, MO) and a kelp fertilizer (Manakelp, Stauffer Chemical Co.) were also tested on nutrient-irrigated 'H' farms.

A wave-powered pump was designed to supply fertilizer to 'H' farms. The pump consisted of a 2 l plastic accordion bottle connected on the closed end to a buoy and on the open end to a fertilizer-containing canister fitted with two one-way valves. The canister-pump apparatus was mounted on the side of a 'H' farm (Fig. 2a). A hose connected the canister to the planting substrate, on which emitter tubes (6 cm long, ¼ inch diameter) were inserted into holes drilled at 30 cm intervals along the pipe (Fig. 2d). As a wave passed by the floating buoy attached to the accordion bottle, it would pull up, extending the bottle and pulling in water. After the wave passed, the buoy would slacken and the bottle would be collapsed by rubber tubes attached to it, expelling the seawater into the canister through the second one-way valve, where it would take up dissolved nutrients and carry them through the feed line and out of the emitters. The efficiency of the wave pump was measured in the laboratory prior to its use in the ocean. A spring scale was attached to the bottle to measure the force needed to extend it. Extension forces of 2.3, 4.5 and 9 kg were tested and all produced water uptake and expulsion. In the sea, fluorescent dye was placed in the canister to show how nutrient-enriched water was ejected in 'puffs' from the emitters as each wave passed. The nutrient irrigation system was also tested by sampling the seawater at increasing distances from the emitters on collapsed plastic bags.

Fronds on the 'Z' accordin farms were supplied with nutrients from four 100 l mixing tanks installed on the adjacent Ellwood Pier. Ammonium nitrate and ambient seawater were added to one pair of tanks to make a 0.1 M ammonium nitrate solution, while the other pair contained only unenriched seawater. The tanks emptied by gravity flow at a rate of 1 l min^{-1} through feed pipes to the horizontal planting substrates. Dye tests were used to determine flow rates and here also seawater was sampled from the emitters to show that fertilizer was being delivered. In addition to making measurements of plant growth, tissue samples were collected and analyzed for carbon, hydrogen and nitrogen content (data not shown).

Farm hydrodynamics. The hydrodynamic observations made during the present study are similar to those previously reported by Charters *et al.* (1968), except that a Eumig Nautica 8 mm underwater movie camera, attached to experimental farm structures or to the natural substrate (Fig. 4), was used. It recorded the movement of *Gelidium* fronds and plants on the different types of experimental farms and in natural populations. The films were projected on a paper tablet and, by the use of a stop-watch and freeze-frame capability, the position of the frond at specific times could be traced, to produce a motion diagram. A digital planimeter was then used to determine the area 'swept' by the frond as it was moved by the waves passing it. Motion diagrams were obtained from two natural populations, Naples Reef and

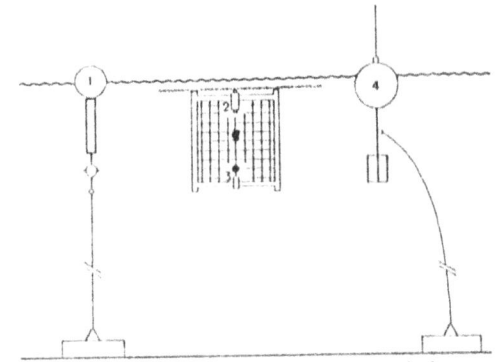

Fig. 4. Ambient water motion, plant-water and plant-water-farm interactions were measured using a large Marsh McBirney 355B ECM (1) moored or attached to the sea floor as a reference, a smaller MMI 512 ECM (3) attached to test farms and an Endeco 949 wave track buoy (4) equipped with a radio telemetry system. Plant movement (motion diagrams) were made from motion pictures taken with a Eumig Nautica Super 8 camera (2) either operated by a diver, or attached to a farm panel.

Ellwood Pier, and from long-line and 'Z' farm structures.

Direct measurements of water motion in natural habitats and on and around experimental farm structures were made using electromagnetic current meters (ECM's) (Marsh-McBirney MMI 512 and MMI 355B) and a wave-track buoy (Endeco 949) (Fig. 4). The wave-track buoy's inertial sensor measured wave frequency and height at the sea surface and the ECM measured flow velocities in two perpendicular directions in the same plane. One ECM attached to a rigid sub-surface spar and the wave-track buoy measured the 'sea state' water motion (Fig. 4). A second ECM was attached to culture panels on the farms where the fronds were growing. Data from these hydrodynamic sensors were stored on floppy disks, using a Vector MZ microcomputer on board the boat and returned to the laboratory for spectral analysis and plotting.

Results

The results of our test-farming of *Gelidium* at three farm sites, on different kinds of farm structures, with and without nutrient irrigation, may be evaluated by comparing the survival and growth of farmed fronds with those of nearby wild plants. The farmed fronds served as a 'bio-assay' of farming efficacy. When the fronds did not grow, we interpreted this to mean that some factor was missing and changes were then made in the design of the test farms, followed by new experiments.

Measurements of sea conditions made from October 1979 to September 1980 at Goleta Pier and in Goleta Bay, showed it to be calm, with a mean wave height of less than 0.5 m most of the year, with occasional wave heights of 1.5 m. Current speeds doubled in winter over the summer rates. Water temperatures were 16–19 °C during the summer and 13–16 °C in winter. Upwelling was noticed during three periods in April and May 1980, when temperatures dropped below 13 °C and water samples showed nitrogen levels in excess of 5 μg at l^{-1}. A severe storm occurred in January 1980 with peak current speeds of

3 m s^{-1}. Storm turbulence produced suspended bottom sediments in excess of 3.75 g cm^{-2} d^{-1}. Sediment levels at the three tet farm sites greatly reduced vertical and horizontal Secchi readings. In 1980–81 there was a windy cool winter and a warm calm summer. Nutrient-level measurements showed nitrogen in excess of 5 μg at l^{-1} in surface and mid-water on five or more occasions at each year at Ellwood, with upwelling in April, May, June, October and December. Goleta Bay and more turbid than Ellwood, with Secchi depth in the former remaining at around 2 m most of the year, while it was 6 m at Ellwood Pier.

Farm seeding and site selection

Results suggest that farmed *Gelidium* growth may be correlated to the initial size of the fronds used to seed the farms. This is shown in Figure 5, where the growth rates of different sized fronds of both *G. robustum* and *G. nudifrons* grown on 'H' farms in two different sites are plotted against their starting weights. The growth rates of *G. robustum* in both Goleta Bay and Campus Point sites were significantly ($P < 0.05$) negatively correlated to the starting weight of the fronds (Fig. 5), suggesting that farms should be seeded with younger, smaller fronds in order to maximize production. The correlation for the case of *G. nudifrons* was not significant (Fig. 5), suggesting that in this species the growth rate may be size independent.

The growth rates of fronds of *G. robustum*, *G. nudifrons* and, for reference, of young sporophytes of the giant kelp *Macrocystis pyrifera*, at 2, 4 and 6 m depths can be compared in Table 1, for two experiments carried out in the spring of 1980 on 'H' farms in Goleta Bay. Cultivation was found to be possible over the depth range with growth rates for both *Gelidium* species ranging between 1% d^{-1} and almost 3% d^{-1}. *Macrocystis* young plants grew faster reaching growth rates as high as 8.7% d^{-1}.

A comparison of test-farms at three different sites showed that site selection was very important. This is shown in Table 2 where growth data

98

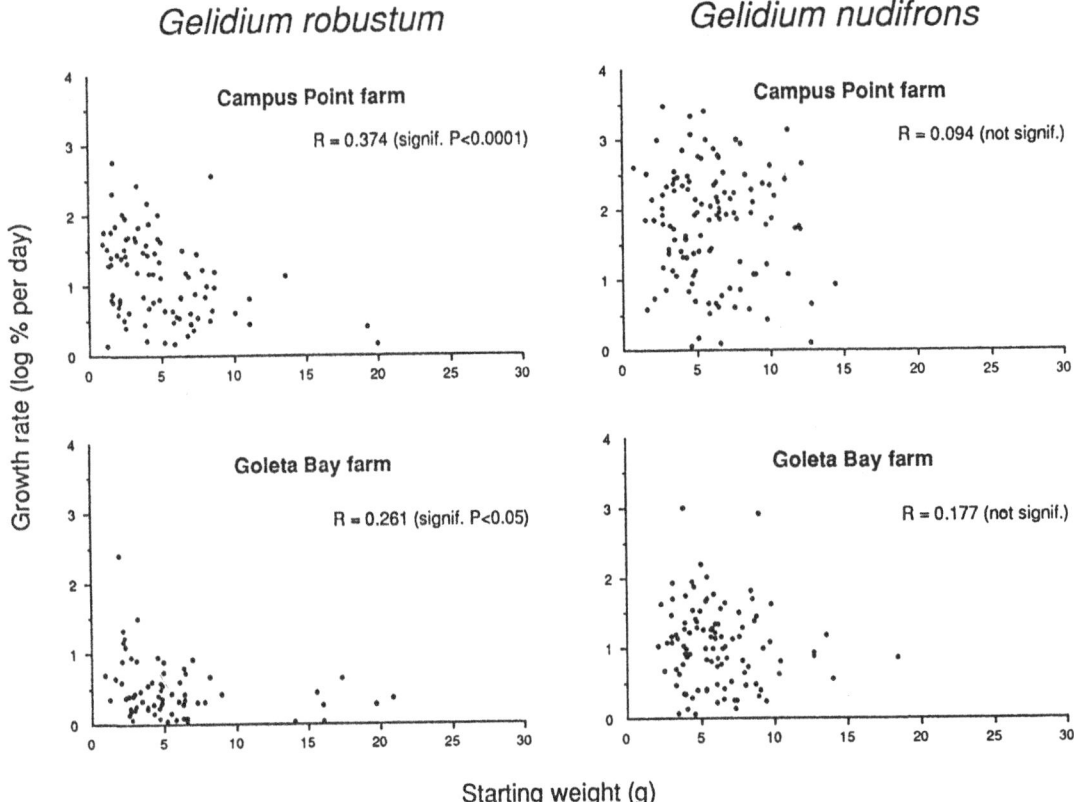

Fig. 5. Comparison of the growth rates of farmed *G. robustum* and *G. nudifrons* fronds in Goleta Bay and Campus Point, on 'H' farms, as a function of frond starting weight.

Table 1. Growth of fronds of *G. robustum* and *G. nudifrons* and of young plants of *Macrocystis pyrifera*, over a 6 m depth range on 'H' farms in Goleta Bay in the spring of 1980.

	Depth	1st experiment			2nd experiment		
		GR	SE	N	GR*	SE	N
Gelidium nudifrons	2 m	2.400	0.260	11	1.586	0.490	15
	4 m	2.300	0.392	27	1.645	0.153	31
	6 m	2.105	0.241	36	2.013	0.175	15
Gelidium robustum	2 m	1.640	0.147	22	2.815	0.483	14
	4 m	1.374	0.169	31	1.662	0.339	24
	6 m	0.935	0.146	37	2.185	0.139	14
Macrocystis pyrifera	2 m	6.442	0.357	7	5.080	1.260	5
	4 m	4.533	2.123	3	6.545	0.496	11
	6 m	8.700	–	1	5.520	1.054	5

GR = mean specific growth rate ($\% \ d^{-1}$); SE = standard error of the mean; N = sample size; * All include fronds with negative growth rates.

Table 2. A comparison of mean growth rates (GR) of *G. robustum* and *G. nudifrons* in Goleta Bay versus Campus Point on 'H' farms using ANOVA.

	Treatment	Mean GR (% d⁻¹)	var	N	F	Probability
Gelidium robustum	Bay	0.042	0.71*	74	68.5	< 0.00001
	Point	0.101	0.86*	95		
Gelidium nudifrons	Bay	0.82	0.12**	104	32.7	< 0.00001
	Point	1.49	0.10**	111		

* Square root transformed;
** Log$_{10}$ transformed.

for fronds grown in two different sites on 'H' farms with nutrient irrigation are compared. The growth rates observed during this experiment were significantly higher ($P << 0.001$) at the more exposed Campus Point farm for both *G. robustum* (GR = 0.10% d⁻¹) and *G. nudifrons* (GR = 1.49% d⁻¹) than at the sheltered Goleta Bay farm (GR = 0.04% d⁻¹ and GR = 0.82% d⁻¹, respectively).

Nutrient irrigation

Seawater samples were taken at increasing distances from the emitters on a 'H' farm horizontal planting substrate (Fig. 2d). As can be seen in Fig. 6, ambient nitrate levels in the seawater surrounding the plants were around 1 μmol L⁻¹ at Campus Point and 0.4 μmol L⁻¹ at Goleta Bay while on the farms, within 0.3 m of the emitters, nitrate concentration was at least one order of magnitude higher. At increasing distances from the emitters, on both farm locations, the nitrate concentrations measured were either unchanged, increased or decreased to close to ambient levels (Fig. 6).

A quantitative assessment of the efficiency of nutrient uptake in the sea showed that only 0.005 to 0.28% of the nutrients supplied by the nutrient irrigation system was used by the plants (unpubl. data). From laboratory experiments we knew that the efficiency can be 70% under controlled conditions, when the plants were exposed to high nitrogen concentrations (80 μmol) and long residence times (5 h) (unpublished results). Experiments comparing the efficacy of tube fertilizers versus pump fertilizers were carried out on 'H' farms at Goleta Bay and campus Point in the summer of 1980 (Table 3). In the case of *G. nudifrons*, the higher growth rates obtained with tube fertilizers

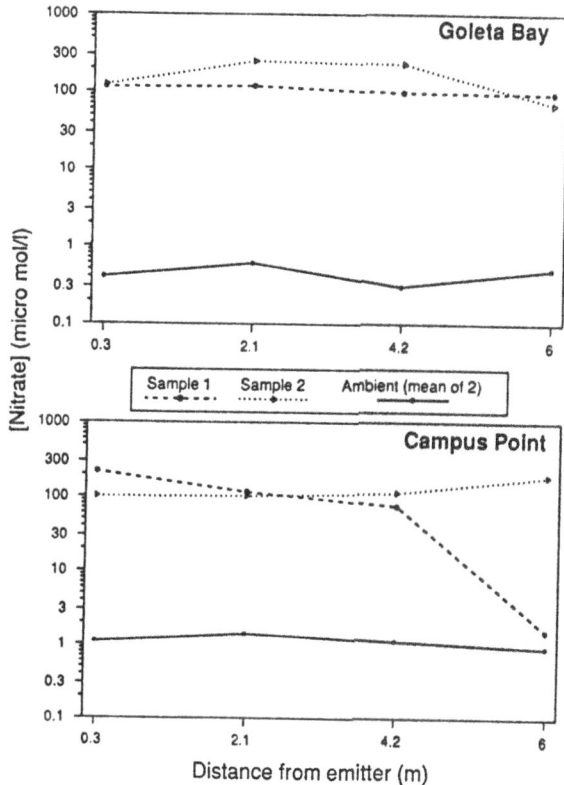

Fig. 6. Tests of fertilizer delivery, as nitrate concentration, from emitters on 'H' farms (see Fig. 2) in Goleta Bay and at Campus Point, from two sets of samples, illustrates that nutrient levels can remain high 6 m from the emitter, or can drop to ambient levels.

Table 3. A comparison of growth rates of *G. robustum* and *G. nudifrons* growing on 'H' farms supplied with nutrients. Two systems for nutrient delivery are compared: a wave-powered pump (see Fig. 2a) and fertilizer tubes attached to the frond-holding collars (Fig. 1).

Treatment	Mean GR ($\% \, d^{-1}$)	SE	N	DF	t-test	Probability
Gelidium nudifrons	0.69	0.127	31	45	3.544	0.0013(*)
pump fertilized	1.26	0.270	16			
tube fertilized						
Gelidium robustum	−0.06	0.109	26	38	2.458	0.0280(ns)
pump fertilized	0.36	0.107	14			
tube fertilized						

GR = specific growth rate; SE = standard error of the mean; N = sample number; DF = degrees of freedom; (ns) = not significant; (*) = significant, $P < 0.01$.

were significantly different ($P < 0.01$) from those obtained with the wave pump (Table 3). In the case of *G. robustum* the same trend was observed, but the mean growth rates obtained were not significantly different (Table 3). By drying and weighing the fertilizer pellets before and after being used in the sea, it was found that the pumps dispersed 1.65 g of fertilizer per plant per day, while the tubes only dispersed 0.18 g per plant per day. As wave-pump irrigation did not produce significantly higher growth rates (Table 3), use of

the more economical tube fertilizers is preferable.

Both nutrient-irrigated and unfertilized fronds grew at rates over $2\% \, d^{-1}$ on 'Z' farms (Table 4, experiment 4) in April–June when natural nutrient levels were 4.0 μmol $NO_3 \, l^{-1}$ at the surface and 10.0 μmol $NO_3 \, l^{-1}$ at the bottom (unpubl. data). In the second growth period (June–July), ambient nutrient levels were low, as is frequent in Santa Barbara coastal waters in the summer months, and the growth rates of unfertilized test fronds dropped to $0.72\% \, d^{-1}$ (Table 4,

Table 4. A comparison of growth rates of *Gelidium robustum* on 'H' and 'I' farms at Ellwood Pier, comparing nutrient irrigated versus unfertilized fronds. The results from 5 different experiments are summarized.

Experiment	Growth period	Treatment	Mean GR ($\% \, d^{-1}$)**	SE	N	DF	t-test	Probability
1. 'H' farm	91 d	Unfertilized	0.48	0.048	59	–	–	–
2. 'I' farm- single fronds	64 d	Fertilized	0.43	0.091	98	195	1.9352	0.0539(ns)
		Unfertilized	0.67	0.086	99			
3. 'I' farm- fronds in lines	50 d	Fertilized	1.63	0.080	8	14	1.4427	0.1765(ns)
		Unfertilized	1.47	0.073	8			
4. 'Z' farm- 1st period	49 d	Fertilized	2.28	0.085	75	149	1.7340	0.0858(ns)
		Unfertilized	2.11	0.053	76			
5. 'Z' farm- 2nd period	40 d	Fertilized	1.42	0.144	68	136	2.9863	0.0037(*)
		Unfertilized	0.72	0.184	70			

GR = specific growth rate; SE = standard error of the mean; N = sample number; DF = degrees of freedom; (ns) = not significant; (*) = significant, $P < 0.01$; ** All include fronds for which weight decreased.

experiment 5). In the same period, growth rates for the fertilized fronds were significantly greater than those for the unfertilized fronds (Table 4, experiment 5). The *G. robustum* fronds on 'Z' farms responded to fertilizer application by more frequent branching and by turning a deeper red. In all the other experiments reported in Table 4, the growth rates of both nutrient-irrigated and unfertilized fronds were not significantly different in any of the different farm types tested.

Epiphytism

Growth of epiphytic animals and plants on fronds of *Gelidium* planted on test farms was monitored. The growth rates of *G. robustum* and *G. nudifrons* fronds planted on 'H' farms at the Goleta Bay and Campus Point sites, without nutrient irrigation,

were not significantly correlated to epiphyte loads on the fronds, (Fig. 7), suggesting that growth of *Gelidium* was not affected by the epiphytic animals and plants; however, the growth rates at the more exposed Point site were higher (Fig. 7) as noted in a different experiment (Fig. 5).

Measurements of the epiphytic loads on *G. robustum* fronds farmed at Ellwood Pier showed that 12–13% of the final weight of the frond was owing to epiphytic encrustation (Table 5). Wild plants on adjacent pier pilings had annual mean epiphyte loads of 56.8%, comprising mostly animals, while on the nearby Naples Reef population the annual mean epiphyte load was 35.8%, consisting mostly of other algae (unpublished results). On 'Z' farms at Ellwood Pier, nutrient-irrigated plants had greater epiphyte loads than unfertilized plants, although the difference was not statistically significant

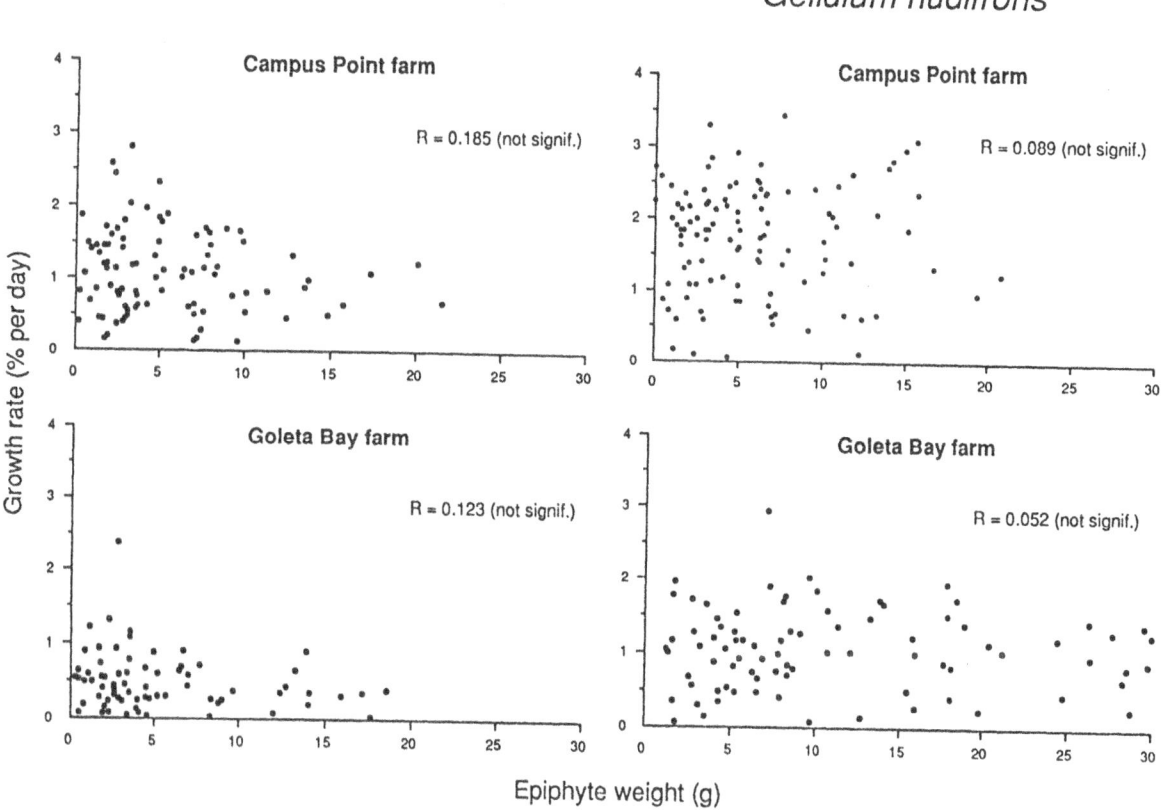

Fig. 7. Comparison of the growth rates of farmed *G. robustum* and *G. nudifrons* on 'H' farms at Campus Point and in Goleta Bay versus epiphyte load (in weight).

Table 5. Comparison of epiphyte loads (in weight) on nutrient supplied versus unfertilized *G. robustum* fronds grown on 'Z' farms at Elwood Pier.

Treatment	Epiphyte % load	Epiphyte mean wt (g)	SE	N	DF	t-test	Probability
Fertilized	12%	12.6	4.33	8	20	0.3124	0.20(ns)
Unfertilized	13%	11.1	2.617	14			

Epiphyte % load = mean epiphyte weight as a percent of mean *G. robustum* frond final weight; Epiphyte mean wt (g) = mean epiphyte weight per *G. robustum* frond; GR = specific growth rate; SE = standard error of the mean; N = sample number; DF = degrees of freedom; (ns) = not significant.

(Table 5). The most common epiphytes were the red algae *Acrosorium uncinatum* (Turn.) Kyl. and *Callophyllis* sp. and the green alga *Ulva* sp. Animals encrusting *Gelidium* thalli included the bryozoan *Membranipora* sp. and other genera, the hydroid *Aglaophenia* sp. and occasionally sponges, tunicates and molluscs.

Harvesting

Individual *G. robustum* fronds, cut with scissors at the base, in the middle and near the tips (Table 6) suggest that harvesting enhances *Gelidium* growth. All harvested treatments had mean growth rates significantly higher ($P < 0.01$)

Table 6. Comparison of re-growth of harvested versus unharvested *G. robustum* fronds grown for 32 days on 'I' farms at Elwood Pier. Fronds were hand cut with scissors at the base, in the middle or close to the tips. Unharvested control fronds were grown under the same conditions.

Treatment	Mean GR (% d^{-1})**	SE	N	DF	t-test	Probability
Base cut harvest	1.61	0.103	5			
Mid cut harvest	1.78	0.154	5			
Tip cut harvest	1.76	0.164	5			
All harvested fronds	1.72	0.079	15	18	3.4216	0.0037(*)
Non harvested fronds	1.08	0.228	5			

GR = specific growth rate; SE = standard error of the mean; N = sample number; DF = degrees of freedom; (*) = significant, $P < 0.01$; ** All include fronds with negative growth rates.

Table 7. The effects of base, mid and tip cut harvests on regrowth of the crop fronds. Duration of regrowth period was 32 days.

Treatment	MHW (g)	Percent harvested	MFG (g)	HW:FG ratio	Mean GR (% d^{-1})
Control unharvested	0	0%	46.2		1.14
Base cut harvest	61.9	75%	13.5	4.6:1	1.57
Mid cut harvest	46.8	50%	35.6	1.3:1	1.78
Tip cut harvest	30.2	29%	60.6	1:2	1.84

MHW = mean harvested weight per frond; MFG = mean frond growth (in weight) after harvest; HW:FG ratio = ratio between mean harvested weight and mean frond growth after harvest; GR = specific growth rate.

Table 8. Excursion areas swept by *G. robustum* fronds in response to ambient water motion on natural habitats and on different farm types.

Substrate	Area (cm²)
Natural habitats:	
Naples Reef	1397.8
Ellwood Pier	740.4
Farm structures:	
'Z' farm	722.3
Long line: parallel	748.9
	722.7
perpendicular	678.1
oblique	703.4

than that of the uncut control fronds (Table 6). Harvesting only at the tips, average of 29% of initial frond weight, resulted in a subsequent growth rate of 1.84% d^{-1} (Table 7); other harvest treatments gave lower growth rates and the control uncut fronds grew the most slowly. The ratio

of 1:2 between the weight harvested and regrowth (Table 7) for the tip harvest similarly shows the enhancement of frond growth possible with this harvest method. When the Naruse harvester was used in long-line farms (Fig. 3), a mean growth rate of 1.42% d^{-1} was observed for mechanically harvested fronds.

Farm assessment

Motion diagrams of *Gelidium* fronds in natural habitats and on test farms were used to compare their excursion areas (Table 8). The excursion area of 1397.8 cm² for wild plants at the offshore Naples Reef was the largest, about twice that of fronds planted on both a 'Z' farm and a long-line farm, oriented parallel to the prevailing swell. Wild plants adjacent to the test farms, on the pilings at Ellwood Pier, showed about the same excursion areas (Table 8). When oriented per-

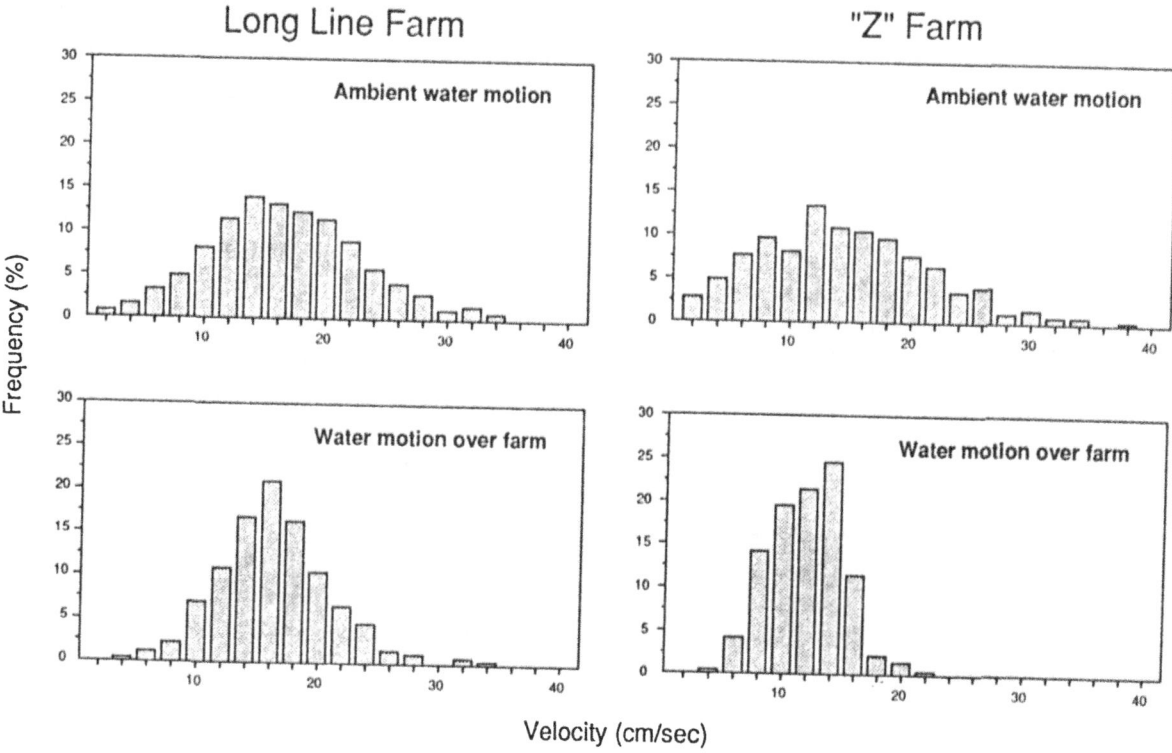

Fig. 8. Ambient water velocity spectra are compared to water motion spectra observed on a long-line and on a 'Z' farm. Compare with frond excursion areas in Table 8.

104

pendicular to the prevailing swell, the long-line farm damped the ambient water motion and the frond excursion areas observed were the lowest (Table 8). ECM measurements of ambient water motion over farm structures in Chinese long-line farms, in parallel orientation, peaked at 16, ranging from 4–34 cm s^{-1} (Fig. 8); under similar conditions the 'Z' farm spectrum peaked at 14 cm s^{-1}, ranging from 4–22 cm s^{-1}, showing a greater damping of ambient water motion (Fig. 8).

Discussion

We assumed that healthy, seemingly identical *Gelidium* fronds planted in similar situations would result in similar growth rates. While there were differences between species and farm sites, the large variation among fronds of the same species at the same location was unexpected. In assuming that the fronds are nearly identical, we have not taken into account possible antecedent events (Doty, 1971) that might influence nutrient uptake rates, branching and other aspects of growth and morphogenesis. It would be logical to consider the terete basal-branch system of *Gelidium* as a possible source of seed stock and to study their growth and the rates of formation of upright fronds.

We have not answered the question of whether *Gelidium* is inherently slow growing (Stewart, 1984). Some of the high growth rates we have observed and some of those reported elsewhere (Santelices, this volume) suggest otherwise. It seems that the 'inherent slow growth' problem can be described as the 'inherent growth variability' problem; however, some of the variability might be explained by considering the upright fronds structures of determinate growth, adapted to serve several functions. The distichously branched or clumped upright fronds of *G. robustum* and *G. nudifrons*, besides functioning in nutrient uptake and light harvesting, may also play a role in fertilization via spermatia-bearing slime strands (Melo, unpublished results) analogous to the smaller fronds and branches of *Tiffaniella* (Fetter & Neushul, 1981); also after

fertilization or spore formation or both, these branches and their branchlets may serve in distribution (Neushul *et al.*, 1976). The heavy epiphyte loads observed (Table 5) may be advantageous by increasing the weight and drag of the fronds, and facilitating their dislodgement. All of these possible adaptations could have a bearing on the growth rate of fronds.

We assumed that in carefully selecting clean, epiphyte-free plants as sources of seed stock, they would remain that way; consequently, the heavy load of epiphytes and epizoans that grew on them was discouraging. The results of an experiment designed to study the effect of epiphyte load on the growth rates of *G. robustum* and *G. nudifrons* suggest that, at least under the conditions tested, growth was not affected by the weight of the epiphytes. Other workers have found that bryozoans of the genus *Membranipora*, although reducing the photosynthetic rate of *Gelidium rex* Santel. et Abb. growing subtidally in central Chile, did not have a net effect on algal growth (Cancino *et al.*, 1987b). In another study with *Gracilaria verrucosa* (Huds.) Papenf. from southern Chile, the heavy load (up to 70% in weight) of epiphytic bivalves negatively affected algal growth while, at the same time, increasing the gel strength of its agar (Cancino *et al.*, 1987a). Increased nitrogen availability, from the epifaunal excretion products, and slow growth were thought to be the mediators of this increased gel strength (Cancino *et al.*, 1987a). The carbon : nitrogen (C : N) ratio in *G. robustum* on the pier pilings at Ellwood and adjacently on a nutrient irrigated 'Z' farm (unpublished results) showed that farmed plants, although fertilized, had a lower ratio (20.0) than wild plants (24.4). It seems likely that the heavy populations of caprellids on the piling plants provided tight recycling of nutrients, which may be the case for some of the epiphytes as well. The effect of reduced exposure to light of Pier plants, because of shading from the Pier, should be considered.

Harvested fronds showed a remarkable ability to initiate regeneration at the cuts, resulting in increased branching, and this is well documented in the literature (Santelices, 1988). The harvest weight-to-plant growth ratios, show a relationship between the amount harvested and the time required for a new crop to grow. Two additional factors determine frequency and severity of harvest: relative cost of harvesting and relative cost of farming structures.

It was assumed that it would be simple to design effective farming structures and nutrient-irrigation methods. In the absence of any principle of marine farm engineering, we had few design constraints and the temptation to try yet another farm design was not resisted. Our 'damped' and rigid pipe farm structures did not take important plant-water interactions into account and, hence, were not effective. By adding water motion measurements to our 'bio-assay' approach, farm effectiveness could be measured rapidly. When these measurements were combined with those of pumping effectiveness and rates of delivery of nutrient enriched water to emitters, the systems operated according to plan, even though the fronds on the farms did not grow as expected.

By moving from 'bio-assay' to 'hydrodynamic' assessments of farm effectiveness, we soon realized that different kinds of farms resulted in 'new' habitats that differed from the natural ones; nutrient irrigation presented even more possibilities. It became obvious that design principles were needed. By comparing functional morphology and hydrodynamic interactions (Neushul, 1972), and by making motion diagrams and hydrodynamic measurements, farm performance could be evaluated without having to grow plants for long time periods. Hydrodynamic measurements are useful to evaluate designs for enhancement or reduction of ambient water motions. Farms that simply support the plant in the water column until they die from nutrient drought, epiphytism, solarization, grazing, or a combination of these, are obviously not effective. It became obvious that similar ambient conditions can produce different water-motion environments for fronds growing on the farms. Much can be learned from small, inexpensive and easily modified test farms, the simplest being a spar buoy. There is also a tendency, which must be resisted, to make farms of ever increasing complexity, instead of making careful measurements

of plants in their natural environments and comparing these with what occurs on farm structures. With instruments for remote sensing, telemetry and computer-based data acquisition systems, the 'environmental complex' can be monitored in considerable detail.

In view of the complexity of the problems facing an agarophyte mariculturalist, a trial-and-error or 'bio-assay' approach to the development of commercial farms is likely to be very slow. In order to speed up this process, we advocate a 'hydrodynamic' approach with test farms where frond and farm performance can be controlled. In the use of these farms, it is important to relate ambient hydrodynamic measurements, farm tension and plant motion and growth. The tools used here for measuring the different frond and farm performances were effective and we are optimistic about the possibility of progressing from test- to commercial-scale *Gelidium* mariculture in the near future.

Acknowledgements

This research was supported by the National Science Foundation, USA (NSF) research awards to MN, BWW and Neushul Mariculture Inc., Goleta (NMI) (PFR-7911715, CPE-8110034 and CBT 8407060) and by the Junta Nacional de Investigação Científica e Tecnológica, Instituto Nacional de Investigação Científica and Luso American Educational Committee, Portugal doctoral fellowships and research awards to RAM. NSF staff members (O. Zaborsky, S. Gunsher, D. Sennich) took the time and trouble to personally visit during the course of this project. Encouragement by the California Department of Fish and Game (E. Smith) was also appreciated. Thanks are due to employees at NMI who installed and operated the first test farms (D. Carlsen, S. Claybusch, S. Ettman, S. Goldstein, J. Woessner) or worked on the laboratory (M. Polne, W. Wheeler, R. Lewis). Mr. H. Fay, Aquaflow Corp, Goleta kindly provided pipe samples and advice on setting up nutrient irrigation systems. Consultants at the University of

California Santa Barbara (UCSB) were particularly helpful (A. Charters, D. Coon, A. Gibor, R. Petty, F. DeWitt). Visitors from other institutions (A.C. Mathieson, X.G. Fei) made significant contributions. In the final stages of the project, help and assistance from colleagues at UCSB were appreciated (D. Reed, C. Amsler, I. Pinto). Thanks are also due to J. Serôdio for doing the illustrations. A modified version of this paper was included in a dissertation submitted by RAM in partial satisfaction of the requirements for the PhD degree at UCSB.

References

Akatsuka, I., 1986. Japanese Gelidiales (Rhodophyta) especially *Gelidium*. Oceanogr. Mar. Biol. annu. Rev. 24: 171–263.

Brinkhuis, B. H., 1985. Growth patterns and rates. In: Littler M. M., Littler D. S. (eds), Handbook of Phycological Methods. Ecological Field Methods: Macroalgae. Cambridge Univ. Press, Cambridge, 461–477.

Cancino, J. M., M. Muñoz & M. C. Orellana, 1987a. Effects of epifauna on algal growth and quality of the agar produced by *Gracilaria verrucosa* (Hudson) Papenfuss. Hydrobiologia 151/152: 233–237.

Cancino, J. M., J. Muñoz, M. Muñoz & M. C. Orellana, 1987b. Effects of the byrozoan *Membranipora tuberculata* (Bosc.) on the photosynthesis and growth of *Gelidium rex* Santelices et Abbott. J. exp. mar. Biol. Ecol. 113: 105–112.

Charters, A. C. & M. Neushul, 1979. A hydrodynamically defined culture system for benthic seaweeds. Aquat. Bot. 6: 67–78.

Charters, A. C., M. Neushul & C. Barilotti, 1968. The functional morphology of *Eisenia arborea*. Proc. Int. Seaweed Symp. 6: 89–105.

Doty, M. S., 1971. Antecedent event influence on benthic marine algal standing crops in Hawaii. J. exp. mar. Biol. Ecol. 6: 161–166.

Doty, M. S., 1986. The production and use of *Eucheuma*. FAO Fish. Tech. Pap. 281: 123–161.

Fetter, R. & M. Neushul, 1981. Studies on developing and released spermatia in the red alga, *Tiffaniella snyderae* (Rhodophyta). J. Phycol. 17: 141–159.

Fredriksen, S. & J. Rueness, 1989. Culture studies of *Gelidium latifolium* (Grev.) Born. et Thur. (Rhodophyta) from Norway. Growth and nitrogen storage in response to varying photon flux density, temperature and nitrogen availability. Bot. mar. 32: 539–546.

Friedlander, M. & Y. Lipkin, 1982. Growing agarophytes and carrageenophytes under field conditions in the eastern Mediterranean. Bot. mar. 25: 101–105.

Friedlander, M. & N. Zelikovitch, 1984. Growth rates, phycocolloid yield and quantity of the red seaweeds, *Gracilaria*

sp., *Pterocladia capillacea*, *Hypnea musciformis* and *Hypnea cornuta* in field studies in Israel. Aquaculture 40: 57–66.

Hansen, J. E., 1980. Physiological considerations in the mariculture of red algae. In: Abbott I. A., M. S. Foster & L. Eklund (eds), Pacific Seaweed Aquaculture. California Sea Grant College Program, La Jolla, 80–91.

Mairh, O. P. & P. Sreenivasa Rao, 1978. Culture studies on *Gelidium pusillum* (Stackh.) Le Jollis. Bot. mar. 21: 169–174.

Matsumoto, F., 1959. Studies on the effect of environmental factors on the growth of nori (*Porphyra tenera* Kjellm.), with special reference to the water current. Hiroshima Univ., Facul. Fish. An. Husbandry J. 2: 249–333.

Miura, A., 1975. *Porphyra* cultivation in Japan. In: Tokida J. & H. Hirose (eds), Advances of Phycology in Japan. Dr W. Junk, The Hague: 273–304.

Neushul, M., 1972. Functional interpretation of benthic marine algal morphology. In: Abbott I. A. & M. Kurogi (eds) Contributions to the Systematics of Benthic Marine Algae of the North Pacific. Univ. Tokyo Press, Tokyo: 47–74.

Neushul, M., 1981. The domestication and cultivation of Californian macroalgae. Proc. Int. Seaweed Symp. 10: 71–69.

Neushul, M., M. S. Foster, D. A. Coon, J. W. Woessner, B. W. W. Harger, 1976. An *in situ* study of recruitment, growth and survival of subtidal marine algae: techniques and preliminary results. J. Phycol. 12: 397–408.

Neushul, M. & B. W. W. Harger, 1987. Nearshore kelp cultivation, yield and genetics. In: Bird K. T. & P. H. Benson (eds), Seaweed Cultivation for Renewable Resources. Elsevier, New York: 69–93.

Oohusa, T., 1984. Technical aspects of nori (*Porphyra*) cultivation and quality preservation of nori products in Japan today. Hydrobiologia 116/117: 95–114.

Santelices, B., 1976. Nota sobre cultivo masivo de algunas especies de Gelidiales (Rhodophyta). Rev. Biol. mar. Dep. Oceanol. Univ. Chile. 16: 27–33.

Santelices, B., 1987. Métodos alternativos para la propagación y el cultivo de *Gelidium* en Chile central. In: Verreth, J. A. J., M. Carillo, S. Zanuy & E. A. Huisman (eds), Investigación Acuícula en América Latina. PUDOC Wageningen: 349–366.

Santelices, B., 1988. Synopsis of biological data on the seaweed genera *Gelidium* and *Pterocladia* (Rhodophyta). FAO Fish. Synop. 145: 1–55.

Santelices, B. & M. S. Doty, 1989. A review of *Gracilaria* farming. Aquaculture 78: 95–133.

Stewart, J. G., 1984. Vegetative growth rates of *Pterocladia capillacea* (Gelidiaceae, Rhodophyta). Bot. mar. 27: 85–94.

Strickland, J. D. H. & T. R. Parsons, 1972. A Practical Handbook of Seawater Analysis, 2nd ed. Fisheries Research Board of Canada, Bulletin no. 167, Ottawa 310 pp.

Tseng, C. K., 1987. Some remarks on the kelp cultivation industry of China. In: Bird K. T. & P. H. Benson (eds), Seaweed Cultivation for Renewable Resources. Elsevier, New York: 147–153.

Hydrobiologia **221**: 107–117, 1991.
J. A. Juanes, B. Santelices & J. L. McLachlan (eds), International Workshop on Gelidium.
© 1991 *Kluwer Academic Publishers.*

Spray system for re-attachment of *Gelidium sesquipedale* (Clem.) Born. et Thur. (Gelidiales: Rhodophyta)

J.M. Salinas
Instituto Español de Oceanografía, Apartado 240, 39080 Santander, Spain

Key words: cultivation, *Gelidium sesquipedale*, rhizoidal cluster, spray system

Abstract

A system is described for rapid re-attachment of the rhodophycean alga *Gelidium sesquipedale* (Clem.) Born. et Thur. on artificial or natural substrata. This method is applicable to industrial cultivation of this species. The function of rhizoidal clusters and the origin of germlings from the apical portion of the thalli are analyzed in relation to the re-attachment process. The role that re-attachment might play in the maintenance and spreading of natural populations and in the observed anomalies of the life-history of this species is discussed.

Introduction

Gelidium sesquipedale (Clem.) Born. et Thur. is a rhodophycean alga (Gelidiales) with a *Polysiphonia*-type life history and isomorphic generations; the gametophytes are of diverse morphology. Thalli can attain lengths greater than 450 mm and have a unique axial structure consisting of small branches which are grouped into bunches. These arise from the encrusting rhizoid system attached to the substratum by thin haptera (Fritsch, 1952; Gayral, 1966; Dixon & Irvine, 1977). The thalli are surrounded by a complicated network of filaments formed from the apical cell. These filaments, referred to as hyphae or rhizoidal clusters, reach their greatest development in the basal zones adjacent to the rhizoid and haptera. In *G. sesquipedale*, the rhizoidal clusters occupy an intermediate position below the cortical layer, although they can reach the medulla in the lower portions of the thalli.

The apparent life history of *Gelidium ses-*

quipedale is not clear, and several distinct genetic forms exist in nature. It might seem logical that gametophytes and sporophytes should alternate and occur in equal proportions in natural populations; however, a strong imbalance is always noted. In the Cantabrian Sea, 92% of the individuals are sporophytic, while of the remaining 8%, only 2% bear cystocarps (*i.e.* female gametophytes). As the common mode of spore production is tetraspores, which occur annually during autumn, a failure in the viability of these asexual spores may explain this imbalance in the proportions of life-history generations; this has also been noted in other species of *Gelidium* (Abbott, 1980). This apparent anomaly may also be explained by frequent occurrences of apomeiosis (Magne, 1987) or other more complex processes, such as apomixis-specific mortalities. Heterozygotic alleles, linked to sex or population distribution, controlled by adaptations of one of the generations, may be important (Barilotti, 1980). These speculations, far from clarifying the mechanisms of maintenance and spreading of the species, are frequently confusing and make it difficult to develop techniques for management of resources, such as through cultivation.

Worldwide interest in species of *Gelidium* continues to increase (Akatsuka, 1986) owing to the high value of its main polysaccharide (agar) and growing demands for its derivatives, such as agarose, which have numerous industrial applications. Scientific interest is sustained by the relative fragility of natural populations and large oscillations which periodically occur in resources of Gelidiales and the production of agar (Moss, 1977). There is, in addition, a need to devise cultivation techniques as an alternative means to stabilize production.

Among the various activities of the Instituto Español de Oceanografía initiated during the 1980's, a programme for the enhancement of natural populations of *G. sesquipedale* was included. During studies of attachment and development of tetraspores, it was observed, in the winter of 1984–85, that the apices of *Gelidium* can re-attach to certain substrata under specific environmental conditions – the air-water interface of calcareous substrata, and in the absence of water movement. Akatsuka (1986), in his review, mentioned the work of Japanese authors from the beginning of the present century who describe processes of re-attachment of fragments of thalli of *G. amansii* (Lamour.) Lamour. This ability to re-attach can also be seen in the Cantabrian Sea, with the appearance, for example, of thalli of *G. sesquipedale* attached by their apices to colonies of the polychaete, *Sabella*.

In the system described below, I have attempted to reproduce certain conditions which favour re-attachment of *G. sesquipedale* and which can be used to study the process in detail. It can also be used for production of clones of this species in the laboratory, permitting studies of growth and spreading of the species, applicable to in-field cultivation, with an outstanding advantage over traditional methods that have depended on spores. The method described is based on a spray system which promotes the appearance of rhizoidal clusters and encourages the re-attach-

ment of *Gelidium* on calcareous substrata. D'Antonio and Gibor (1985) studied the influence of irradiances and photoperiod on germlings and rhizoidal clusters of *G. robustum* (Gardn.) Hollenberg and Abott, and Moeller *et al.* (1982, 1984) and Lignell and Pedérsen (1986) used a spray technique as a method for mass culture of seaweeds; in none of these cases, however, was a functional relation between spraying and *Gelidium* re-attachment sought.

Materials and methods

The re-attachment studies were done in a system consisting of 4 spray chambers in which either apical fragments, about 4 cm long, or entire thalli were exposed. Continuous spray was delivered through nozzles similar to those used in horticulture. Each spray chamber consisted of a semi-cylindrical polyester tank, 2.0×1.0 m, covered with a translucent glass-fiber reinforced polyester sheet. Each tank had a drain to discharge continuously the water being sprayed. The chambers also had individual static ventilation systems of holes in the upper and lower parts, and three 'Matavi' nozzles, placed to deliver a uniform distribution of water. Each nozzle discharged water at a rate of 920 ml min^{-1} at a pressure of 3 kg cm^{-2}. In total, each chamber received a spray of 166 l h^{-1}. A pump supplied water (working pressure, 3 kg cm^{-2}) at a flow rate of 644 l h^{-1} (16986 l day^{-1}) to the 4 chambers. The pumps used were 1.15 kw 'Espa' with a delivery capacity of $25–90$ l min^{-1} at pressures of $50–15$ kg cm^{-2}. The water intake supplied the general network in the greenhouse culture system of the Oceanographic Center of Santander from an intermediate tank; this allowed the pump to operate continuously and permitted intermittent disinfections. Between the pump and the spray chambers, there was a 'Cuno' cartridge filter (110 μm pore size) to control the pressure of the supply to the tanks, and a regulatory manometer. The water was distributed among the tanks by a ring, to avoid pressure gradients between them, and entered the tanks through valves, which allowed the spraying to be interrupted individually; Fig. 1 illustrates a spray chamber and its components. Additional components of the system consisted of an arrangement for different lighting levels for each tank. Light was provided by daylight fluorescent lamps arranged on the transparent overhead panel of the chamber, and light levels were adjusted to 15, 21, 32 and 40 μmol m^{-2} s^{-1} (Chambers I, II, III and IV respectively) on a 12 : 12-h Light : Dark cycle controlled by a timer. A thermograph in one of the chambers recorded the temperature variations during the week.

The nature of the substrata and the spraying technique were fundamental to the experiment. The principal feature of the substrata was their chemical compositions, rich in calcium carbonate or magnesium carbonate or both. A wrinkled, semi-cylindrical polythene base was used, obtained from a drainpipe to which a thin layer (5 mm) of seawater-resistant calcified mortar was applied. Adherent and salt-resistant paint ('Estolite'), containing calcium and magnesium carbonates, was applied to this mortar layer. Although this layer was not strictly necessary, a mortar of this kind enhanced the subsequent re-attachment of G. sesquipedale. The substrata were tile-shaped with dimensions 40 × 10 × 4 cm at

their highest point, and this prevented accumulation of stagnant water which would hamper growth of the plants in tanks or in the sea. Natural substrata such as dolomite, or cement covered with calcite gravel, have also been successfully employed. Although these materials are heavy and difficult to manipulate, they are acceptable because of their carbonate composition. The best method for re-attachment and subsequent development of the plants is shown in Fig. 2.

The G. sesquipedale samples were collected on the Isle of Mouro (43° 28' N, 3° 45' W) and at Cabo Mayor (43° 29' N, 3° 47' W), both on the coast of Santander, by divers and taken to the laboratory in darkened containers. After the more abundant epiphytes (Plocamium, Dictyota, Sphaerococcus, etc.) had been removed, selected thalli were washed in a disinfectant solution of sodium hypochlorite (domestic quality) made up in seawater (40 mg Cl$_2$ l^{-1}). This treatment was sufficient to eliminate the remaining fauna and

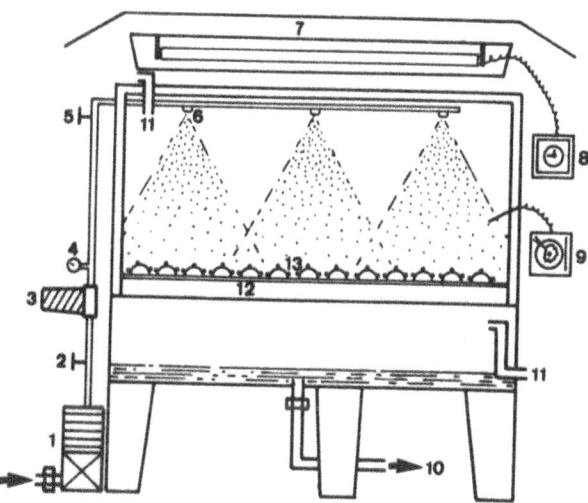

Fig. 1. Diagrammatic scheme of the spray chamber: 1 pump, 2 & 5 valves, 3 filter, 4 manometer, 6 nozzles, 7 lamps, 8 timer, 9 thermograph, 10 drain, 11 ventilation, 12 and 13 substrata & algae.

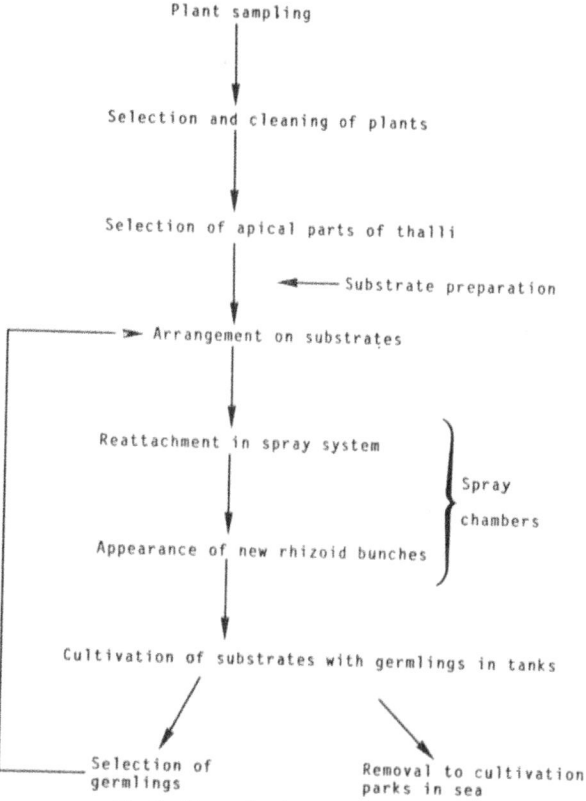

Fig. 2. General scheme of operation.

110

juvenile phases of most epiphytic algae, without significantly damaging the meristematic cells of *Gelidium*. The terminal portions of the principal axes and primary branches of *G. sesquipedale*, 5–10 cm in length, cut without tearing the cortical layer, were used in the re-attachment experiments; it is also possible to use complete parts of the thallus cut from the attachment haptera. Regardless of the parts used, these did not contain reproductive elements, such as tetrasporangia, as tetraspore-bearing pinnules, once empty, decay and create putrefaction foci. After being liberally washed in seawater, the portions of thalli were placed in such a way so that all apices were in contact with the substrata. This was conveniently done by arranging the basal part of the apices after the latter had been placed in contact with the substrata. A semi-open seawater circulation system with a volume of 50000 l and a turnover rate of 10% was used. It was unnecessary to include nutrients in the culture medium. Occasional nutrient deficiency in summer was corrected by an intermittent application of A-G medium (ammonia nitrate and urea-total N: 5.46 mg l^{-1}; total P: 0.50 mg l^{-1}; minerals, vitamin B, and indolacetic acid). The germlings obtained were grown initially in transparent cylindrical tanks of 200 l capacity in which the water was changed daily and periodically aerated. Once the germlings were 3 to 4 cm long, the substrata were placed in the sea and fixed to rocks at 8 to 10 m depth by screws (Fig. 3). This permitted qualitative assessment of the behaviour of the

plant-substrate unit, and was relevant to subsequent cultivation programmes.

Results

Figure 4 shows changes which occur in the thallus of *Gelidium sesquipedale* during spring: (a) production of rhizoidal clusters, (b) attachment of rhizoids to substrate by rhizoidal clusters and (c) production of new germlings from rhizoids. After 2 days of spraying, the apices became typically flattened and curved slightly upwards (positive phototropism), especially when the irradiance was increased in the chamber. After 5 days of spraying, the rhizoidal clusters appeared on the lower apical face of the curved part and also sprouted, to a lesser extent, from the upper surface. This sprouting was most intense on the apices of the principal axes, but it also occurred on primary and secondary branches (Fig. 5). There was a clear relationship between the size of

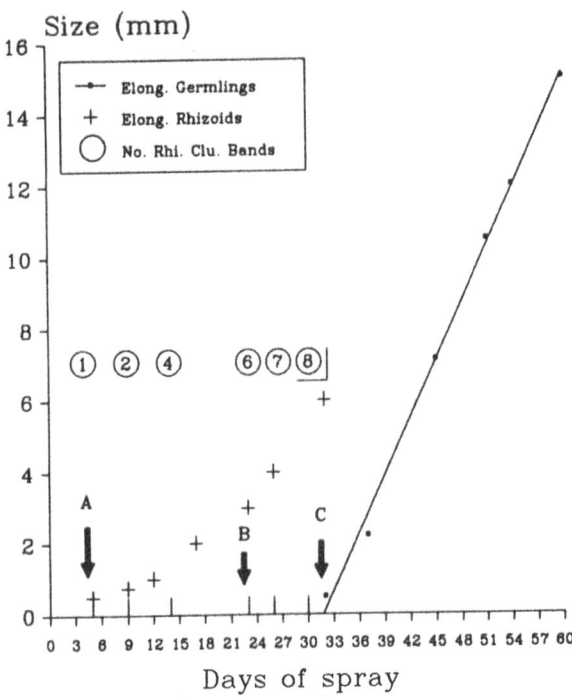

Fig. 4. Re-attachment sequence: (+) apical growth, (line) germlings, (A) appearance of rhizoidal clusters, (B) appearance of primary buds on shoots, (C) clumps of germlings with apical growth.

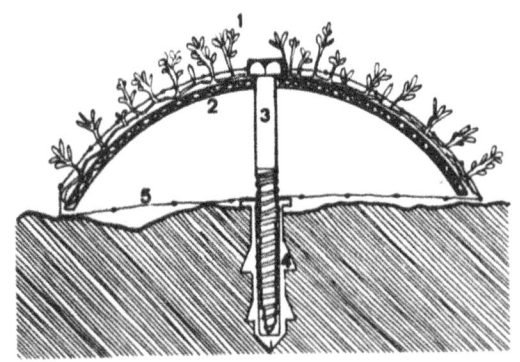

Fig. 3. Substratum positioned in the sea: 1 plant, 2 substratum, 3 & 4 bolt and washer, 5 protective netting.

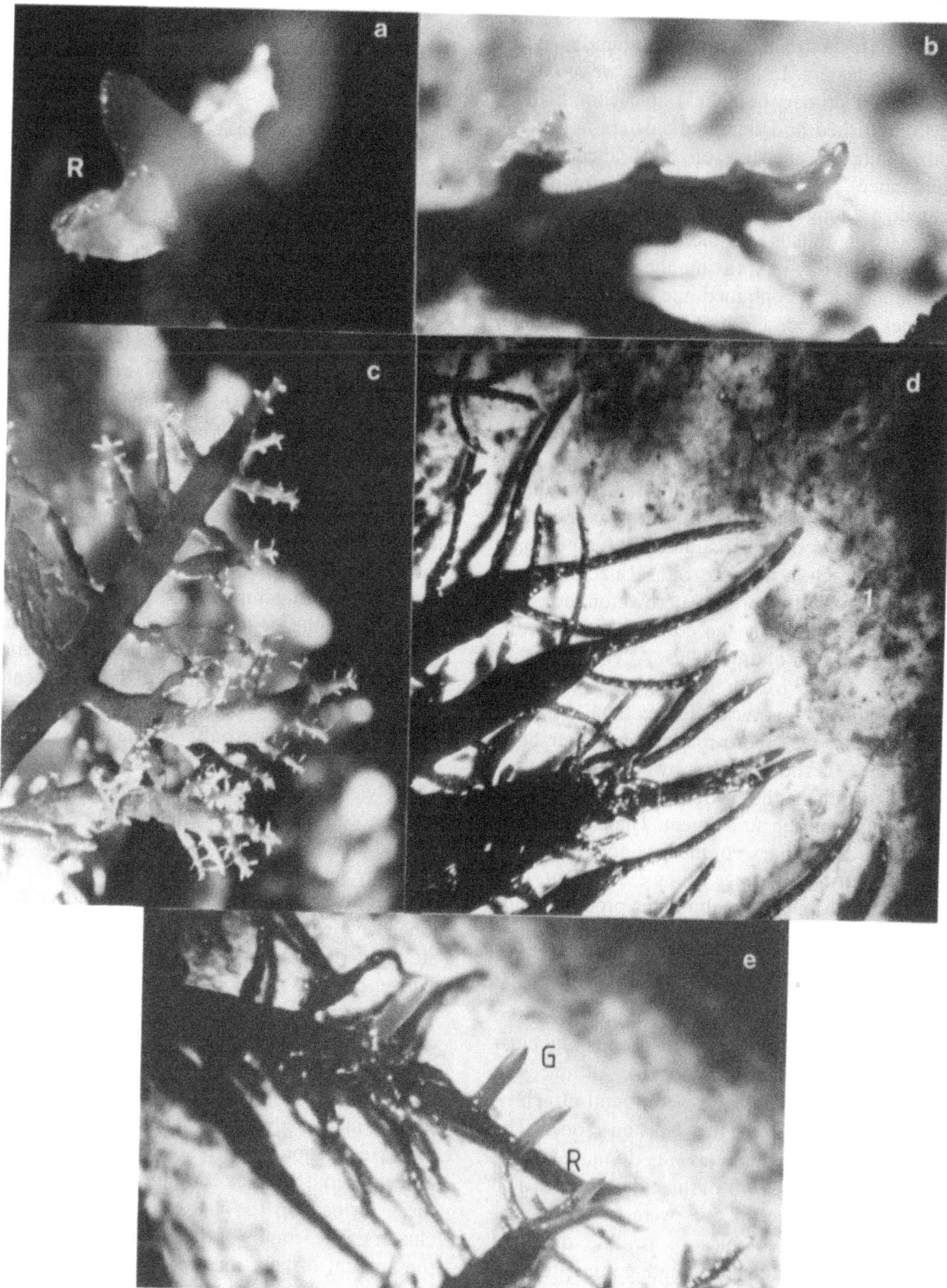

Fig. 5. Early developmental stages in the reattachment of *Gelidium sesquipedale* under the spray-system conditions: (a) apex of
G. sesquipedale with rhizoidal clusters (**R**); (b) initial stage of apical attachment in *G. sesquipedale*; (c) thallus of *G. sesquipedale*
obtained by spraying; covered with rhizoidal clusters; (d) *G. sesquipedale* rhizoids expanding across the substratum; (e) sequential
production of germlings (**G**) from rhizoid (**R**) re-attached by haptera (**A**).

112

the re-attached thallus and the abundance of rhizoidal clusters. In thalli less than, or equal to, 5 cm long (8 primary branches), apical dominance was diminished and rhizoidal clusters appeared with similar frequencies on all branches (co-dominance), indicating that initial fragments should be longer than 6 mm. Co-dominance decreased as thallus size increased.

On the 9th day of spraying, the appearance of rhizoidal clusters continued, the original curve of the apex increased, and its end began to become cylindrical with a diameter of 0.7 mm (rhizoid initiation) and bore 2–3 crowns of rhizoidal clusters. The lower ones, which were in contact with the substrate, began to attach (Fig. 5). After 14 days of spraying, the rhizoidal cluster extended to the secondary branches and terminal pinnules. The main axis already had four crowns of rhizoidal clusters from which numerous apices had begun to attach to the substrate. Rhizoidal clusters appeared in the form of a crown on incipient rhizoids, generated by the cortical layer on the posterior incision of the thallus. After 17 days of spraying, apical growth had attained 2 mm in length, although this varied among thalli. Positive phototropism was more marked at higher light intensities. In the chambers with less light, it seemed as though phototropism of the apices was inhibited, favouring attachment. Thalli with more abundant secondary branches and pinnules showed, during this stage of the experiment, less apical growth but more abundant production of rhizoidal clusters; this was indicated by the presence of secondary branches attached to the substrate (Fig. 5).

On the 23rd day of spraying, apices were 3–3.5 mm long with six crowns of rhizoidal clusters separated by 0.5 mm (Fig. 6). All apices in contact with the substrate were found attached, including those from the incision zone. The rhizoid of the principal axis was dominated by a larger number of crowns of rhizoidal clusters. The regions around the rhizoids were free from epiphytes, such as colonies of benthic diatoms, which otherwise may have inhibited the progress and penetration of the rhizoidal clusters into the substrate. The first signs of sprouting of new

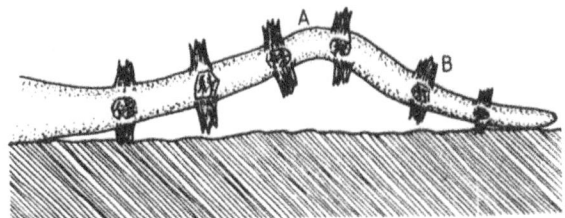

Fig. 6. Rhizoid not responding to phototaxis (A), and showing crown of rhizoidal clusters (B).

germlings were noted on the oldest parts of the rhizoids, in the form of microscopic swellings. After 26 days of spraying, the plants were clearly attached to the substrate and had apical rhizoids of variable length (about 4 mm), with 7–8 crowns of rhizoidal clusters. Phototropism was no longer effective and the rhizoids began to curve towards the substrate, independent of light intensity. At the same time, the first crowns of the rhizoidal clusters on the principal apex (the oldest) began to decay after attachment. A general sprouting of germlings on the most developed rhizoids became apparent as slight swellings (buds) in the cortical layer, with more refringent and less pigmented cells. The first new germlings, with perfectly developed apices, appeared after 32 days of spraying on rhizoids of variable length (about 6 mm). These were cylindrical and showed strong positive phototropism. The thallus had abundant rhizoidal clusters and was attached to the substrate by several haptera (Fig. 7).

Positive phototropism of the rhizoids was no longer evident after 37 days of spraying, and all attached rhizoidal clusters continued to decay. Those rhizoids in contact with the substrate actively expanded across the surface, forming a complex network adapted to the texture of the substrate (Fig. 5). Although new germlings were not noted in chambers with low light levels, those present increased in size when the irradiance was increased (Table I). After 45 days of spraying, only the aerial rhizoids continued to produce, and these formed up to 11–12 crowns of rhizoidal clusters. The lengths of the germlings varied with the amount of light received (Table I), and their abundance was notably greater in the chamber receiving the highest level of irradiance; never-

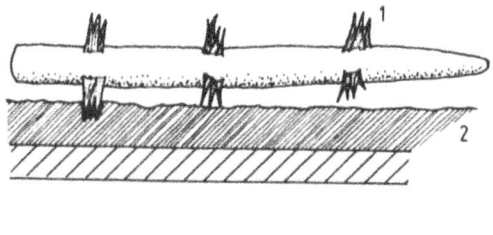

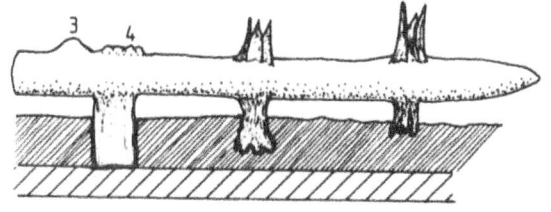

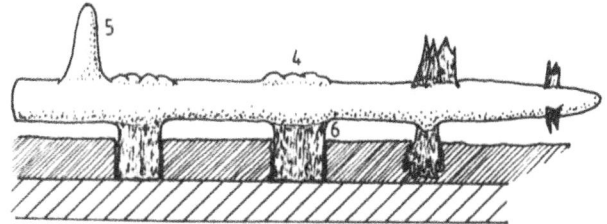

Fig. 7. Development of re-attachment rhizoid: 1, rhizoidal clusters; 2, caleareous substrata; 3, buds of shoots; 4, degenerate rhizoidal clusters; 5, germling; 6, attachment haptera.

of spraying (Table I), although successive series of germlings were produced along the rhizoid. Only those not in contact with the substrate produced rhizoidal clusters; those which were already attached and producing germlings stopped growing, bringing an end to the reattachment phase. Finally, after 60 days of spraying, some of the germlings, about 15 mm in size, began to show rhizoidal clusters at their apices, having previously become tapered apical rhizoids. This process was very active in the chambers with higher irradiances, and marked the end of the reattachment process of thalli and production of germlings of *G. sesquipedale*.

In the course of 2 months, 60–70 germlings with lengths greater than 10 mm were produced by cuttings from an apex 7 cm long with 6 primary branches. The real number was much greater, however, because successive growth sequences on the thallus and rhizoids were incorporated into the generation of clonal germlings. Studies of rhizoidal clusters during recent years have shown that some aspects of re-attachment are linked to taxonomic characters within the Gelidiales; for example *Pterocladia* reattaches only to a limited degree, as rhizoidal clusters grew poorly under the same conditions used for *G. sesquipedale*. In *Gelidium*, I have seen marked differences both in the intensity of growth and in the uniformity of the rhizoidal clusters. *Gelidium pulchellum* (Turn.) Kützing showed the greatest growth intensity of rhizoidal clusters, slightly superior to *G. sesquipedale*, while in *G. pusillum* (Stackh.) Le Jol. the process is more moderate and re-attachment is poor. Results obtained with *G. latifolium* (Grev.) Born. et Thur. have shown the lowest growth

theless, germlings must have been continuously produced, as there was a range of sizes of the rhizoids (Fig. 5). The generalized sprouting after 50 days of spraying led to practical cessation of rhizoidal growth. In this instance, the first shoot coincided with the base of the apical rhizoid; subsequent shoots appeared along the rhizoid towards the apex, provided that the rhizoid was attached to the substrate. The size of the bunches continued to be related to irradiance after 54 days

Table 1. Length (mm) of first germlings on principal rhizoids obtained by spraying at different light levels.

Chamber	$\mu E\ m^{-2}\ s^{-1}$	Days of culture			
		32	37	45	54
type I	15	–	+	2.0	5.0
type II	21	+	0.5	3.0 – 4.0	8.0
type III	32	+	1.0	4.0 – 5.0	10.0
type IV	40	+	1.5 – 2.0	5.0 – 7.0	12.0

114

intensity of rhizoidal clusters (all samples of *Gelidium* species were collected on the coast of Santander).

Rhizoidal clusters in contact with calcium and magnesium carbonate substrata formed a much more compact cylindrical structure than the aerial

Fig. 8. Final stages in the re-attachment process: (a) Attachment haptera (A) with active rhizoidal clusters in substrate. (b) *G. sesquipedale* plant with re-attachment on calcium and magnesium carbonate substrate; obtained by spraying. (c) *G. sesquipedale* plants grown in the sea to commercial size.

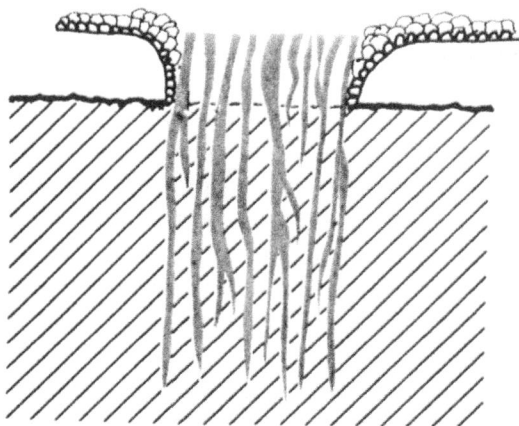

Fig. 9. Rhizoidal clusters perforating substrate.

forms and penetrated the substrata by dissolution. This process seems to occur through the apical part of the tubules because the pit made is perfectly cylindrical. Microscopic observation of these attachment haptera showed that, even after months of attachment, the tubes of the rhizoidal clusters remained active, dissolving and penetrating the substrate (Fig. 8). Attachment was possible as dissolution of the substrate was incomplete, consisting of a microscopic network of substrate and perforating filaments, in the form of a chimney (Fig. 9). As the attachment haptera consolidated, they acquired a layer of epidermal cells and took on a definitive form. No relationship has been observed between the appearance of germlings and the presence of rhizoidal clusters. There was a direct relationship between light level and the abundance and size of the shoots; best results were obtained at $40\,\mu E\,m^{-2}\,s^{-1}$. Temperature seems to play a secondary role, with an optimum of 22–24 °C; above 26 °C, the thalli developed necrotic patches.

Discussion

The process of reattachment and germling production in *Gelidium sesquipedale*, brought about by spraying, is a consequence of the initiation and transformation of those structures referred to as rhizoidal clusters. Analysis by scanning electron microscopy has indicated that sulphur is, qualita-

tively, the most important element in the composition of rhizoidal clusters, 145 times more abundant than in the cortical layer (unpubl. results). Agaropectin (sulphated fraction of agar) may be the most important constituent of the rhizoidal clusters. The role of sulphate groups in the rhizoidal clusters is not known, although it may be related to the dissolution of carbonates and the attachment of the haptera (Seoane-Camba, 1989). The development of groups of rhizoidal clusters depends mainly on the rhizoid. If the latter is attached to the substrate, rhizoidal clusters, 20–22 days old, begin to decay, progressing from the oldest at the base to the youngest at the apex. A relation between attachment and the end of rhizoidal growth indicates a regulatory role of re-attachment in rhizoidal development. If cluster development occurs and the rhizoids do not become attached to the substrate, the banded appearance and necrosis of rhizoidal clusters last more than 2 months. The same chronological appearance of rhizoidal clusters and the need to reach a minimum size of 15 mm for each germling indicate an intricate physiological regulation of this process. Many recent laboratory observations suggest that the capacity to produce germlings is affected by several factors.

Intrinsic character
Although germlings are always produced, at times under the same conditions, some plants produced thrice the number of shoots as others. This observation can be also related to plant size; the greater the size of the re-attached thallus, the greater the number of apical shoots. This seems logical, assuming that the reattached thallus uses translocated materials to supply growth of germlings. The variable production capacity provides an interesting means of selecting individuals with a special capacity to produce shoots.

Life-history
Female gametophytes with cystocarps do not readily reattach and produce germlings. These thalli remain inactive during almost the entire re-attachment process and remain rigid and darkened. Their apices do not respond phototropi-

cally, nor do they produce classical rhizoids after months of spraying under various light regimes. This does not provide serious limitation to the reattachment process in the Cantabrian Sea as 92% of the population is comprised of sporophytic individuals of *G. sesquipedale*. It does, though, raise new questions about the role of different genetic types in the life-history of *G. sesquipedale* populations and supports the hypothesis that in natural populations reproduction is exclusively vegetative or apomeiotic (Magne, 1987).

Germlings become individual plants when growth buds appear on the primary branches. Strictly speaking, the shoots cannot be considered as separate individuals (genets) as they form part of the same thallus; they are, in fact, clones derived from the same parent. Once the germlings reach 15 mm in length and, if spraying continues, the apical cylinder tapers rapidly (pseudo-rhizoids) and begins to produce bands of rhizoidal clusters in the same way as the apices of adult thalli. This change, the opposite of germling development, can be prevented by transferring the substrata to liquid culture with directional currents. The response originates from the variable arrangement of the apical meristematic cells during mitosis. This is easily confirmed in the laboratory, as the apices can be changed from normal to rhizoidal form, and visa versa, with or without water currents. The absence of currents may be the main stimulus of rhizoid initiation and, subsequently, of rhizoidal clusters. The penetration of the substrate by the rhizoids would ultimately prevent rhizoidal development, even in the absence of water motion. When a flow exists (tank culture), the cylindrical plants gradually become flattened and rounded. When the germlings are more than 20 mm in length, the first buds begin to emerge, opposite the apices, in bunches which correspond to the primary branches. From an evolutionary viewpoint, this marks the beginning of the 'plant stage', with vigorous and continuous growth of new primary branches on the principal axis and leaflets on the primaries; subsequently, these give rise to secondary branches and pinnules, thus completing the typi-

cal morphology of the plant. Densities of about 8 plants per cm^2 can be achieved after 4 months of spraying, much higher than those which can be obtained in the natural environment (Fig. 8). The production of plants in the sea does not need a lengthy initiation period of laboratory cultivation; rhizoids attached to substrata placed in the sea have subsequently developed normally. The sea, qualitatively and quentitatively, provides the best and most economical conditions for the germlings.

Although this study was focussed on growth of rhizoidal clusters, complementary studies of plants cultured in the sea to the adult stage have provided information on the relationship between plant and substrata. Perfectly developed adult thalli have been obtained, although their morphology differed from that in nature in having less characteristic juvenile cylindrical forms. This may be because when plants have a greater surface available for rhizoidal development, they occupy it almost completely and prevent competition. The more rapid growth seen in cultivated plants, as a result of cutting and sprouting (11 cm y^{-1}), favours this hypothesis. The maximum growth-rate obtained on substrata with 15 mm long germlings, and without a stage of cultivation in the laboratory, was 17 cm y^{-1} (Fig. 8). Many aspects of the development and growth of *Gelidium* remain unknown; nevertheless, the role of the rhizoidal clusters in re-attachment provides a new alternative towards understanding such important aspects as population dynamics, reproduction, genetics and population grouping.

Conclusions

The spray technique is an highly efficient method for obtaining rapid and predictable re-attachment of apical fragments and germling of *Gelidium sesquipedale* at any time of the year. The rhizoidal clusters of *G. sesquipedale* are the means of re-attachment to organic or mineral substrata. The observed links between the existence of growth of *Gelidium* and calcareous substrata may account for population structures. It may also clarify the difficulties which this species has in recolonizing habitats in which calcareous organic substrata (*Litophyllum* and *Mesophyllum*) have been destroyed. Many other unknown aspects of *Gelidium* biology can be explained through the rhizoidal-cluster reattachment process, such as vegetative reproduction and the great difference in frequencies of gametophytic and sporophytic individuals in natural populations.

The re-attachment method of producing clones opns new avenues in the culture of Gelidiales. It is more reliable and more rapid than systems based on the use of spores, and it allows the selection of germlings through successive series of reattachment from the same individual. It also resolves the problem of the reproductive parts being seasonal, as it is effective throughout the year; substrata with germlings in different stages of development can easily be transported to the sea. The reattachment capacity of *G. sesquipedale* provides a new interpretation of dispersion and colonization of rocky shores, and it lends credence to the model of intense aggregation of *Gelidium* populations. This study provides new information on the response of different *Gelidium* species to the re-attachment process; we are now in a better position to understand specific requirements and populations dynamics, and to consider the possibilities of restoring and cultivating the more interesting species.

Acknowledgements

My sincere thanks to Drs B. Santelices and J.L. McLachlan for their critical comments on the manuscript, which have clarified and improved this work. I also thank my many colleagues, and in particular L. Valdes, who have made this study possible.

References

Abbott, I. A., 1980. Seasonal population biology of some carragenophytes and agarophytes. In Abbott I. A., M. S. Foster & I. F. Eklund (eds), Symposium on Useful Algae, Seagrant Program, La Jolla, CA, 45–53.

Akatsuka, I., 1986. Japanese Gelidiales (Rhodophyta) especially *Gelidium*. Oceanogr. Mar. Biol. Ann. Rev. 24: 171–263.

Barilotti, D. C., 1980. Genetic considerations and experimental design of outplanting studies. In Abbott, I. A., M. S. Foster & I. F. Eklund (eds), Symposium on Useful Algae, Seagrant Program, La Jolla, CA, 10–18.

D'Antonio, C. M. & A. Gibor, 1985. A note on the influence of photon flux density on the morphology of germlings of *Gelidium robustum* (Gelidiales, Rhodophyta) in culture. Bot. Mar. 28: 313–316.

Dixon, P. S. & L. M. Irvine, 1977. Seaweeds of the British Isles. Vol. I., Part I, Rhodophyta. British Museum (Natural History) London, 252.

Fritsch, F. E., 1952. The Structure and Reproduction of the Algae, Vol. I. Cambridge University Press, Cambridge, 939.

Gayral, P., 1966. Les Algues des Cotes Français (Manche & Atlantique). Doin Deren & Cie, Paris, 632.

Lignell, A. & M. Pedérsen, 1986. Spray cultivation of seaweeds with emphasis on their light requirements. Bot. mar. 29: 509–516.

Magne, F., 1987. Is the frequency of apomeiosis in the Rhodophyta a genetic character? Hydrobiologia 151/152: 221–232.

Moeller, H. W., G. Griffin & V. Lee, 1982. Aquatic biomass production on sand using seawater spray. In Symposium Papers, Energy from Biomass and Waste VI, Lake Buena Vista, FL. Institute for Gas Technology, Chicago 299–302.

Moeller, H. W., S. M. Gaber & G. Griffin, 1984. Biology and economics of growing seaweeds on land in a film culture. Hydrobiologia 116/117: 299–302.

Moss, J. R., 1977. Essential consideration for establishing seaweed extraction factories. In Krauss R. W. (ed) Marine Plant Biomass of the Pacific Northwest Coast: A Potential Economic Resource. Oregon State University Press, Corvallis, 301–314.

Seoane-Camba, J. A., 1989. On the possibility of culturing *Gelidium sesquipedale* by vegetative propagation. In Kain J. M., J. W. Andrews & B. J. McGregor (eds) Proceedings of the Second Workshop of COST 48, Subgroup 1: 59–68.

Hydrobiologia **221**: 119–124, 1991.
J. A. Juanes, B. Santelices & J. L. McLachlan (eds), International Workshop on Gelidium.
© *1991 Kluwer Academic Publishers.*

Artificial sporeling and field cultivation of *Gelidium* in China

X.G. Fei & L.J. Huang*
EMBL, Institute of Oceanology, Academia Sinica, 7 Nan-hai Road, Qingdao, 266071 P.R. China.
*Shandong Marine Cultivation Institute, 29 Jinkou Road, Qingdao, 266003 P.R. China

Key words: agar, cultivation, *Gelidium amansii*, *Gelidium pacificum*, raft culture, tank culture

Abstract

The supply of *Gelidium* resources is dependent on collection from natural habitats. As these resources are very limited, one possible means of augmenting production is by implementation of artificial cultivation as has been done with *Laminaria*, *Porphyra* and *Eucheuma*. Chinese phycologists have conducted research on sporeling and field cultivation of *Gelidium* for many years. Trials have been made using: (1) raft cultivation based on vegetative propogation of thallus fragments, (2) cultivation based on spore-collecting and land-based tank culture to provide seedstock, (3) cultivation based on regeneration from small thallus fragments to provide seedstock. Some of these results have been promising, but development remains at the experimental stage and methods of cultivation need to be improved. This paper is an updated review of the research and development on artificial sporeling production and field cultivation of *Gelidium* in China.

Introduction

Natural resources of *Gelidium*, seaweeds used internationally for agar production, are comparatively limited along the Chinese coast. Research on *Gelidium* is thus a major concern of phycologists in China. The successful commercial production of *Laminaria* and *Porphyra* in China, *Porphyra*, *Undaria* and *Laminaria* in Japan and Korea and *Eucheuma* in the Philippines and Malaysia, has prompted Chinese phycologists and mariculturists to pursue *Gelidium* cultivation since the 1950's. Some progress has indeed been made; however, all efforts are still at the experimental or development stage, and commercial cultivation of *Gelidium* along the Chinese coast has not been achieved. Experiences of Chinese researchers and future developments in *Gelidium* cultivation in China are both presented in our paper.

Species of *Gelidium* in China

There are 13 species of *Gelidium* recorded from China (Zhang & Xia, 1988): *Gelidium amansii* (Lamouroux) Lamouroux, *Gelidium crinale* (Turner) Lamouroux, *Gelidium divaricatum* Martens, *Gelidium japonicum* (Harvey) Okamura, *Gelidium johnstonii* Setchell & Gardner, *Gelidium kintaroi* (Okamura) Yamada, *Gelidium latiusculum* Okamura, *Gelidium pacificum* Okamura, *Gelidium planiusculum* Okamura, *Gelidium pusillum* (Stackhouse) Le Jolis, *Gelidium tsengii* Fan, *Gelidium vagum* Okamura, *Gelidium yamadae* Fan. Among these 13 species, *G. amansii* and *G. pacificum* are the most important raw materials for agar production. Both have large thalli and produce good-quality agar; therefore, these two species have been used for the cultivation experiments on *Gelidium* in China.

120

Raft culture based on vegetative propagation of thallus fragments

Fragments of thallus obtained from sporophytes or gametophytes of *Gelidium* have been used for vegetative propagation and have become the basis for mariculture of *Gelidium*. Okamura (1911) pioneered research on propagation of *Gelidium* in Japan, while the first growth experiment with thallus fragments in China was made in 1956 in an artificial shallow pool on the Qingdao coast (Li, 1981). *Gelidium amansii* fragments (45 g), taken from natural beds by divers, were planted at the bottom of a 1 × 0.2-m area. After 90 days, the wet weight had increased to 180 g, with a daily-growth rate of 1.54%.

Successful mariculture of *Laminaria*, by the raft-culture method, stimulated experimental raft culture of *Gelidium* in shallow, nearshore Chinese waters. These trials showed that *Gelidium* fragments grew much better when positioned near the surface than at deeper positions (Li *et al.*, 1989). Horizontal strings of thallus fragments are therefore usually positioned immediately below the surface. Experimental results from small- and large-scale cultivation of *Gelidium*, undertaken in Qingdoa in 1956, 1981, 1983, 1984, and 1986, have been reported by Huang (1989) and Li *et al.* (1989). Growth-rates ranged from 1.23–4.58% day^{-1} (avg. 2.5% day^{-1}); culture period was from 56–128 days (usually 60–90 days).

In seaweed cultivation the following simple relationship holds:

$$G = ln\ (W_t - W_o)\ t^{-1} \times 100$$

where G is growth rate (% day^{-1}), t is cultivation period (days), W_o is weight at the start of the experiment W_t is weight at time t.

An average growth rate of > 3% day^{-1} is usually acceptable for a cultivation period of 120 days or more. In seaweed cultivation, the 'Multiple value, M' (W_t/W_o and therefore dependent on G and t) is an important parameter. Longer cultivation periods and higher daily growth rates result in higher values for M. M must be at least 20 for the cultivation to be profitable. In the Qingdao trial cultivation above, M was < 10,

because of the slow growth rates and shorter cultivation periods, and cultivation was not commercially viable. In these experiments, fragments of *Gelidium* from local habitats were usually collected by divers for cultivation at a density of 100 g m^{-1} on parallel strings normally set 0.30–0.4 m apart on the rafts. The resulting cultivation density for this kind of raft culture was thus 167–222 kg per cultivation unit (1600 m of seedstock strings). This method required too much valuable natural *Gelidium*, and therefore, cultivation starting from mature thalli was not practical.

Cultivation based on spore-collecting and land-based tank culture to provide seedstocks

Methods of seaweed cultivation include starting the culture from spores (*e.g. Laminaria, Porphyra*) and starting from thallus fragments (*e.g. Eucheuma, Gracilaria*). Our research efforts on *Gelidium* cultivation have been based on spore culture (Huang, 1982, 1988; Huang *et al.*, 1986, 1989). Four steps have been distinguished in the method:

Spore-collecting

Gelidium amansii thalli were collected by divers from nearshore habitats in Qingdao. Mature tetrasporophytic and cystocarpic thalli were selected and washed. Tanks filled to 0.1–0.2 m depth with sterilized seawater were used as containers. Substrata of glass slides, clam shells and small screens made from 3-mm diameter strings were used for spore-collecting and placed on the tank bottom. Clean and reproductive *Gelidium* fronds were positioned above the substrata. Spore-collecting started in the morning, when the water temperature was 22–25 °C, in indoor natural light > 50 μE m^{-2} s^{-1}. *Gelidium* fronds were removed from the container in the evening and all substrata the next morning.

Tank culture of juvenile Gelidium

Substrata with attached *Gelidium* spores were transferred to culture tanks filled with clean N- and P-enriched seawater 0.2 m in depth. Temperature and light conditions were similar to those

Table 1. Development of juvenile creeping branches of *Gelidium amansii* in indoor cultures (1981).

Days	n	Average length
5–6	30	36.3 ± 5.7
9–12	30	54.3 ± 12.0
15–16	30	96.5 ± 24.5
20	40	143.4 ± 43.8
25–26	35	255.4 ± 89.3
31–31	30	384.2 ± 144.3
34–35	40	518.3 ± 195.5
53–56	30	1286.3 ± 506.5

Table 3. Results of experiment on adult *Gelidium* cultivation.

Date	Wet weight	Multiple	Period	Growth rate
27 Aug.	86.4 ± 27.0	1	0	
11 Sep.	130.9 ± 37.7	1.52	15	2.79
26 Sep.	180.0 ± 48.9	2.08	29	2.25
11 Oct.	329.0 ± 75.7	3.81	45	3.78
23 Oct.	439.5 ± 91.1	5.09	57	2.44
8 Nov.	8700.9 ± 188.0	9.27	71	4.28
23 Nov.	1077.7 ± 226.7	12.48	98	1.11
17 Dec.	1513.6 ± 252.1	17.58	112	2.45

Average growth rate = 2.56% day

in the spore-collecting room during tank cultivation of juveniles. Substrata were cleaned and seawater was changed once or twice weekly. Under these conditions, 0.3-mm juvenile creeping branches were observable after 30–40 days and these elongated to about 1.0 mm after 50 days (Table 1).

Mariculture of Gelidium juvenile seedstocks

In our experiments, juvenile seedstocks of *Gelidium* did not usually become fully developed under indoor culture conditions. Transplanting to the sea was, therefore, necessary when creeping branches reached 5–10 mm in length. After transplantation, erect and creeping branches grew faster, and juvenile shoots of *Gelidium* were produced from creeping branches. Table 2 and Fig. 1 show the results of experiments in 1980–1981, and these experiments continued until the following summer, when wet weight of a single string was about 80 g m⁻¹. Seedstock produced by cul-

tivation can be used for cultivation of *Gelidium* thalli.

Adult Gelidium cultivation

With a string of *Gelidium* seedstock, cultivation of *Gelidium* thalli can be started. Table 3 shows the results of an experiment begun in August and extended to December 1984. This experiment included 11, 3.3-m long strings of seedstock (produced from spores collected in July 1983), cultivated on a raft from 1983 to 1984.

These experimental results (Table 3) show that, with raft-culture production from seedstock to adult, there was more than a 17-fold increase in weight within 112 days. This is nearly acceptable for commercial production, even though further improvements must be made. For cultivation based on spores only, collection of the few mature *Gelidium* thalli will not deplete or affect the natural resources. This is therefore the direction which should be encouraged. The main problem in this

Table 2. Development and growth of juvenile *Gelidium* after transplantation to the sea. In all experiments *n* = 10.

Date	15 Oct./80	5 Nov.	5 Dec.	5 Jan./81	5 Feb.	5 April
Days	10	30	60	90	120	180
Length mm	1.65 (± 0.47)	1.85 (± 0.58)	4.50 (± 0.71)	6.20 (± 1.55)	6.25 (± 1.65)	6.40 (± 1.07)
Stage	CB*	JP⁺	JP	JP	JP	JP

*CB = creeping branches
⁺JP = juvenile plantlets

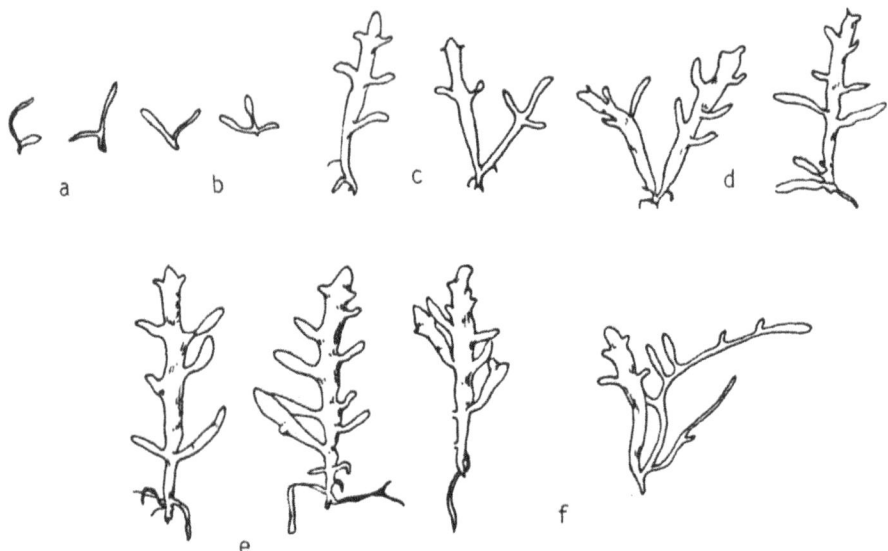

Fig. 1. Development and growth of juvenile *Gelidium* after transplantation to the sea a. 10 days; b. 30 days; c. 60 days; d. 90 days; e. 120 days; f. 180 days.

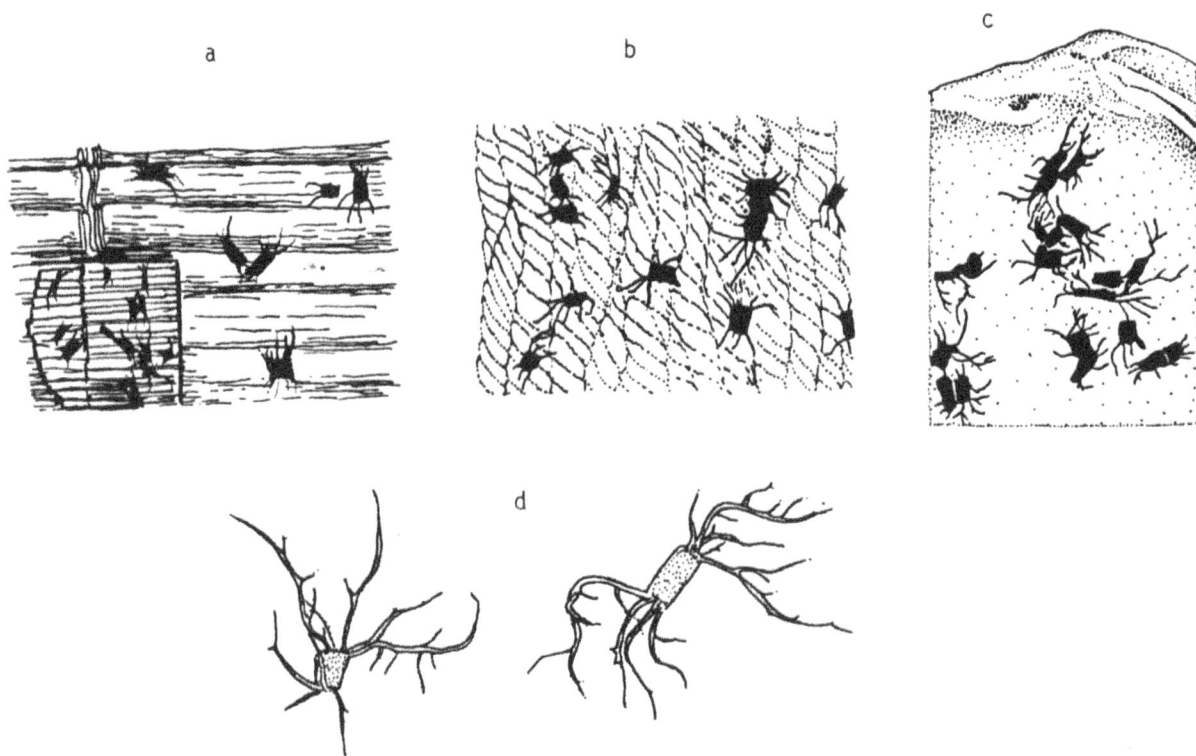

Fig. 2. Creeping branches from small thallus fragments attached to substrata. (a) bamboo slide; (b) vinyl screen; (c) clam shell; (d) small fragments and creeping branchlets.

type of cultivation is that it takes about a year from preparation of seedstock strings indoors to cultivation in the sea, to commercial-size thalli. This procedure is obviously complicated, difficult and expensive, although the results have indicated the potential for *Gelidium* cultivation in the future.

Cultivation based on regeneration of small thallus fragments to provide seedstocks

Pei *et al.* (1988) in Zhejiang produced seedstock by inducing regeneration from small fragments of *Gelidium pacificum* thalli. Following are the main experimental process and results:

On 29 January 1988, 2–5 mm thallus fragments of *G. pacificum* were cut, sowed on substrata made of clam shells, vinyl string screens and thin bamboo slides, and cultured indoors at room temperature in shallow tanks filled with N- and P-enriched seawater; light of about $50 \, \mu E \, m^{-2} \, s^{-1}$ was from a north-facing window. Regenerative buds of 0.5–1 mm were produced after 30 days; 75 days after inoculation, these buds developed into about 7-mm long creeping branches which attached firmly to substrata (Fig. 2).

On 10 June 1988, these materials were transplanted for cultivation on rafts in the sea. Within 35 days, all the minute thallus fragments on screens developed into 20–40 mm long creeping branches with erect (1–2 mm) juveniles. After another 17 days, these juveniles reached 25 mm in length and became suitable seedstock.

The above cultivation procedure took about 130 days of indoor-tank culture and 53 days of raft cultivation in the sea to obtain suitable seedstock strings of *Gelidium*. Compared with a 12-month period needed to produce seedstock strings based on the spore-collecting method, regeneration of tiny thallus fragments required only about one-third of this itme. The technique is promising for seedstock production; however, more research and experiments are needed in this field.

Discussion

Cultivation of *Gelidium* on a commercial scale might be the way to augment limited raw materials for agar production based on *Gelidium*. The provision of sufficient and cheap sources of seedstock is the key factor in *Gelidium* cultivation. Among the three ways outlined above for providing seedstocks for cultivation, the one based on propagation of thallus fragments from natural habitats is not encouraging. The technique of spore-collecting is promising even though further studies and improvements are needed before it is applicable. The third way, based on the regenerative capacity of thallus fragments, shows greater potential and viability, and future research efforts should be focussed in this direction.

Gelidium cultivation on rafts at sea has usually been acceptable, although growth rates are comparatively lower than for other seaweed species. From cultivation examples cited in this paper, we obtained an average growth rate of 2.5% day^{-1} for *Gelidium* sp. which is only acceptable if the cultivation period is of 4 to 5 months' duration. The highest growth rate we achieved (4.6% day^{-1}) indicated a potential for field cultivation if further improvements can be made.

References

Huang, L. J., 1982. Preliminary observations on the growth of *Gelidium amansii* Lamx. in the sporlings stage. Acta Oceanol. Sinica 4: 223–230. (in Chinese, English abstract).

Huang, L. J., 1988. *Gelidium* sporeling and cultivation. Ocean Press, Beijing. 130 pp. (in Chinese).

Huang, L. J. & S. Z. Zhao, 1986. Preliminary research on seedstock culture based on spores. Trans. Oceanol. Limnol. 1986: 49–55. (in Chinese).

Huang, L. J., Z. T. Xi, S. Z. Zhao, L. C. Wang & S. Sun, 1989. Research on the propagation ways of *Gelidium amansii*. Acta Oceanol. Sinica 11: 481–485 (in Chinese).

Li, H. J., 1981. *Gelidium* cultivation research and its problems. Mariculture 1981: 1–7. (in Chinese).

Li, H. J., Q. Y. Li & B. Y. Zhuang, 1988. Experiment on technical cultivation of *Gelidium amansii* Lamx. by raft culture method. Trans. Oceanol. Limnol. 1988: 98–103. (in Chinese, English abstract).

Li, H. J., Q. Y. Li & B. Y. Zhuang, 1983. The effect of temperature and water depth on the growth of *Gelidium*

124

amansii Lamx. J. Fish. China 7: 373–383. (in Chinese, English abstract).

Pei, L. S., Z. Q. Fei, G. M. Ma, J. M. Zhou & Y. F. Zhu, 1988. A preliminary study on the raising of seedlings of *Gelidium pacificum* Okam. by regeneration of thallus fragments. J. Zheijiang College Fish. 7: 99–105. (in Chinese, English abstract).

Zhang, J. F. & E. Z. Zia, 1988. Chinese species of *Gelidium* Lamouroux and other *Gelidials* (Rhodophyta), with key, list, and distribution of the common species. In: Abbott, I. A. (ed), Taxonomy of economic seaweeds Vol. II, California Sea Grant College Program, California: 109–113.

Hydrobiologia **221**: 125–135, 1991.
J. A. Juanes, B. Santelices & J. L. McLachlan (eds), International Workshop on Gelidium.
© 1991 *Kluwer Academic Publishers.*

General principles of on-shore cultivation of seaweeds: effects of light on production

J.L. McLachlan
National Research Council, 1411 Oxford Street, Halifax NS, B3H 3Z1, Canada

Key words: cultivation, efficiencies, growth, light, production, seaweed

Abstract

Numerous species of seaweed have been successfully cultivated in the sea for commercial purposes. Although considerable experimental work has been done on on-shore cultivation systems, none of these has yet proved to be economically viable on a sustained basis; nevertheless, such cultivation systems offer the potential for productivities greater than can be achieved in other systems. In on-shore systems, factors other than light can be controlled and provided at saturation levels. As density of biomass is controllable, all the light entering the cultivation system is absorbed. This results in efficient conversion of light energy to biomass when only light is limiting; moreover, density, rather than growth rate, is the major factor determining productivity. As growth of seaweeds in on-shore systems is only vegetative, there is no interruption for reproduction or maturation of the plants, and all of the net production can be recovered. Seaweeds have, though, relatively low percentage carbon composition, compared with terrestrial plants, and this may result in apparent high productivities based on dry matter.

Introduction

Historically, seaweed resources were obtained by foraging for wild crops, which limited both species available and quantities that could be obtained. It is only relatively recently that, because of increased understanding of the biology and life histories of many seaweed species, attempts to provide resources through cultivation have been possible. Early practices in cultivation of seaweeds were largely limited to providing additional substrata which, in turn, were 'seeded' by natural means. Extensive cultivation of seaweeds began only about four decades ago. Undoubtedly, the single-most important contribution in applied phycology has been that of Drew (1949) who elucidated the life history of a species of *Porphyra*. Availability of this information led, almost imme-

diately, to an explosive development in the cultivation of *Porphyra*. Because of the rather late application of cultivation to seaweeds, advanced technologies are already available, so that relatively high productivities can be expected without the necessity of long periods of development, characteristic of the history of agriculture.

In addition to their consumption as food, seaweed species are used for their cell-wall extractives or phycocolloids. These unique polysaccharides, alginates, agars and carrageenans, have many applications in the hydrocolloid industry, where seaweeds represent a multimillion dollar segment of this industry (McHugh, 1987; McLachlan, 1985). It has only been since the early 1950's, when these demands began to increase exponentially (Moss, 1978), that it was quickly appreciated that insufficient natural resources were available to meet the requirements of the seaweed extractive industry (Woodward, 1952). This was especially true of agarophytes and carrageenophytes, both of which are species of red algae. As the life histories of most commercial colloid-yielding species became reasonably well-known, cultivation was an obvious adjunct to the seaweed industry in providing additional resources. This has, indeed, occurred in the carrageenan industry, where today the majority of resources are derived from cultivation. These resources, though, are largely limited to species of *Kappaphycus* and *Eucheuma*, which account for nearly 80% of the resource base of the carrageenan industry (McHugh, 1991). Presently, demands for carrageenans continue to increase, and it will only be through additional cultivation that these requirements can be met.

There is a high demand for agarophytes in an industry that has been severely resource-limited, and over the past decade or so there have been virtually no increases in the quantities of agars that have been marketed (*cf.* McLachlan, 1985; McHugh, this volume). This is attributable to a lack of development in the cultivation of agarophytes, which has failed to parallel cultivation of *Kappaphycus*; however, recent reports on cultivation of species of *Gracilaria* suggest that the situation is greatly improving (Santelices & Ugarte, 1990). Parallel developments have apparently not occurred in the *Gelidium* sector, although on-shore cultivation trials of *Gelidium* species are being undertaken on Vancouver Island, British Columbia, by Agar Technologies Inc.

Successful commercial cultivation of seaweeds has been done in the sea. In some instances, for example with species of *Porphyra* and *Laminaria*, cultivation is done through spores; the crop is annual or biannual, with a sequence of planting and harvesting similar to that of the farming of many agricultural crops. These resources have, therefore, a relatively high cost of production, particularly for labour and maintenance of life-history stages under artificial conditions. These species are clearly not amenable to vegetative propagation under cultivation conditions. For seaweed resources that are relatively cheap, cultivation has been through vegetative propagation; where practical, this has been done very successfully, employing generally low-level technologies in areas where labour is cheap.

In addition to in-sea and pond cultivation, there have been numerous proposals for commercial cultivation of seaweeds in land-based tanks. Much of the pioneering work with land-based systems was done by Neish (Neish *et al.*, 1978) and Ryther (Ryther *et al.*, 1979). In all instances, vegetative propagation has been used, so that the basic concept of cultivation remains extremely simple, whether for cultivation in the sea, in small laboratory vessels, or in large, outdoor tanks. At present, cultivation in tanks, starting from spores, has apparently not been undertaken, apart from where alternate life-history stages are required for establishment of commercial crops in the field. Spore cultivation has been impractical because of high costs and low productivities.

Considerable experimental work has been done in relatively small tanks. Scaling up of tank cultures, which is essentially increasing the size of a monospecific ecosystem, is conceptually simple, although many technical problems have been encountered. Tank cultivation is capital-intensive, requiring extensive periods of development, which greatly add to the costs. To date commercially-viable operations remain to be proven, although possibilities are that such a system may be realized (Deveau, 1989; Craigie, 1989; Craigie & Shacklock, 1989). This effort has, however, required large inputs of public monies, prolonged backing by a benevolent government department, and recent significant changes in the market and demand for certain seaweed resources; moreover, significant financial backing must be continually available as insurance against unforeseen losses, largely the result of stochastic events; these can be extremely expensive, both in money and time.

On the positive side, on-shore cultivation offers the possibility to control many of the factors affecting growth, so that production approaches the limits of solar radiation that impinge on the cultivation system; consequently, productivities can be expected to be among the highest attainable when the full photosynthetic potential of the system is realized (McLachlan *et al.*, 1986).

In this report, economic factors related to cultivation have not been considered, and many of the technical innovations that are necessary to implement commercial cultivation have been ignored; at the same time, it is emphasized that technological innovations, however ingenious, can never surpass the limitations imposed by solar radiation. In order to obtain maximum yields, large inputs of nutrients and so forth are required to sustain maximum photosynthesis. This is, perhaps, more easily and effectively achieved in an on-shore cultivation system because of the control that is possible; in this respect, such cultivation systems are unique as, in practice, it should be possible to exploit the limitations of light throughout the production cycle. Light, in these circumstances, becomes the central issue; however, as noted previously (Bidwell *et al.*, 1985) commercially-successful systems must be cost-effective, which does not necessarily imply the operation of a cultivation system at maximum biological efficiencies.

In part some of the ideas presented here result from experience in the development of an on-shore system for Irish moss, *Chondrus crispus* Stackh. (Bidwell *et al.*, 1985; McLachlan *et al.*, 1986).

The on-shore cultivation system

A principal advantage of on-shore cultivation is that overall control is possible through integration of the system components. Light is the non-controllable variable; the primary objective becomes, then, to manage a system in which light becomes the single factor limiting production (Bidwell *et al.*, 1985). The major processes are photosynthesis, which is rate limiting, production and maintenance, and a description of light energy is appropriate only when light is the limiting factor.

An on-shore cultivation system is shown in Figs 1 and 2, where only the physical and chemical factors are included. The predetermined di-

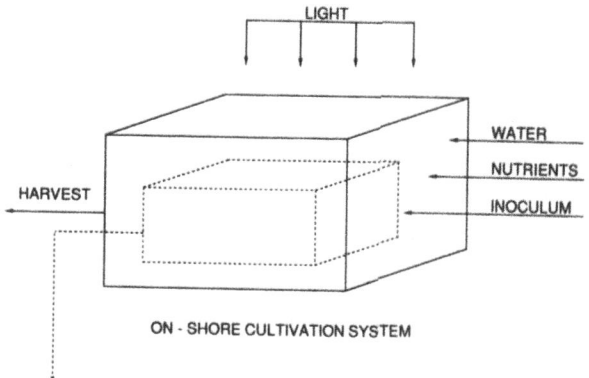

Fig. 1. Diagrammatic illustration of an on-shore cultivation system, showing the major inputs to the system and harvest as the major withdrawal. The box indicated by the dotted lines refers to that in figure 2.

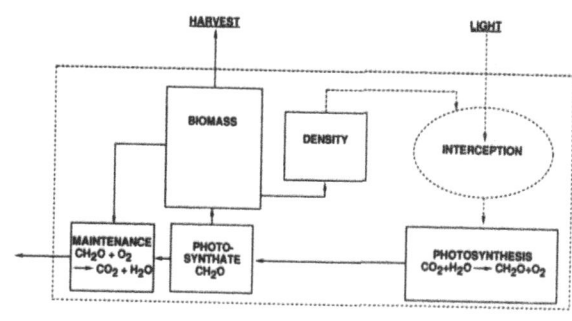

$$E = (Q - W)$$

Fig. 2. Relationships among input of light into the cultivation system, the pathway of production and harvest.

mensions of the system include surface area and depth, which set the volume. The outline of the system, although commonly rectangular, can be of any configuration. Shallow depths are usual ensuring that the system has a relatively large surface-area-to-volume ratio, and, in general, the greater the SA/V ratio the greater the predicted productivities (*e.g.* Khailov & Silkin, 1986). Seawater is provided by pumping, and exchange rates are controlled. Temperature and nutrients are, to an extent, regulated by flushing rates, although nutrients are usually added separately; most frequently, these include nitrogen, phosphorus (Schreiber's medium; Schreiber, 1927; McLachlan, 1973), and perhaps iron; all are provided at saturation levels. In productive systems, carbon dioxide must also be added, as this

128

nutrient is rapidly depleted during plant growth, resulting in unfavourable pH levels (Bidwell & McLachlan, 1985; Bidwell *et al.*, 1985; McLachlan *et al.*, 1986). Concentration and ratios of nutrients, types of salts, timing of additions and pH are easily manipulated.

Temperature of a cultivation system can be regulated by several means, although for larger systems this may be impractical or uneconomical. Flushing rates offer some temperature control, especially if water can be pumped from sources of differing temperatures. In some areas there is little annual change in seawater temperature so that changes or regulation are not major considerations. This does not necessarily imply that ambient temperatures are optimal. In other areas, considerable annual fluctuations in seawater temperatures pertain (*e.g.* Fig. 3), and, if uncontrolled, can affect production, making it largely seasonal (*cf.* Fig. 4); moreover, ambient seawater temperatures usually lag annual radiation fluxes, as shown in Figs. 3 & 4, although in shallow systems, temperature may more closely track ambient air temperature, even undergoing

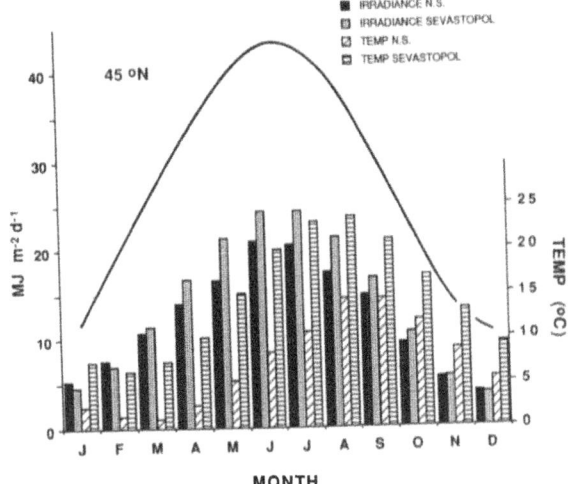

Fig. 3. Mean daily total irradiance (MJ m^{-2} d^{-1}) for two sites at 45 °N and monthly seawater temperatures at these sites. The curve shows the calculated incoming irradiance at the top of the atmosphere at this latitude. (curve redrawn from Kirk, 1983, data for Nova Scotia from Craigie, 1989 and data for Sevastopol provided by V.A. Silkin).

large diurnal fluctuations. Though temperature may be a controllable factor, in many situations it is probably unrealistic or impractical to control it.

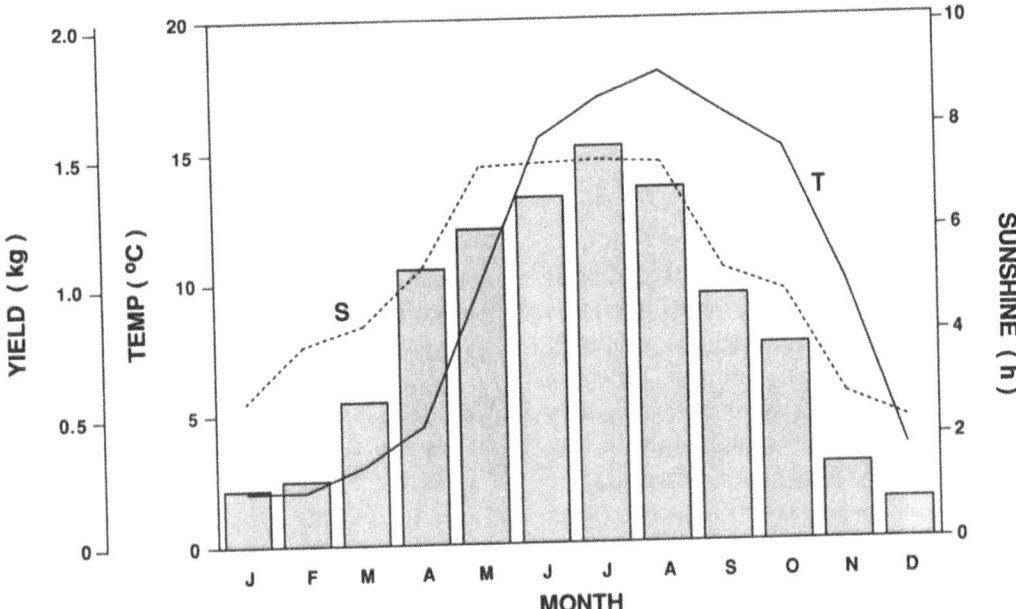

Fig. 4. Monthly production (bars), as mean dry weight per month, of *Chondrus crispus* in an on-shore cultivation system. The solid line is the mean monthly temperature and the broken line the mean daily hours of sunshine for the period shown (data from Bidwell *et al.*, 1985).

Growth and reproduction

In on-shore systems, cultivation of seaweeds is through vegetative propagation, although this is not easily done for all species. Growth interrupted by reproduction is a major disadvantage. It is usual for the plant material to be frown suspended in the medium, rather than attached; normal morphological development is unnecessary, and for species having heteromorphic morphologies, only one of the thalloid types need be cultivated. Clonal material is easily established and maintained through vegetative propagation; at the same time, it is necessary to appreciate that disasters, such as diseases, can quickly devastate an entire crop.

Cultivated seaweeds, particularly where thalli are used, have a distinct advantage over most crop species in that the entire thallus is utilized rather than just an 'economic' portion of the plant. In on-shore systems, 100% of net production can essentially be recovered, and the entire growing season is available for production at that level of economic yield. Even with terrestrial species where much of the above-ground portion of the plant is utilized, little or none of the below-ground productivity may be recovered, and this can represent a considerable portion of the production.

Plants or thalli propagated vegetatively have indeterminate growth affected by irradiance, and irradiances, whether the level, duration or both, should be directly related to production; irradiances and light periods affecting reproduction and morphogenesis are not relevant. Densities of inocula in cultivation systems are controllable, and this is probably the single-most important feature of onshore systems. Incubation times for germination, closing of canopies and maturation are not part of the cultivation strategy, and, with saturation nutrients and favourable temperatures, light should always be limiting. Even where temperature is limiting for part of the year, maximum production can at least be achieved throughout the growing season.

Light

Growth or production of plant material depends upon the conversion of light energy, which is 'free' (Bidwell *et al.*, 1985), to dry matter. Light energy is the driving force of the system; at the same time, it is the uncontrollable factor. Light is photo-energy with the downward flux varying instantaneously. Only a portion of the total irradiance can be used in the conversion of photo-energy to chemical energy as absorption of energy by plant pigments is limited to the wave band 400–700 nm, the photosynthetic active radiation (PAR); even then, only a small fraction of PAR is converted into chemical energy by photosynthetic plants. There is a negative correlation between PAR and total irradiance: at low sun angles, for instance, irradiance may be about 65% PAR and decline to about 55% at solar noon. It is generally accepted, though, that over a wide range of conditions, PAR constitutes about 50% of total irradiance.

Irradiances change throughout the day, with maxima near solar noon, throughout the year and with geographical location (*cf.* Fig. 3). Accordingly, a number of factors need to be taken into account in describing plant growth in terms of light utilization efficiencies, although as they change continuously, these variables are difficult to model. As a consequence, many experimental designs rely on controlled conditions using artificial radiance. In spite of these limitations, the daily incidence of irradiance, over growth periods of finite durations, remains relatively constant, and it may be assumed that the amount of light energy utilized by a culture will be proportional to the duration of plant growth.

At any location on earth, day length and solar elevation reach maximal values in summer and are minimal in winter. While the annual-mean-value of daily irradiances declines as latitude increases, the relative seasonal amplitude increases with increasing latitude. In this way, the summer effects of increased daily irradiances are counteracted by seasonal amplitudes. At higher latitudes, considerable local variations are to be expected, owing to atmospheric losses among other factors (Campbell & Aarup, 1989). Even in tropical or subtropical latitudes, large year-to-year variations can occur (*e.g.* Doty, 1987); such variations are to be expected, and are, to an extent, stochastic. If the cultivation system is light-limited, such variations can have significant effects on production; this must be expected and taken into account in modelling production. On an annual basis, considerably more radiation is to be expected with decreasing latitude, and, other factors being equivalent, production can be related inversely to latitude.

At equivalent latitudes, different geographical sites can receive quite different amounts of annual irradiances (Fig. 3), and, regardless of location, these are always considerably less than the calculated in-coming radiation at the top of the atmosphere. Differences in total irradiances, in the example shown in Fig. 3, result largely from mid-year differences. These differences, together with different temperatures of the seawater at the two sites, predict greater production for the Crimea than for Nova Scotia. Experimental results comparing two sites of comparable latitude have supported such predictions (Bidwell *et al.*, 1984; Lloyd *et al.*, 1981). Choice of sites, based on irradiances is, therefore, a key factor for the success of on-shore cultivation systems, and should be evaluated where maximum production from available light is required.

Conversion of light energy

Photo-energy incident upon the system is incorporated into chemical energy which is withdrawn in the form of plant biomass (Fig. 2). Photo-energy is intercepted by the inoculum, photosynthesis occurs and light energy is converted to chemical energy – the photosynthate. Some of this is used directly in maintenance, while, under favourable conditions, most accumulates as biomass. Excess biomass is harvested and the system continues to self-generate. The general principle of the system can be described by the first law of thermodynamics; its most efficient mode is in steady-state so that:

$$\Delta E = [Q - W] = 0,$$

where ΔE is the change in internal energy state of the system, Q is the quantity of energy assimilated by the system and W is the energy withdrawn as both maintenance and harvest. Thus ΔE is the difference between Q and W and if $Q > W$ the internal system gains energy as biomass, whereas if $Q < W$ energy is lost and the system is deteriorating.

Light entering the cultivation system is absorbed by the plant material or biomass, and the amount depends upon both the daily integral of irradiance and the proportion intercepted by the biomass. As density of biomass in the cultivation system is controllable, virtually all of the light entering the system should be intercepted by the biomass. Under these circumstances, light will be limiting, and these relationships can be stated as

$$J_i = I_o K,$$

where J_i = irradiance intercepted, I_o = mean daily integral of light and K = proportion of light intercepted – a direct function of density of biomass. As K approaches 100%, light becomes limiting and production becomes proportional to irradiance.

Efficiencies of conversion

For a cultivation system to operate at maximum efficiency, biomass density must remain essentially constant; accordingly, biomass must also be continually withdrawn. In other words the system must be in steady state so that $\Delta E = 0$. An on-shore cultivation system can operate at steady-state throughout the growth-cycle, always with light limiting. It is for this reason that highest productions are to be expected from aquatic systems where all of the light is absorbed.

Optimal stocking densities depend upon a number of factors, including morphology of the thallus. With increasing densities, production increases (Fig. 5). For maximum production efficiencies to be achieved, all of the biomass in the cultivation system must be exposed to the light,

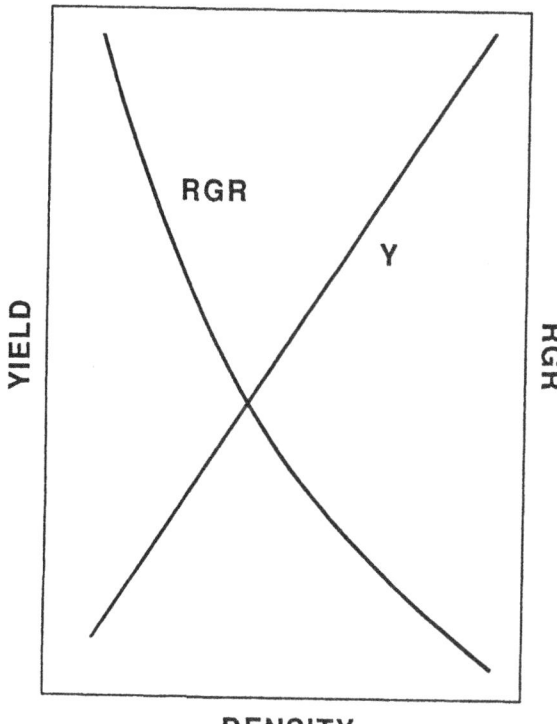

Fig. 5. Relationship between relative growth rate (RGR) and yield (Y) in a cultivation system. RGR declines as production (yield) and density increase. (based on data given in Bidwell *et al.*, 1985).

but exposure need only be intermittent. The light-dark relationships of this phenomenon remain emperical, which may be species dependent, and it is essentially a means of increasing maximum densities allowable under the existing light regime. In the modular system developed by Bidwell *et al.* (1985), light-dark cycles of about one-minute duration increased production significantly, and the general principle of this innovation is probably applicable to most, if not all, cultivation systems.

The photosynthetic process, in converting light energy to chemical energy, involves the reduction of carbon dioxide to the carbohydrate level. This conversion requires, theoretically, 8 photons per molecule of CO_2 reduced, or 4 for each hydrogen atom transferred. Calculations of conversions of CO_2 to starch equivalent suggest that about 25% of the light energy would be transformed to chemical energy as carbohydrate, the maximum theoretical efficiency (Kirk, 1983); however,

experimental data (*e.g.* Bugbee & Salisbury, 1988) have shown that the quantum requirements are 12–13 photons per molecule of CO_2 reduced, rather than the theoretical 8 quanta, so that maximum conversion efficiencies must be less than 20%. Production data based on efficiency = biomass per unit irradiance show that efficiencies are much lower, 2–6% of PAR being more usual.

Efficiencies can be estimated (Charles-Edwards *et al.*, 1986) by the slope of the regression of dry-matter produced upon the cumulated amount of absorbed light-energy:

$$\varepsilon = g\ NP\ \alpha\ Ps_{max}/(\alpha KI_o + Ps_{max}) + g\ MC,$$

where ε = efficiency, g = g CO_2 assimilated and converted to g dry biomass, NP = net production, α = quantum yield, K = proportion of light intercepted, Ps_{max} = rate of light-saturated photosynthesis, I_o = mean daily integral of irradiance, MC = maintenance coefficient.

In instances where light is limiting, K becomes a constant, resulting in high production per-unit-irradiance, as essentially complete absorption of light occurs throughout the production period. For terrestrial plants, high efficiencies have been obtained in short-term experiments. Productions in the long term have, in contrast, been much less than calculated values, and this has been attributed to incomplete interception of PAR (Bugbee & Salisbury, 1988). Over the growing season, only about 50% of available irradiance is absorbed by higher plant systems compared to the nearly 100% that is possible for an on-shore cultivation system. With complete absorption of PAR, photosynthesis is then limited by light, and quantum requirements probably remain constant for a given species.

The maintenance coefficient is extremely difficult to assess, largely because few data are available. For seaweeds it has been estimated that respiratory costs are about 30% of gross production (Bidwell & McLachlan, 1985; Enright & Craigie, 1981). This accords with values obtained for many crop plants (Bugbee & Salisbury, 1988) where it has been suggested that most maintenance respiration is associated with continued

132

synthesis of compounds in response to changing environmental conditions. In on-shore cultivation systems, major environmental changes are largely light fluxes; carbon required for respiration may not be so great as might be expected in other environments.

The conversion of photosynthate to biomass can have a major influence on the calculated efficiency values. These depend to a major extent upon the chemical nature of the dry matter produced. If the dry matter is mainly carbohydrate, high efficiencies can be expected, whereas the synthesis of nitrogenous and unsaturated products results in much lower conversion efficiencies. The organic constituents of most seaweed species, and certainly those used in the extractive industry, are composed largely of polysaccharides. For this reason, relatively high efficiencies are anticipated; moreover, on a dry-weight basis, seaweeds contain around 30% carbon compared with 40–45% for many terrestrial plants. Seaweeds should thus produce a relatively high content of dry matter per unit-carbon-assimilated provided maintenance costs are not exceptionally high. In the above equation, the percentage carbon contributes significantly to higher calculated efficiencies for seaweeds.

Growth and production

The primary objective of a cultivation system is the production of biomass, and it is necessary to have means of measuring growth and production of the seaweed. Growth is a fundamental biological activity which describes gross weights in mass, and rates in loss of mass, so that net changes occurring over finite time are:

$$\Delta W/\Delta t = \text{Gross Weight Gain} - \text{Rate of Loss}.$$

It is necessary to define size in dimensions of mass or other terms. Growth can be related to increases in cell numbers, the outcome resulting from two competing influences, exponential increase and mitigating factors. It is also easier to quantify and define the increasing process than

the limiting processes, which are usually dealt with in retrospective analysis. Growth is correlated with changes in the growth environment, and, contrary to some statements in the literature, is never negative, always involving irreversible enlargement or divisions of cells. There are, though, extensive differences in growth among organisms.

Growth rates: logistic

In organisms such as unicellular algae, increases in weight can be attributed to increases in cell numbers so that there is little change (only transient) in biomass of individuals in a population of cells. Where there are no constraints, growth remains constant over time, and can be calculated as compound interest. Absolute growth is proportional to the number of cells, or biomass, initially present and their rate of increase with time.

$$N = N_o \exp(\mu t),$$

where N_o = number of cell in inoculum, N = total number of cells, t = time, μ = growth constant. Taking the natural logarithm of both sides of the equation:

$$\ln(N) = \ln(N_o) + \mu t$$

so

$$ln(N/N_o) = 0 \text{ when } t = 0 \text{ and slope } \mu,$$

or:

$$\mu = lnN - lnN_o/t.$$

These equations assume no limitation to growth in the population. Limitations, in fact, always occur to the extent that growth is slowed or stopped, and in continuous culture the population, or biomass, is altered, as are other factors in the growth environment. Primary descriptions of growth do not explicitly take into account changes in environment during growth; consequently, more complex models, relating changes in the growth environment, are unavailable.

Growth in multicellular seaweeds (*cf.* Morgan, 1984) can also be expressed as changes in cell numbers, or cell weight, with time as:

$$dN/dt = RpN,$$

where N = cell number, t = time, R = growth of meristematic cells and p = proportion of cells remaining meristematic. The shape of the growth curve is determined by p. In the simplest relationship, all cells in the meristematic region are meristematic, where $p = 1$ with the simple logistic being $p = Nt - N/Nt$. Increases in weight of non-dividing cells (non-meristematic) may be written as:

$$d(1 - p)N/dt = spN,$$

where s = relative growth rate of non-dividing cells or $spNW_m$, W_m = mean final weight of expanding cells. As biomass increases, dW/dt is a summation of increases in cell number and size:

$$dW/dt = pN(sW_n + (r - s) W_m),$$

where W_m = mean weight of non-meristematic cells and W_n = mean weight of meristematic cells and reflects reduction in growth of an organism with increase in its size.

While pertinent to the present discussion, such equations are difficult to apply to seaweeds. A new formulation of the logistic growth equation recently set forth by Thornley (1990) should be assessed for its applicability to seaweeds. There is little information available to quantify seaweeds in this respect, even though these are extremely important considerations in the selection of species for cultivation and in describing their behaviour in cultivation systems. A critical area to be developed in seaweed biology is functional morphology. Qualitative relationships have been described (*e.g.* Littler & Littler, 1980; Wheeler, 1988), and there have been attempts to quantify these using surface area/volume models (Khailov & Kamenir, 1988; Nielsen & Sand-Jensen, 1990).

Growth rates: relative

To approximate growth of multicellular seaweeds, rates of growth are commonly expressed in relative terms, RGR, which is the rate of growth divided by its weight, or the rate of growth per unit of biomass when growth is not restricted and exponential:

$$W(t) = W_o \exp (\mu t),$$

where $W(t)$ = weight of plant at any time, W_o = initial weight at time 0, μ = growth constant. Thus, relative growth rate at time dW/dt is:

$$dW/dt = \mu W_o \, exp \, (\mu t)$$

or

$$(1/W) (dW/dt) = RGR = \mu.$$

$RGR = \mu$ so that RGR is a constant during exponential growth. However, as RGR is a function of thallus weight and as the thallus increases in size, or mass, with growth, RGR decreases (Fig. 5):

$$RGR = (1 - W) (dW/dt) = \mu (1 - \delta),$$

where δ is a constant and a reciprocal of the maximum growth rate; therefore, as W becomes bigger, $(1 - \delta)$ becomes smaller and approaches 0 as W approaches $1/\delta$ so when $W = 1/\delta$, the product dW = unity and RGR = 0.

Production and yield

The primary biological objective of on-shore cultivation is to obtain maximum production, in other words to capture light energy and efficiently convert it to biomass. This requires a unity of growth and productivity (Charles-Edwards *et al.*, 1986). Productivity is net yield of biomass over an increment of time and unit of surface area. The important point is that production is estimated on an areal basis, and can be approximated by summing increments of dry matter produced over time, times the area of the system:

$$W = \sum_{i=0}^{i=t} (\delta J - V) t\, A.$$

J = light intercepted, δ = efficiency of light utilization, V = loss of production, t = duration of production, A = area of the production system. Production, as noted previously, is extremely sensitive to the efficiencies of light utilization, whereas growth rates are not, provided that light is not limiting. Growth rates and production are generally related reciprocally (Fig. 5); when growth rates are high light is not limiting and production is low; when production is high, light is limiting and growth rates are low; thus, different sets of factors must be considered. A primary consideration of on-shore cultivation studies is to be able to predict production with reasonable reliability.

The net assimilation rate (NAR) is the relationship between biomass growth rate and cultivation area:

$$\text{NAR} = (1 - A)\, dW/dt,$$

where A = area, dW/dt = growth rate; accordingly, NAR is related to RGR:

$$\text{NAR} = (A/W)\, \text{RGR},$$

where A/W = area divided by the total dry weight. NAR can thus be logically related to the amount of light intercepted by the biomass of the system. If the inoculum is small, all of the light entering the cultivation system is not intercepted by the biomass. Provided that light is replete, rapid growth rates are expected. With increased growth and density, mutual shading occurs so that the amount of light received per unit surface area of biomass declines, hence RGR declines in a manner shown in Fig. 5. With continuous production, all of the light of the system is eventually intercepted by the biomass. While RGR declines, assimilation of photoenergy by dense biomass results in the continuation of production. Production will continue to increase until losses, principally through maintenance (Fig. 2), offset gains through production and there is no further net assimilation by the system. If A/W can be kept constant, then, provided that the amount of incident light converted to biomass in the system is constant, the cultivation system can be operated at steady state and maximum production can be achieved, limited only by light.

Conclusions

An on-shore cultivation system has the potential for greater productivities than other types of cultivation systems. Maximum production can be achieved only when light is the limiting factor. As density of biomass in an on-shore system is controllable, all of the light entering the cultivation system can be absorbed by the biomass and photosynthesis becomes rate-limiting. This, in turn, results in high efficiencies of light utilization. Maximum biomass densities can be maintained throughout the growing season, and when light is the factor limiting production, efficiencies of conversion of light to biomass can be maintained during this period. The seasonal integral of light will determine total production so that location of the cultivation site becomes an important consideration. It must be appreciated in these considerations that loss of inoculum in high-density systems can be disastrous because of time necessary to recover the density.

On-shore cultivation takes advantage of vegetative growth of the seaweed which is cultivated suspended in the medium. Growth is indeterminate, 'pure' morphological development of the plant is not a consideration, and all of the biomass can be recovered and utilized for economic purposes; in addition, greater biomass densities for maximum production are achievable if plants are not attached to substrata. This is particularly important as density has a major effect on production. In seaweeds, particularly those used for their cell-wall extractives, a large portion of the dry matter occurs as carbohydrate. This, together with a relatively low carbon content of the dry mater, results in high conversion rates of dry matter per unit of carbon synthesized, and further increases potential productivity in an on-shore cultivation system limited by light.

Acknowledgements

I am grateful to Drs R.G.S. Bidwell, A.E. Miller and B. Santelices for their helpful comments on the manuscript which has led to its improvement. Issued as NRC 31968.

References

Bidwell, R. G. S., N. D. H. Lloyd & J. McLachlan, 1984. The performance of *Chondrus crispus* (Irish moss) in laboratory simulations of environments in different locations. Proc. int. Seaweed Symp. 11: 292–294.

Bidwell, R. G. S. & J. McLachlan, 1985. Carbon nutrition of seaweeds: photosynthesis, respiration and photorespiration. J. exp. mar. Biol. Ecol. 86: 15–46.

Bidwell, R. G. S., J. McLachlan & N. D. H. Lloyd, 1985. Tank cultivation of Irish moss, *Chondrus crispus* Stackh. Bot. mar. 28: 87–97.

Bugbee, B. G. & F. B. Salisbury, 1988. Exploring the limits of crop productivity. I. Photosynthetic efficiency of wheat in high irradiance environments. Pl. Physiol. 88: 869–878.

Campbell, J. W. & T. Aarup, 1989. Photosynthetically available radiation at high latitudes. Limnol. Oceanogr. 34: 1490–1499.

Charles-Edwards, D. A., D. Doley & G. M. Rimmington, 1986. Modelling Plant Growth and Development. Academic Press, New York, 235 pp.

Craigie, J. S., 1989. Irish moss cultivation: some reflections. In Yarish C., C. A. Penniman & P. van Patten (eds). Economically Important Plants of the Atlantic: Their Biology and Cultivation. Connecticut Sea Grant Program, Marine Science Institute, University of Connecticut, Groton CT, 37–52.

Craigie, J. S. & P. F. Shacklock, 1989. Culture of Irish moss. In Boghen A. D. (ed.). Cold-water Aquaculture in Atlantic Canada, The Canadian Institute of Research and Regional Development, Moncton NB, 243–270.

Deveau, L., 1989. cultivation of *Chondrus crispus*. XII'th Int. Seaweed Symp. Vancouver, BC. (Abstracts TB5).

Doty, M. S., 1987. The production and use of *Eucheuma*. In Doty M. S., J. F. Caddy & B. Santelices (eds). Case studies of seven commercial seaweed resources. FAO Technical Paper 281, Rome: 123–164.

Drew, K. M., 1949. *Conchocelis*-phase in the life history of *Porphyra umbilicalis* (L.) Kütz. Nature London 164: 748.

Enright, C. T. & J. S. Craigie, 1981. Effects of temperature and irradiance on growth and respiration of *Chondrus crispus* Stackh. Proc. int. Seaweed Symp. 10: 271–276.

Khailov, K. M. & YuG. Kamenir, 1988. Photoassimilation surface of plants and its correlation with the respiratory surface of animals in size rows. J. gen. Biol. 49: 844–853 (in Russian).

Khailov, K. M. & V. A. Silkin, 1986. Ecological classification of aquaculture bioproductive systems with respect to their aquaculture quality. J. gen. Biol. 47: 769–779 (in Russian).

Kirk, J. T. O., 1983. Light and Photosynthesis in Aquatic Ecosystems. Cambridge Univ. Press, Cambridge, 401 pp.

Littler, M. M. & D. S. Littler, 1980. The evolution of thallus form and survival strategies in benthic marine macroalgae: Field and laboratory tests of a functional form model. Am. Nat. 116: 25–44.

Lloyd, N. D. H., J. L. McLachlan & R. G. S. Bidwell, 1981. A rapid infra-red carbon-dioxide analysis screening technique for predicting growth and productivities of marine algae. Proc. int. Seaweed Symp. 10: 461–466.

McHugh, D. J. ed., 1987. Production and utilization of products from commercial seaweeds. FAO Tech. Paper 288, Rome 189 pp.

McHugh, D. J., 1991. Worldwide distribution of commercial resources of seaweeds including *Gelidium*. Hydrobiologia 221: 19–29.

McLachlan, J., 1973. Growth media-marine. In Stein J. R. (ed). Handbook of Phycological Methods: Culture Methods and Growth Measurements. Cambridge Univ. Press, Cambridge: 25–51.

McLachlan, J., 1985. Microalgae (seeweeds): industrial resources and their utilization. Plant & Soil 89: 137–157.

McLachlan, J., R. G. S. Bidwell, R. G. Smith & C. M. Moseley, 1986. Cultivation strategies for seaweeds. In Westermeir R. (ed.) Actas II Congr. Nacional sobre Algas Marinas Chilenas. Univ. Austral de Chile, Valdivia, Chile, 47–62.

Morgan, J. M., 1984. Modelling environmental effects on crop productivity. In Pearson C. J. (ed.) Control of crop productivity. Academic Press, New York: 289–304.

Moss, J. M., 1978. Essential considerations for establishing seaweed extraction factories. In Krauss R. W. (ed.). The Marine Plant Biomass of the Pacific Northwest Coast. Oregon State Univ. Press, Corvallis, OR: 301–314.

Nielsen, S. L. & K. Sand-Jensen, 1990. Allometric scaling of maximal photosynthetic growth rate to surface/volume ratio. Limnol. Oceanogr. 35: 177–181.

Neish, A. C., P. F. Shacklock, C. H. Fox & D. R. Robson, 1978. The cultivation of *Chondrus crispus*. Factors affecting growth under greenhouse conditions. Can. J. Bot. 55: 2263–2271.

Ryther, J. H., J. A. DeBoer & B. E. Lapointe, 1979. Cultivation of seaweeds of hydrocolloids, waste treatment and biomass for energy. Proc. int. Seaweed Symp. 9: 1–16.

Santelices, B. & R. Ugarte, 1990. Ecological differences among Chilean populations of commercial *Gracilaria*. J. appl. Phycol. 2: 17–26.

Schreiber, 1927. Die Reinkultur von marinen Phytoplankton und deren Bedeutung für die Erforschung de Produktionsfähigkeit des Meerwassers. Wiss. Meeresuntersuch. N.F. 16: 1–34.

Thornley, J. H. M., 1990. A new formulation of the logistic growth equation and its application to leaf area growth. Ann. Bot. 66: 309–311.

Wheeler, W. N. 1988. Algal productivity and hydrodynamics – a synthesis. Progr. Phycol. Res. 6: 23–58.

Woodward, J. N., 1952. Forward. Proc. int. Seaweed Symp. 1: iv.

Hydrobiologia **221**: 137–148, 1991.
J. A. Juanes, B. Santelices & J. L. McLachlan (eds), International Workshop on Gelidium.
© 1991 *Kluwer Academic Publishers.*

Chemical structure and physico-chemical properties of agar

Marc Lahaye[1] and Cyrille Rochas[2]
[1] *Institut National de la Recherche Agronomique, Laboratoire de Biochimie et Technologie des Glucides, BP 527, 44026 Nantes Cedex 03, France;* [2] *Centre de Recherche sur les Macromolécules Végétales, Centre National de la Recherche Scientifique, BP 57, 38042 Grenoble Cedex, France*

Key words: agar, agarose, review, chemistry, physico-chemistry

Abstract

Advances in the chemistry and physico-chemical properties of agar since the review of Araki at the Fifth International Seaweed Symposium in 1965 are discussed. These advances are essentially the result of better separation techniques of the heterogeneous family of polysaccharides known as agar, the use of nuclear magnetic resonance spectroscopy, the use of agarases and, particularly, the use of combinations of the three approaches. Although physico-chemical methods have evolved, particularly molecular-weight determinations, X-ray diffraction data and molecular modelling of agar, correlations between chemical and functional properties of agar and agarose and their gelation mechanisms remain to be studied.

Introduction

The many uses of agar and agarose relate to formation of thermoreversible gels at low concentration in water with large hysteresis (Meer, 1980). The physico-chemical and rheological properties of these algal polysaccharides are linked to their chemical structure. In 1965, at the Fifth International Seaweed Symposium, Araki (1966) reviewed the chemistry of agar. Since then, the development of methodology has revealed fine structural features of these polysaccharides. The aim of this review is to summarize recent advances made in the field of chemistry and physico-chemistry of agar.

Discussion

Heterogeneity of agar

Because understanding of the chemical structure of agar had advanced through fractionation studies based on different specific physico-chemical properties of these polysaccharides, these features will be reviewed first.

Probably the oldest used physico-chemical fractionation of agar is the classical 'gel and thaw' method. The gelling polysaccharides are separated from non-gelling ones simply by freezing and thawing or by pressing an agar gel. Agar structure of some of the non-gelling polysaccharides has been recognized (Usov *et al.*, 1979; Friedlander *et al.*, 1981; Wen & Craigie, 1984; Lahaye, 1986).

Very early on, differential solubility of native or substituted agar polysaccharides was exploited to

138

(1) demonstrate the heterogeneity of agar and (2) separate agarose, the polysaccharide fraction having the highest gelling potential, from the so-called 'agaropectin', the poor or non-gelling charged polysaccharides (Samec & Isajevic, 1922; Araki, 1937). In particular, Yanagawa (1936, in Yanagawa, 1946) extracted *Gelidium amansii* Lamour. with water at 30, 50–55, and 75–80 °C and obtained fractions with different sulfate contents. He also obtained different yields in fractions of agar from *Gelidium*, *Acanthopeltis*, *Ceramium*, *Gracilaria*, and *Gloiopeltis* species extracted by boiling in 70, 60, 50% alcohol and water. Guiseley (1970) found that methylated agar was soluble in hot ethanol used to precipitate the polysaccharides, a property that was exploited in the development of a sequential solvent extraction of agar involving water at different temperatures and different concentrations of boiling ethanol-water solutions (Lahaye *et al.*, 1986).

Differences in charge densities of agar polysaccharides were exploited in anion exchange chromatographic techniques introduced by Izumi (1970, 1971, 1972) and Yaphe's group (Duckworth & Yaphe, 1971a, 1971b; Duckworth *et al.*, 1971).

Thus, agar is composed of an heterogeneous populations of molecules differing in their physico-chemical properties.

Physico-chemistry of agar and agarose

The gel-forming ability and solubility of agar polysaccharides rely on the relative hydrophobicity of the basic repeating unit, the alternating 1,3-linked β-D-galactopyranose and 1,4-linked 3,6-anhydro-α-L-galactopyranose or agarobiose (1) (Araki, 1966), and its substitution by hydrophobic (methoxyl) and polar (sulfate, pyruvate) groups.

Agar gels are formed if a helical conformation of agar polysaccharides is possible and if these helices can aggregate (Rees & Welsh, 1977). Agarose ordered conformation was considered to be two intertwined left-handed helices with a 3-fold symmetry of pitch 1.90 nm, axial advance of 0.634 nm, translation between strands of 0.95 nm

and internal cavity of 0.45 nm. Hydrogen bonds between water molecules and O_2 of galactose and O_5 anhydrogalactose would stabilize this structure (Arnott *et al.*, 1974). Recent X-ray diffraction studies show that extended single helices of axial advances of 0.888–0.973 nm are formed in particular drying and/or gelation conditions (Foord & Atkins, 1989). Such structure was corroborated by molecular modelling studies from agaro-oligosaccharides giving a 3-fold left-handed single helix of axial advance of 0.95 nm and pitch of 2.85 nm (Barbero *et al.*, 1989). Thus, clustering of single helices may also be a possible step in agarose gelation.

Gel formation is believed to occur through the aggregation of hexagonal fibers of 6 double helices (Djabourov *et al.*, 1989). Replacement of the anhydro-sugar by its biological precursor, L-galactose 6-sulfate (2) (Rees, 1961a) or by L-galactose having both a $^1C_4$ conformation instead of the $^4C_1$ of 3,6-anhydrogalactose, are known to prevent

(1)

(2)

(3)

(4)

helix formation or to introduce interruptions or 'kinks' along agar helices (Rees & Welsh, 1977). These 'kinks' would contribute to the formation of the three-dimensional network of the gel; furthermore, if chemical substitution on O_2 of anhydrogalactose and O_6 and O_4 galactose do not appear to affect markedly the helical conformation of agar (Arnott et al., 1974), they can contribute to a decrease in, or prevention of aggregation of these helices and thus prevent the gel formation, or increase the temperature at which the ordered conformation will occur (Arnott et al., 1974; Guiseley, 1970). Guiseley (1970) discovered that the higher the methoxyl content was in native agar, the higher the gelling temperature; however, when agar was synthetically methylated, an inverse relationship was obtained, demonstrating the site-specificity of methylation of this physico-chemical property (Guiseley, 1987). Indeed, most methylated derivatives in agar occur on O_6 of galactose and/or O_2 of 3,6-anhydrogalactose, whereas synthetic methylation probably occurs at random.

Average molecular weight of agar, ranging between 35 700 and 144 000 for commercial preparations, does not appear to determine the differential solubility of agar when the sequential solvent extraction method is used (Rochas & Lahaye, 1989a). Solubility of agar depends, like other polymers, on the ability of the solvent to disrupt and melt the ordered conformations (gel and helices). In aqueous ethanol solutions, agar solubility reflects the degree and strength of aggregation of the polysaccharides and their affinity for different concentrations of ethanol, once the aggregation and helical conformation have been melted at high temperatures. Thus, high concentration of methoxyl and 3,6-anhydrogalactose in agar increase the hydrophobic properties of the molecules, allowing for their solubility in hot solutions of ethanol-water (40–80%); by contrast, agar substituted by charged groups and/or of low 3,6-anhydrogalactose content will have increased hydrophobic properties with concomitant solubility in polar solvents (rich in water) at more or less lower temperatures (Duckworth & Yaphe, 1971a; Lahaye et al., 1986; Lahaye & Yaphe,

1989). It is therefore essential to know precisely the fine chemical structures of agar which determine the physico-chemical and rheological properties of the molecules.

Chemistry of agar

Since the discovery of the basic chemical structure of agarose, agarobiose (1), by the pioneering work of Araki (1966), numerous deviations from that basic structure have been described. Hirase (1957) discovered a pyruvic acid acetal in agar from *Gelidium amansii* as 4,6-O-(1-carboxyethylidene)-D-galactose (3), latter found in *Gracilaria* agar (Duckworth & Yaphe, 1971b; Duckworth et al., 1971; Young et al., 1971; Lahaye & Yaphe, 1989). The biological precursor of 3,6-anhydrogalactose, L-galactose 6-sulfate (2) was proposed by Rees (1961a). Methylated galactose units: 6-O-methyl-D-galactose (4) and 4-O-methyl-L-galactose, L-galactose, methyl-pentose, and xylose were described (Hirase & Araki, 1961; Peat et al., 1961; Araki et al., 1967). Although the exact location of methyl-pentose remains unknown, recent studies established that 4-O-methyl-L-galactose occurs as a branch on galactose in the polymer backbone (5) (Craigie & Jurgens, 1989; Karamanos et al., 1989). A similar location of xylose in agar from *Curdiea flabellata* Chapman and *Melanthalia abscissa* (Turn.) Hook f. & Harvey has been proposed (Furneaux et al., 1990). *Ceramium rubrum* (Huds.) J. Agardh agar contains an unusual neutral repeating unit in which the 3,6-anhydrogalactose was replaced by L-galactose (6) (Turvey & Williams, 976). Bowker and Turvey (1968) identified 2-O-methyl-3,6-anhydrogalactose, 2-O-methyl-L-galactose 6-sulfate, and D-galactose 2-sulfate from agar extracted from *Laurencia pinnatifida* Lamour. The 2-O-methylated anhydro-sugar was later found in the agar of *Rhodomela larix* (Turn.) C. Agardh (7) (Shashkov et al., 1978) and is the major sugar in agar from *Gracilaria eucheumoides* Harvey where it coexists with 6-O-methyl-D-galactose and galactose 4-sulfate (8) (Ji et al., 1985; Lahaye et al., 1986, 1989). Such di-methylated agarose

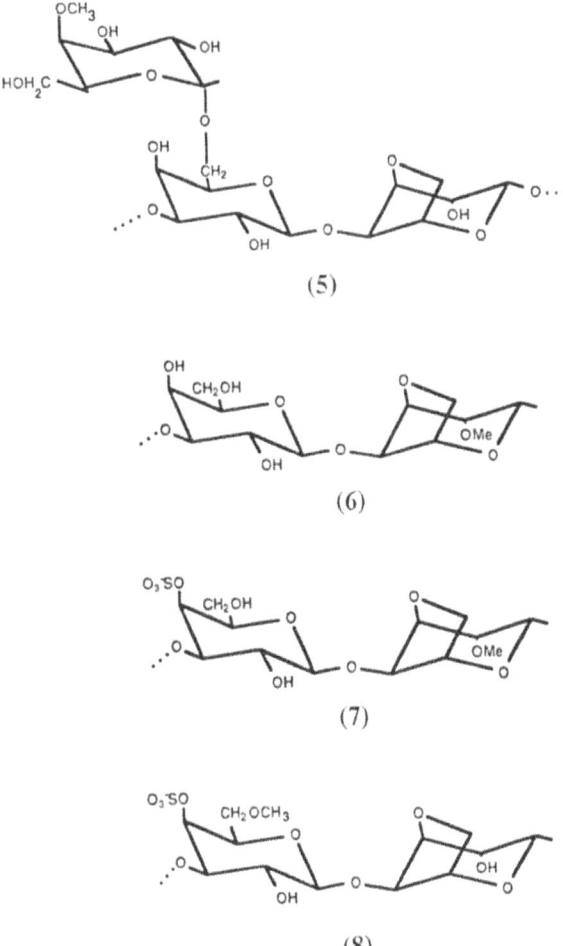

(5)

(6)

(7)

(8)

(9)

(10)

(11)

repeating units hav also been recently described in the agar from *Curdiea coriacea* (Hook f. & Hanv.) Chapman, *C. flabellata*, and *M. abscissa* (Furneaux *et al.*, 1990). Natural methylation of agar is common; in fact, most (if not all) commercial agarose contain some 6-*O*- and/or 2-*O*-methylated repeating units. The chemical analysis of the agar extract from *Polysiphonia lanosa* (L.) Tandy (Batey & Turvey, 1975) introduced 6-*O*-methyl-D-galactose 4-sulfate (9). It was also found in the agar extract of *Odonthalia corymbifera* (Gmel.) J. Agardh and *Gracilaria tikvahiae* McLachlan (Shashkov *et al.*, 1978; Craigie & Jurgens, 1989), and the unmethylated sulfated derivative was identified in *Gracilaria* agars (10) (Lahaye & Yaphe, 1988, 1989). Other derivatives were identified from the agar-like polysaccharide extracts

from *Gloiopeltis furcata* (Post. & Rupr.) J. Agardh: D-galactose 6-sulfate (11) and 3,6-anhydrogalactose 2-sulfate (Hirase & Watanabe, 1971). The former sulfated residue was described in agars from other algae (Usov *et al.*, 1983; Usov & Ivanova, 1987; Ji *et al.*, 1988; Lahaye & Yaphe, 1989; ochas & Lahaye, 1989b).

It is noteworthy that most of these natural chemical modifications, except the biological precursor, occur on sites that do not affect an helical conformation of the polysaccharides, that is on O_6 and O_4 of galactose and O_2 of 3,6-anhydrogalactose; however, they may affect aggregation of helices and thus gelation.

Factors affecting the chemistry and the physicochemical properties of agar and agarose

Many applications of agar and agarose require the formation of firm gels. Sulfate alkyl-transferase in algae (Rees, 1961a) and alkali treatment of agar convert L-galactose 6-sulfate to 3,6-anhydrogalactose (Rees, 1961b) and was first shown by Yanagawa (1936, in Yanagawa, 1946) to increase the gel strength of algal extracts. The effects of such chemical treatment on the rheological properties of agar have been described and can be interpreted in relation to the gel network theory proposed by Rees (1969), Matsuhashi and Hayashi (1971) and Watase & Nishinari (1981a, 1981b, 1982).

The synthesis, and thus the yield and chemistry of agar, is affected by factors modifying the physiology of the algae. Seasonality of agar quality has been observed by several authors (John & Asare, 1975; Hoyle, 1978b; Oza, 1978; Asare, 1980; Whyte *et al.*, 1981; Yang *et al.*, 1981; Nelson *et al.*, 1983; Onraët & Robertson, 1987; Lahaye & Yaphe, 1988) while others reported no variation (Kim & Humm, 1965; Duraitnam & Santos, 1981; Doty & Santos, 1983). Ji *et al.* (1985) observed a different fractionation pattern by anion-exchange chromatography and different 6-*O*-methyl-D-galactose contents in agar from *Gracilaria verrucosa* (Huds.) Papenfuss (*asiatica*) collected in north and south China. These differences most probably reflected seasonal variations between the two locations, although genetic variations cannot be ruled out. Environmental conditions were correlated with the 3,6-anhydrogalactose content in *Porphyra* agars (Rees & Conway, 1962).

From these results, it is, in fact, difficult to evaluate the effects of season, location and biological factors on the quality of agar, as a variety of extraction and purification methods have been used; furthermore, a complex set of factors involved in seasonal effects are likely to be different from one locale to another and to have different effects depending on the algal species. A wiser approach is to evaluate the effect of defined parameters, those altering the physiology of the algae, on the quality of the agar synthesized using extraction and purification protocols that avoid losses.

Single parameters, or combinations of different physical and nutritional parameters have been tested to model natural-variation effects on the yield and quality of agar. The effect of nutrients on agar was investigated by varying the nitrogen content of cultivated *Gracilaria* species (Bird *et al.*, 1981; Craigie *et al.*, 1984; Bird, 1988; Chiles *et al.*, 1989). Generally, an increase in nitrogen

content in the culture medium leads to a higher thallus nitrogen content and thus, to a decrease of agar yield on the algal dry-weight basis but a higher gel strength (after alkali treatment); however, Chiles *et al.* (1989) questioned the validity of the general observation that increased agar yields are obtained in nitrogen-deficient conditions since these agar extracts have high starch contents; furthermore, such starch contamination may not be totally removed using usual purification schemes and thus may interfere with the mechanical properties of the gel. The use of a heat-stable amylase like Termamyl or amyloglucosidase during agar extraction or recovery may avoid such contamination. Although modifying the nitrogen status in the culture medium may have most likely acted as limiting factor in growth, modification of growth rates were also obtained either by changing light or biomass density or, by changing water temperature of *Gracilaria* cultures. These were shown to affect agar populations and chemistry (Craigie & Wen, 1984; Guerin & Bird, 1987; Christiaen *et al.*, 1987; Ondarza *et al.*, 1987; Bird, 1988; Chiles *et al.*, 1989; Lignell & Pedersén, 1989); for example, high water temperature (27 °C) resulted in low gel-strength and 3,6-anhydrogalactose content, and in an increase in concentration of sulfate and 4-*O*-methyl-L-galactose in alkali modified agar from *G. tikvahiae*.

Cellular factors, in particular the age of the tissues, were related to the quality of *G. tikvahiae* agar (Craigie & Wen, 1984). The content of 3,6-anhydrogalactose was high in agar from young tissue, whereas sulfate and methyl contents were higher in agar extracted from old tissue. Changes in the chemical structure and particularly in the distribution of agar populations from *Gracilaria pseudoverrucosa* collected in non-growing and growing seasons were attributed to reflect different ratios of old and young algal tissues. Modification of agar chemistry with age was proposed to arise from a type of 'secondarization' of the algal cell-wall (Lahaye & Yaphe, 1988). Polysaccharides in the 'primary' (young) cell-wall consist of agar enriched in precursor-repeating units, with physico-chemical properties that would decrease mechanical constraints for the elongating and dividing cells of actively-growing tissues. As the algal tissues age, fewer cell divisions occur and the cell-walls thicken to act as a flexible but resistant skeleton. Synthesis and/or *in vitro* substitution of agar polysaccharides with chemical groups and/or formation of cross-links would increase cohesiveness and rigidity of the 'secondarized' cell-wall in addition to performing other biological functions such as increasing resistance toward pathogens, for example.

The life-history stage of an alga was reported by Kim & Henriquez (1978) and Whyte *et al.* (1981) to affect the quality of agars. According to the former authors, cystoscarpic *G. verrucosa* yielded a higher amount of agar but of a poorer quality than that of tetrasporic plants; by contrast, the latter authors observed that agar quality of *Gracilaria* (*verrucosa* type) decreased according to the order: cystoscarpic, tetrasporic, vegetative, and male gametophyte. Usov *et al.* (1979) found no major difference in the agar synthesized by gametophytic male or female, or tetrasporophytic *G. verrucosa*, but noticed that vegetative immature young algae contained 'unfinished' agar enriched in L-galactose 6-sulfate; similarly Hoyle (1978a), Sheng *et al.* (1984), Ji *et al.* (1985), and Onraët & Robertson (1987) did not find major variations in the quality of agar synthesized by different life-history stages of *Gracilaria bursa-pastoris* (Gmelin) Silva, *G. coronopifolia* J. Agardh, *G. verrucosa* and *Onikusa pristoides* (Turn.) Akatsuka. It appears that the structure of agar is not markedly affected by the stage in the algal life history as are certain members of the carrageenophytes (McCandless, 1981).

The specific genotype of various algal strains within a species has been shown by Patwary & van der Meer (1983), Craigie *et al.* (1984) and Lignell & Pedersén (1989) to affect the chemical and physico-chemical properties of agar.

Agar is thus a generic name for a continuum of related molecules based on agarobiose repeating units. Such polysaccharides vary in the extent and types of substituent groups and thus in physico-chemical properties. Agarose is defined as the least substituted agar molecule having, as a conse-

quence, the highest gelling potential. This family of polysaccharides is dynamic in that agar populations defined by particular chemical and physico-chemical properties change in relation to environmental, physiological and cellular factors. The biochemical implications of these changes need to be elucidated.

Methods in the chemical analysis of agar and agarose

Information on the fine structure of agar has advanced from the development of combinations of a variety of methods. Physico-chemical fractionations (see above), specific enzymatic hydrolysis and physical methods – in particular nuclear magnetic resonance spectroscopy – have supplemented the basic chemical techniques.

Agarases have been of great help in determining the fine chemical structure of agar. First used by Araki and Arai (1956, 1957) to confirm the structure of agarobiose (1), several agarases have been purified (Yaphe, 1957; Groleau & Yaphe, 1977; Usov & Miroshnikova, 1975; oung *et al.*, 1978; Morrice *et al.*, 1983a; Aoki *et al.*, 1990) and used in the investigation of agar structures from different origins (Turvey & Christison, 1967a, 1967b; Duckworth & Turvey, 1969a, 1969b; Hong *et al.*, 1969; Duckworth & Yaphe, 1971c; Usov & Lvanova, 1981; Morrice *et al.*, 1983b; Lahaye *et al.*, 1989).

Among the physical methods, infared (IR) spectroscopy has been applied to the analysis of *Gracilaria verrucosa* agar (Christiaen & Bodard, 1983), in the determination of sulfate content in agars (Rochas *et al.*, 1986b) and to follow variations in sulfate and 3,6-anhydrogalactose contents between agar fractions (Ji *et al.*, 1988; Lahaye & Yaphe, 1988; Lahaye *et al.*, 1988); however, this analytical method is insufficiently precise to assign definitive polysaccharide structures. Whyte *et al.* (1985) found the combination of IR analysis with high performance liquid chromatography of methanolysates from algal galactans helpful for their classification into agar or carrageenan structures.

Major advances in the chemistry of agar were made this past decade using nuclear magnetic resonance spectroscopy (NMR). ^{13}C and ^{1}H NMR spectroscopy are non-destructive ways to quantify and analyse linkages and substitution patterns in agar (Usov, 1984). Although introduced first, ^{1}H NMR spectroscopy (Izumi, 1973; Welti, 1977) is not generally used because of the complexity of the spectra obtained; however, quantitative data can be obtained and detection of minor concentrations of particular repeating units is possible (Izumi, 1973; Lahaye, 1986); on the other hand, although less sensitive than ^{1}H NMR with today's techniques, ^{13}C NMR spectroscopy

Table 1. ^{13}C N.M.R. chemical shifts (ppm, relative to dimethyl sulfoxyde H_6, 39.6 ppm) assignment at 80 °C of agarobiose and substituted agarobiose repeating units in agar.

Structure	Ref.	Unit[1]	Carbon							
			1	2	3	4	5	6	CH$_3$	COOH
1	1	G	102.4	70.2	82.2	68.8	75.3	61.4		
		A	98.3	69.9	80.1	77.4	75.7	69.4		
2	2	G	103.7	69.8	81.2	69.1	75.9	61.6		
		A	101.3	69.2	71.0	79.0 / 78.7[2]	70.2	67.7 / 67.5[2]		
3[3]	3	G	102.2	70.0	79.5	71.6	66.7	65.3	25.7	176.3
		A	98.4	69.9	80.1	77.5	75.7	69.4		
4	3	G	102.4	70.2	82.2	69.0	73.6	71.8	59.1	
		A	98.3	69.9	80.2	77.4	75.7	69.4		
5	4	G	102.5	70.2	82.2	69.0	73.6	71.8		
		B	98.5	69.9[4]	70.7	79.4	72.7	61.4	61.9	
		A	98.3	69.9	80.2	77.4	75.7	69.4		
6	2	G	103.7	70.0	81.1	68.9	75.7	61.4		
		A	100.9	69.4	71.0	79.3	72.2	61.2		
7	3	G	102.7	70.2	82.7	68.8	75.3	61.4		
		A	98.7	78.9	78.5	77.6	75.7	69.5	59.2	
8	3	G	102.5	70.5	79.8	76.6	75.2	61.4		
		A	98.2	79.1	78.5	77.7	75.7	69.9	59.1	
9	5	G	102.4	70.0	80.0	71.5	70.8	70.0	59.0	
		A	96.2	70.0	80.0	77.4	75.6	69.0		
10	3	G	102.4	70.8 / 71.0[2]	80.0	77.0	75.0	61.4		
		A	96.8	69.9	80.1	77.4	75.7	69.4		
11	6,7	G	102.6 / 102.4[2]	70.2	82.2	68.4	73.0	67.4		
		A	98.3	69.9	80.1	77.8 / 77.6[2]	75.7	69.4		

[1] G, A, and B refer to the 3-linked, 4-linked, and branched sugars, respectively; [2] split signals result from the arrangement of the repeating unit in the polysaccharide; [3] the signal for the quaternary carbon of the pyruvic acid group was not observed under the NMR conditions used for the polysaccharides, a signal at 101.7 ppm was seen for the oligosaccharides (ref. # 3 for details); [4] this signal is overlapped by others and may be at 70.2 or 69.4 ppm.

1. Rochas *et al.*, 1986; 2. Lahaye *et al.*, 1985; 3. Lahaye *et al.*, 1989; 4. Lahaye *et al.*, 1988; 5. Usov *et al.*, 1980; 6. Usov *et al.*, 1983; 7. Lahaye, 1986.

yields easier spectra to interpret, with usually one well-resolved signal per carbon. Several characteristic signal patterns (Table 1) have been attributed to peculiar agar chemical structures (Bhattacharjee et al., 1978, 1979; Shashkov et al., 1978; Nicolaisen et al., 1980; Usov et al., 1980, 1983; Brasch et al., 1981; Miller et al., 1982; Lahaye et al., 1985, 1989; Lahaye & Yaphe, 1989; Craigie & Jurgens, 1989; Karamanos et al., 1989).

The combination of special extraction and fractionation techniques, enzymatic hydrolysis and ^{13}C and/or ^{1}H NMR spectroscopy has been particularly useful in more precise studies of the basic chemical structure and distribution of the repeating units in various agars (Hamer et al., 1977; Young et al., 1978; Morrice et al., 1983b; Lahaye, 1986; Rochas et al., 1986a; Usov & Ivanova, 1987; Lahaye et al., 1989).

Until recently, the use of gas chromatography for analysis of agar was hampered by the extreme lability of the 3,6-anhydrogalactose residue to acid hydrolysis and, thus to its loss. Stevenson and Furneaux (1991) devised two-step hydrolysis and reductive-hydrolysis methods which are two promising means of allowing conservation of the anhydro-sugar and, thus of its observation and quantification by conventional capillary gas chromatography.

Ideally, the direct characterization of the cell-wall polysaccharides in the algal would avoid artifacts associated with long and tedious extraction and purification procedures. The following techniques have already given some interesting results, although these have to be further investigated.

Pyrolysis-gas chromatography, with or without mass spectrometry, has been used to determine the basic chemical structure of algal cell-wall polysaccharides either after extraction or in situ (Helleur et al., 1985a, 1985b; Bird et al., 1987).

Solid state ^{13}C NMR spectroscopic analysis of agarose powders demonstrated that spectra of larger line-width than, but of similar peak chemical shifts to, those observed by conventional high resolution ^{13}C NMR spectroscopy were obtained (Rochas & Lahaye, 1989b); furthermore, analysis of algal fragments yielded spectra of their major

cell-wall polysaccharides allowing the characterization of some substituents such as methoxyl and pyruvate groups. Because of the similarity between agar and algal powders and solution spectra, the conformation of agar is not markedly different in the ordered and random coil state, thus supporting the idea that agarose in solution is a relatively-rigid molecule (Rochas & Lahaye, 1989a).

Conclusions and perspectives

Although Gelidium is an economically-important genus for agar production, the fine chemistry of the agar synthesized by members of this genus has been less studied (Araki, 1966; Araki et al., 1967; Izumi, 1971; Whyte & Englar, 1981; Matsuhiro, 1990 and reference herein) than that of Gracilaria species. Future work may deal with such agars, as recent developments in chemical analysis and especially in the possibility of obtaining chemically well-defined agar fractions should provide grounds for in-depth investigations of structural/functional properties of these polysaccharides. Future studies should be able to correlate the physico-chemical and rheological properties of agar with molecular weight, type and location of substituent groups leading to a better understanding of agar gelation. Indeed, the relationships between mechanical properties and molecular weight of regular polysaccharides such as κ-carrageenan (Rochas et al., 1990) or chemically-defined alginate (Smidsrod, 1974) are known. The relationships between the elastic modulus and gel strength with molecular weight of agarose are unknown because: (1) until recently, agar molecular-weight measurements were difficult and (2) agar samples are required to have similar substitution levels, types and locations (distributions) which are not realistic for samples of different origins. Quantification of methyl groups, and particularly 2-O-methyl 3,6-anhydrogalactose, is only recent and was made possible through the use of NMR spectroscopy. Although sequencing of substituted repeating units in agar has been approached (Morrice et al., 1983b;

Lahaye *et al.*, 1989) work remains in this area. There is also need of information about the biosynthesis of these polysaccharides so that ultimately the chemical structure of agar can be tailored at the biosynthetic level or *in vitro* using specific enzymes to obtain desired rheological and physico-chemical properties.

Thus, the understanding of the functional properties of agar and agarose is still a challenge to the biochemist, physico-chemist and rheologist. Advances in this field will be possible only with the close cooperation of biologists (taxonomy, culture) and industrialists.

References

Aoki, T., T. Araki & M. Kitamikado, 1990. Purification and characterization of a novel β-agarase from *Vibrio* sp. AP-2. Eur. J. Biochem. 187: 461–465.

Araki, C., 1937. Chemical studies of agar-agar. III. Acetylation of agar-like substance of *Gelidium amansii*. J. Chem. Soc. Japan 58: 1338–1350.

Araki, C., 1966. Some recent studies on the polysaccharides of agarophytes. Proc. int. Seaweed Symp. 5: 3–17.

Araki, C. & K. Arai, 1956). The chemical constitution of agar-agar. XVIII. Isolation of a new crystalline disaccharide by enzymatic hydrolysis of agar-agar. Bull. Chem. Soc. Japan 29: 339–345.

Araki, C. & K. Arai, 1957. The chemical constitution of agar-agar. XX. Isolation of a tetrasaccharide by enzymatic hydrolysis of agar-agar. Bull. Chem. Soc. Japan 30: 287–293.

Araki, C., K. Arai & S. Hirase, 1967. Studies on the chemical constitution of agar-agar. XXIII. Isolation of D-xylose, 6-O-methyl-D-galactose, 4-O-methyl-L-galactose and O-methylpentose. Bull. Chem. Soc. Japan 40: 959–962.

Arnott, S., A. Fulmer, W. E. Scott, I. C. M. Dea, R. Moorhouse & D. A. Rees, 1974. Agarose double helix and its function in agarose gel structure. J. Mol. Biol. 90: 269–284.

Asare, O., 1980. Seasonal changes in sulphate and 3,6-anhydrogalactose content of phycocolloids from two red algae. Bot. Mar. 23: 595–598.

Barbero, J. J., C. Bouffar-Roupe, C. Rochas & S. Perez, 1989. Modelling studies of solvent effects on the conformational stability of agarobiose and neoagarobiose and their relationship to agarose. Int. J. Biol. Macromol. 11: 265–272.

Batey, J. F. & J. R. Turvey, 1975. The galactan sulfate of the red alga *Polysiphonia lanosa*. Carbohydr. Res. 43: 133–143.

Bhattacharjee, S. S., W. Yaphe & G. K. Hamer, 1978. ¹³C N.m.r. spectroscopic analysis of agar, κ-carrageenan and ι-carrageenan. Carbohydr. Res. 60: C1–C3.

Bhattacharjee, S. S., W. Yaphe & G. K. Hamer, 1979. Study of agar and carrageenan by ¹³C nuclear magnetic resonance spectroscopy. Proc. int. Seaweed Symp. 9: 379–385.

Bird, C. J., R. J. Helleur, E. R. Hayes & J. McLachlan, 1987. Analytical pyrolysis as a taxonomic tool in *Gracilaria* (Rhodophyta: Gigartinales). Proc. int. Seaweed Symp. 12: 207–212.

Bird, K. T., 1988. Agar production and quality from *Gracilaria* sp. strain G-16: Effects of environmental factors. Bot. Mar. 31: 33–39.

Bird, K. T., M. D. Hanisak & J. Ryther, 1981. Chemical quality and production of agars extracted from *Gracilaria tikvahiae* grown in different nitrogen enrichment conditions. Bot. Mar. 24: 441–444.

Bowker, D. M. & J. R. Turvey, 1968. Water-soluble polysaccharides of the red alga *Laurencia pinnatifida*. Part I. Constituent units. J. Chem. Soc. C: 983–988.

Brasch, D. J., C. T. Chuah & L. D. Melton, 1981. A ¹³C N.M.R. study on some agar-related polysaccharides from New-Zealand seaweeds. Austr. J. Chem. 34: 1095–1105.

Chiles, T. C., K. T. Bird & F. E. Koehn, 1989. Influence of nitrogen availability on agar-polysaccharides from *Gracilaria verrucosa* strain G-16: structural analysis by NMR spectroscopy. J. Appl. Phycol. 1: 53–58.

Christiaen, D. & M. Bodard, 1983. Spectroscopie infrarouge de films d'agar de *Gracilaria verrucosa* (Huds.) Papenfuss. Bot. mar. 26: 425–427.

Christiaen, D., T. Stadler, M. Ondarza & M. C. Verdus, 1987. Structures and functions of the polysaccharides from the cell wall of *Gracilaria verrucosa* (Rhodophycae, Gigartinales). Proc. int. Seaweed Symp. 12: 139–146.

Craigie, J. S. & A. Jurgens, 1989. Structure of agars from *Gracilaria tikvahiae* Rhodophyta: location of 4-O-methyl-L-galactose and sulfate. Carbohydr. Polymers 11: 265–278.

Craigie, J. S. & Z. C. Wen, 1984. Effects of temperature and tissue age on gel strength and composition of agar from *Gracilaria tikvahiae* (Rhodophyceae). Can. J. Bot. 62: 1665–1670.

Craigie, J. S., Z. C. Wen & J. P. van der Meer, 1984. Interspecific, intraspecific and nutritionally-determined variations in the compositions of agars from *Gracilaria* spp. Bot. mar. 27: 55–61.

Djabourov, M., A. H. Clark, D. W. Rowlands & S. B. Ross-Murphy, 1989. Small-angle X-ray scattering characterization of agarose sols and gels. Macromolecules 22: 180–188.

Doty, M. & G. A. Santos, 1983. Agar from *Gracilaria cylindrica*. Aquatic Bot. 15: 299–306.

Duckworth, M., K. C. Hong & W. Yaphe, 1971. The agar polysaccharides of *Gracilaria* species. Carbohydr. Res. 18: 1–9.

Duckworth, M. & J. R. Turvey, 1969a. An extracellular agarase from a *Cytophaga* species. Biochem. J. 113: 139–142.

Duckworth, M. & J. R. Turvey, 1969b. The action of a bacterial agarase on agarose, porphyran and alkali-treated porphyran. Biochem. J. 113: 687–692.

146

Duckworth, M. & W. Yaphe, 1971a. Preparation of agarose by fractionation from the spectrum of polysaccharides in agar. Anal. Chem. 44: 636–641.

Duckworth, M. & W. Yaphe, 1971b. The structure of agar. Part I. Fractionation of a complex mixture of polysaccharides. Carbohydr. Res. 16: 189–197.

Duckworth, M. & W. Yaphe, 1971c. The structure of agar. Part II. The use of a bacterial agarase to elucidate structural features of the charged polysaccharides in agar. Carbohydr. Res. 16: 435–445.

Duraitnam, M. & N. Q. Santos, 1981. Agar from *Gracilaria verrucosa* (Hudson) Papenfuss and *Gracilaria sjoestedtii* Kylin from northeast Brazil. Proc. int. Seaweed Symp. 10: 669–674.

Foord, S. A. & E. D. T. Atkins, 1989. New X-ray diffraction results from agarose: Extended single helix structure and implications for gelation mechanism. Biopolymers 28: 1345–1365.

Friedlander, M., Y. Lipkin & W. Yaphe, 1981. Composition of agars from *Gracilaria* cf. *verrucosa* and *Pterocladia capillacea*. Bot. Mar. 24: 595–598.

Furneaux, R. H., I. J. Miller & T. T. Stevenson, 1990. Agaroids from New Zealand members of the Gracilariaceae. A novel dimethylated agar. Hydrobiologia 204/205: 454–654.

Guerin, J. M. & K. T. Bird, 1987. Effects of aeration period on the productivity and agar quality of *Gracilaria* sp. Aquaculture 64: 105–110.

Groleau, D. & W. Yaphe, 1977. Enzymatic hydrolysis of agar: purification and characterization of β-neoagarotetraose-hydrolase from *Pseudomonas atlantica*. Can. J. Microbiol. 23: 672–679.

Guiseley, K. B., 1970. The relationship between methoxy content and gelling temperature of agarose. Carbohydr. Res. 13: 247–256.

Guiseley, K. B., 1987. Natural and synthetic derivatives of agarose and their use in biochemical separations. In M. Yalpani (ed.) Industrial Polysaccharides: Genetic Engineering, Structure/Property Relations and Applications. Elsevier Science Publishers B.V., Amsterdam, 139–147.

Hamer, G. K., S. S. Bhattacharjee & W. Yaphe, 1977. Analysis of the enzymatic hydrolysis products of agarose by ^{13}C-n.m.r. spectroscopy. Carbohydr. Res. 54: C7–C10.

Helleur, R. J., E. R. Hayes, J. S. Craigie & J. L. McLachlan, 1985a. Characterization of polysaccharides of red algae by pyrolysis-capillary gas chromatography. J. Anal. appl. Pyr. 8: 349–357.

Helleur, R. J., E. R. Hayes, W. D. Jamieson & J. S. Craigie, 1985b. Analysis of polysaccharide pyrolysate of red algae by capillarly gas chromatography-mass spectrometry. J. Anal. appl. Pyrolysis 8: 333–347.

Hirase, S., 1957. Studies on the chemical constitution of agar-agar. XIX. Pyruvic acid as a constituent of agar-agar (Part 3). Structure of the pyruvic acid-linking disaccharide derivative isolated from methanolysis products of agar. Bull. Chem. Soc. Japan 30: 75–79.

Hirase, S. & C. Araki, 1961. Isolation of 6-*O*-methyl-D-galactose from the agar of *Ceramium boydenii*. Bull. Chem. Soc. Japan 34: 1048.

Hirase, S. & K. Watanabe, 1971. Fractionation and structural investigation of funoran. Proc. int. Seaweed Symp. 7: 451–454.

Hong, K. C., M. E. Goldstein & W. Yaphe, 1969. A chemical and enzymic analysis of the polysaccharides from *Gracilaria*. Proc. int. Seaweed Symp. 6: 473–482.

Hoyle, M., 1978a. Agar studies in two *Gracilaria* species (*G. bursapastoris* (Gmelin) Silva and *G. coronopifolia*. Ag.) from Hawaii. I. Yield and gel strength in the gametophyte and tetrasporophyte generations. Bot. mar. 21: 343–345.

Hoyle, M., 1978b. Agar studies in two *Gracilaria* species (*G. bursapastoris* (Gmelin) Silva and *G. coronopifolia*. Ag.) from Hawaii. II. Seasonal Aspects. Bot. mar. 21: 347–352.

Izumi, K., 1970. A new method for fractionation of agar. Agr. Biol. Chem. 34: 1739–1740.

Izumi, K., 1971. Chemical heterogeneity of the agar from *Gelidium amansii*. Carbohydr. Res. 17: 227–230.

Izumi, K., 1972. Chemical heterogeneity of the agar from *Gracilaria verrucosa*. J. Biochem. 72: 135–140.

Izumi, K., 1973. Structural analysis of agar-type polysaccharides by NMR spectroscopy. Biochim. Biophys. Acta 320: 311–317.

Ji, M., M. Lahaye & W. Yaphe, 1985. Structure of agar from *Gracilaria* spp. (Rhodophyta) collected in People's Republic of China. Bot. mar. 28: 521–528.

Ji, M., M. Lahaye & W. Yaphe, 1988. Structural studies on agar fractions extracted sequentially from Chinese red seaweeds: *Gracilaria sjeostedtii*, *G. textorii* and *G. salicornia* using ^{13}C-NMR and IR spectroscopy. Chin. J. Oceanol. Limnol. 6: 87–103.

John, D. M. & S. O. Asare, 1975. A preliminary study of the variations in yield and properties of phycocolloids from Ghanaian seaweeds. Mar. Biol. 30: 325–330.

Karamanos, Y., M. Ondarza, F. Bellanger, D. Christiaen & S. Moreau, 1989. The linkage of 4-*O*-methyl-L-galactopyranose in the agar polymers from *Gracilaria verrucosa*. Carbohydr. Res. 187: 93–101.

Kim, D. H. & N. P. Henriquez, 1978. Yields and gel strengths of agar from cystocarpic and tetrasporic plants of *Gracilaria verrucosa* (Florideophyceae). Proc. int. Seaweed Symp. 9: 257–262.

Kim, C. S. & H. J. Humm, 1965. The red alga, *Gracilaria foliifera* with special reference to the cell wall polysaccharides. Bull. Mar. Sci. 15: 1036–1050.

Lahaye, M., 1986. Agar from *Gracilaria* spp. PhD Thesis, McGill University, Montréal, Québec, Canada: pp 330.

Lahaye, M., J. F. Revol, C. Rochas, J. McLachlan & W. Yaphe, 1988. The chemical structure of *Gracilaria crassissima* (P. & H. Crouan in Schramm & Mazé) P. & H. Crouan in Schramm & Mazé and *G. tikvahiae* McLachlan (Gigartinales, Rhodophyta) cell-wall polysaccharides. Bot. mar. 31: 491–501.

Lahaye, M., C. Rochans & W. Yaphe, 1986. A new procedure

for determining the heterogeneity of agar polymers in the cell walls of *Gracilaria* spp. (Gracilariaceae, Rhodophyta). Can. J. Bot. 64: 579–585.

Lahaye, M. & W. Yaphe, 1988. Effects of seasons on the chemical structure and gel strength of *Gracilaria pseudoverrucosa* agar (Gracilariaceae, Rhodophyta). Carbohydr. Polymers 8: 285–301.

Lahaye, M. & W. Yaphe, 1989. The chemical structure of agar from *Gracilaria compressa* (C. Agardh) Greville, *G. cervicornis* (Turner) J. Agardh, *G. damaecornis* J. Agardh and *G. domingensis* Sonder *ex* Kützing (Gigartinales, Rhodophyta). Bot. Mar. 32: 369–377.

Lahaye, M., W. Yaphe, M. T. Phan Viet & C. Rochas, 1989. ^{13}C-N.M.R. spectroscopic investigation of methylated and charged agarose oligosaccharides and polysaccharides. Carbohydr. Res. 190: 249–265.

Lahaye, M., W. Yaphe & C. Rochas, 1985. ^{13}C-N.m.r. spectral analysis of sulfated and desulfated polysaccharides of the agar type. Carbohydr. Res. 143: 240–245.

Lignell, Å. & M. Pedersén, 1989. Agar composition as a function of morphology and growth rate. Studies on some morphological strains of *Gracilaria secundata* and *Gracilaria verrucosa* (Rhodophyta). Bot. Mar. 32: 219–227.

Matsuhashi, T. & K. Hayashi, 1971. Rheological behavior of agar gels processed from *Gracilaria foliifera* of Florida. Proc. Int. Seaweed Symp. 7: 464–468.

Matsuhiro, B. & C. C. Urzúa, 1991. Agars from Chilean Gelidiaceae. Hydrobiologia 221: 149–156.

McCandless, E. L., 1981. Polysaccharides of the Seaweeds. In Lobban, C. S. & M. J. Wynne (eds.) The Biology of Seaweeds. Blackwell Scientific Publications: pp. 559–588.

Meer, W., 1980. In Davidson, R. S. (ed.) Handbook of Water Soluble Gums and Resins. McGraw Hill, 7: 17–19.

Miller, I. J., H. Wong & R. H. Newman, 1982. A carbon-13 NMR study of some disaccharides from algal polysaccharides. Austr. J. Chem. 35: 853–856.

Morice, L. M., M. W. McLean, F. B. Williamson & W. F. Long, 1983a. β-agarase I and II from *Pseudomonas atlantica*. Purifications and some properties. Eur. J. Biochem. 135: 553–558.

Morrice, L. M., M. W. McLean, W. F. Long & F. B. Williamson, 1983b. Prophyran primary structure. An investigation using β-agarase I from *Pseudomonas atlantica* and ^{13}C-NMR spectroscopy. Eur. J. Biochem. 133: 673–684.

Nelson, S. G., S. S. Yang, C. Y. Yang & Y. M. Chiang, 1983. Yield and quality of agar from species of *Gracilaria* (Rhodophyta) collected from Taiwan and Micronesia. Bot. mar. 26: 361–366.

Nicolaissen, F. M., I. Meyland & K. Schaumburg, 1980. ^{13}C NMR spectra at 67.9 MHz of agarose solutions and partly 6-*O*-methylated agarose at 95 °C. Acta Chem. Scand. Ser. B 34: 103–107.

Ondarza, M., Y. Karamanos, D. Christiaen & T. Stadler, 1987. Variations in the composition of agar polysaccharides from *Gracilaria verrucosa*, cultivated under controlled conditions. Food Hydrocolloids 5/6: 507–509.

Onraët, A. C. & B. L. Robertson, 1987. Seasonal variation in yield and properties of agar from sporophytic and gametophytic phases of *Onikusa pristoides* (Turner) Akatsuka (Gelidiaceae, Rhodophyta). Bot. mar. 30: 491–495.

Oza, R. M., 1978. Studies on Indian *Gracilaria*. IV. Seasonal variation in agar and gel strength of *Gracilaria corticata* J. Ag. occurring on the coats of Veraval. Bot. mar. 21: 165–167.

Patwary, M. U. & J. P. van der Meer, 1983. Genetics of *Gracilaria tikvahiae* (Rhodophyceae) IX: Some properties of agars extracted from morphological mutants. Bot. mar. 26: 295–299.

Peat, S., J. R. Turvey & D. A. Rees, 1961. Carbohydrate of the red alga, *Porphyra umbillicalis*. J. Chem. Soc.: 1590–1595.

Rees, D. A., 1961a. Enzymatic synthesis of the 3,6-anhydro-L-galactose with porphyran from L-galactose 6-sulphate units. Biochem. J. 81: 347–352.

Rees, D. A., 1961b. Estimation of the relative amounts of isomeric sulphate esters in some sulphated polysaccharides. J. Chem. Soc.: 5168–6171.

Rees, D. A., 1969. Structural, conformation and mechanism in the formation of polysaccharide gels and networks. Adv. Carbohydr. Chem. Biochem. 24: 267–332.

Rees, D. A. & E. Conway, 1962. The structure and biosynthesis of porphyran: a comparison of some samples. Biochem. J. 84: 411–416.

Rees, D. A. & E. J. Welsh, 1977. Secondary and tertiary structure of polysaccharides in solution and gels. Angew. Chem. Int. Ed. Eng. 16: 214–224.

Rochas, C. & M. Lahaye, 1989a. Average molecular weight and molecular weight distribution of agarose and agarose-type polysaccharides. Carbohydr. Polymers 10: 289–298.

Rochas, C. & M. Lahaye, 1989b. Solid state ^{13}C-NMR spectroscopy of red seaweeds, agars and carrageenans. Carbohydr. Polymers 10: 189–204.

Rochas, C., M. Lahaye & W. Yaphe, 1986a. Sulfate content of carrageenan and agar determined by infrared spectroscopy. Bot. mar. 29: 335–340.

Rochas, C., M. Lahaye, W. Yaphe & M. T. Phan Viet, 1986b. ^{13}C NMR-spectroscopic investigation of agarose oligomers. Carbohydr. Res. 148: 199–207.

Rochas, C., M. Rinaudo & S. Landry, 1990. Role of molecular weight on the mechanical properties of kappa-carrageenans gels. Carbohydr. Polymers 12: 255–266.

Samec, M. von, & V. Isajevic, 1922. Studien uber pipanzen kolloide. XIV. Physico-chemische analyse der agargallerte. Kolloidchem. Beih. 16: 285–300.

Shashkov, A. S., A. I. Usov & S. V. Yarotskii, 1978. Polysaccharides of algae. XXIV. The application of ^{13}C NMR spectroscopy to the analysis of the structure of polysaccharides of the agar group. Bioorg. Khim. 4: 74–81 (in Russian).

Sheng, S. Y., Z. Y. Xia, L. Z. En & L. W. Qing, 1984. The yield and properties of agar extracted from different life stages of *Gracilaria verrucosa*. Proc. int. Seaweed Symp. 11: 551–553.

148

Smidsrod, O., 1974. Molecular basis for some physical properties of alginates in the gel state. Faraday Dis. Chem. Soc. 57: 263–274.

Stevenson, T. T. & R. H. Furneaux, 1991. Chemical methods for the analysis of sulphated galactans from red algae. Carbohydr. Res. (in press).

Turvey, J. R. & J. Christison, 1967a. The hydrolysis of algal galactans by enzymes from a *Cytophaga* species. Biochem. J. 105: 311–316.

Turvey, J. R. & J. Christison, 1967b. The enzymic degradation of porphyran. Biochem. J. 105: 317–321.

Turvey, J. R. & E. L. Williams, 1976. The agar-type polysaccharide from the red alga *Ceramium rubrum*. Carbohydr. Res. 49: 419–425.

Usov, A. I., 1984. NMR spectroscopy of red seaweed polysaccharides: Agars, carrageenans, and xylans. Bot. mar. 27: 189–202.

Usov, A. I. & E. G. Ivanova, 1981. Polysaccharides of algae. XXXI: Enzymatic cleavage of an agar-like polysaccharide from the red alga *Rhodomela larix* (Turn.) C. Ag. Bioorg. Khim. 7: 1060–1068 (in Russian).

Usov, A. I. & E. G. Ivanova, 1987. Polysaccharides of algae. XXXVII. Characterization of hybrid structure of substituted agarose from *Polysiphonia morrowii* (Rhodophyta, Rhodomelaceae) using β-agarase and ¹³C-NMR spectroscopy. Bot. mar. 30: 365–370.

Usov, A. I., E. G. Ivanova & V. F. Makienko, 1989. Polysaccharides of algae. XXIX: Comparison of samples of agar from different generations of *Gracilaria verrucosa* (Huds.) Papenf. Bioorg. Khim. 5: 1647–1653 (in Russian).

Usov, A. I., E. G. Ivanova & A. S. Shashkov, 1983. Polysaccharides of algae. XXXIII: Isolation and ¹³C NMR spectral study of some new gel-forming polysaccharides from Japan Sea red seaweeds. Bot. mar. 26: 285–294.

Usov, A. I., L. I. Miroshnikova, 1975. Isolation of agarase from *Littorina mandshurica* by affinity chromatography on Biogel A. Carbohydr. Res. 43: 204–207.

Usov, A. I., S. V. Yarotsky & A. S. Shashkov, 1980. ¹³C NMR spectroscopy of red algal galactans. Biopolymers 19: 977–990.

Watase, M. & K. Nishinari, 1981a. Effect of alkali metal ions on the rheological properties of κ-carrageenan and agarose gels. J. Text. Stud. 12: 427–445.

Watase, M. & K. Nishinari, 1981b. Effect of sodium hydroxide pretreatment on the relaxation spectrum of concentrated agar-agar gels. Rheol. Acta 20: 155–162.

Watase, M. & K. Nishinari, 1982. Effect of alkali metal ions on the viscoelasticity of concentrated kappa-carrageenan and agarose gels. Rheol. Acta 21: 318–324.

Welti, D., 1977. The 300 MHz proton magnetic resonance spectra of methyl β-D-galactopyranoside, methyl 3,6-anhydro-α-D-galactopyranoside, agarose, κ-carrageenan, and segments of ι-carrageenan and agarose sulphate. J. Chem. Res. (S): 312–313, (M): 3566–3587.

Wen, Z. C. & J. S. Craigie, 1984. Composition and properties of agar-type polysaccharides from *Gracilaria sjoestedtii* Kylin. Chin. J. Oceanol. Limnol. 2: 88–91.

Whyte, J. N. C. & J. R. Englar, 1981. The agar component of the red seaweed *Gelidium purpurascens*. Phytochem. 20: 237–240.

Whyte, J. N. C., J. R. Englar, R. G. Saunders & J. C. Lindsay, 1981. Seasonal variations in the biomass, quantity and quality of agar from the reproductive and vegetative stages of *Gracilaria* (*verrucossa* type). Bot. mar. 24: 493–501.

Whyte, J. N. C., S. P. C. Hosford & J. R. Englar, 1985. Assignment of agar or carrageenan structures to red algal polysaccharides. Carbohydr. Res. 140: 336–341.

Yanagawa, T., 1946. Kanten (Agar), 2nd edn (in Japanese), Sangiotosho Co. LTD Tokyo., Japan: 352 pp.

Yang, S. S., C. H. Yang & H. H. Wang, 1981. Seasonal variation of agar-agar produced in Taiwan area. Proc. int. Seaweed Symp. 10: 737–742.

Yaphe, W., 1957. The use of agarase from *Pseudomonas atlantica* in the identification of agar in marine algae (Rhodophyceae). Can. J. Microbiol. 3: 987–993.

Young, K. S., S. S. Battacharjee & W. Yaphe, 1978. Enzymic cleavage of the α-linkages in agarose to yield agarooligosaccharides. Carbohydr. Res. 66: 207–212.

Young, K. S., M. Duckworth & W. Yaphe, 1971. The structure of agar. Part III. Pyruvic acid, a common feature of agars from different agarophytes. Carbohydr. Res. 16: 446–448.

Hydrobiologia **221**: 149–156, 1991.
J. A. Juanes, B. Santelices & J. L. McLachlan (eds), International Workshop on Gelidium.
© 1991 *Kluwer Academic Publishers.*

Agars from Chilean Gelidiaceae

Betty Matsuhiro & Carlos C. Urzúa
Departamento de Química, Facultad de Ciencia, Universidad de Santiago de Chile, Casilla 5659, Santiago 2, Chile

Key words: agar, cystocarpic thalli, *Gelidium chilense*, *Gelidium lingulatum*, *Gelidium rex*, tetrasporic thalli

Abstract

Cystocarpic, tetrasporic and vegetative thalli of *Gelidium chilense* were extracted with water at 95 °C. The contents of 3,6-anhydro-galactose and sulfate group of the hydrocolloids correspond to those of an agar-type polysaccharide. The percentages of 6-*O*-methyl-galactose and of pyruvic acid ranged between 5.7–6.2% and 0.42–0.54%, respectively. The gel melting and gelation temperatures of *Gelidium chilense*, *G. rex* and *G. lingulatum* agars were determined. A correlation between 6-*O*-methylgalactose content and gelation temperatures was not observed. It was found by anion-exchange chromatography that 19.8% of tetrasporic and 4.9% of vegetative *G. chilense* agars are unsulfated polymers. Structural studies on the neutral fraction from tetrasporic *G. chilense* agar by partial hydrolysis and ^{1}H NMR spectroscopy have shown that it is mainly composed of agarose. Methylation analysis, oxidative hydrolysis and partial hydrolysis, followed by ^{1}H NMR spectroscopy, have shown that the neutral fraction of the agar from tetrasporic *Gelidium rex* is agarose. The results obtained in this work are compared with previously reported data on studies of agars from Chilean *Gelidium* species.

Introduction

Species of the genus *Gelidium* are among the world's most important agar resources. While most food-grade and industrial-grade agar is extracted from *Gracilaria* species, bacteriological-grade agar is obtained from species of the genus *Gelidium* and minor amounts from *Pterocladia* (McLachlan & Bird, 1983).

The pioneering studies by Araki's group on agar from *Gelidium amansii* Lamouroux showed that this hydrocolloid contained D-galactose, 3,6-L-anhydrogalactose and 0.64% sulfur (Mori, 1953). Further studies (Araki & Arai, 1957) confirmed that the main polysaccharide of agar, agarose, is composed of 1,3 linked β-D-galactopyranose and 1,4 linked 3,6-anhydro-α-L-galac-

topyranose. From the hydrolysis products of *G. amansii* agar, D-xylose, 6-*O*-methyl-D-galactose, 4-*O*-methy-L-galactose and *O*-methylpentose were isolated and identified (Araki *et al.*, 1967). Izumi (1971) reported on the fractionation of agar from *G. amansii* by anion-exchange chromatography. He found that this agar consisted of a family of polydispersed polysaccharides which range from neutral agarose to polymers of similar structure with variable proportions of acidic substituents such as sulfate, pyruvic acid and uronic acids.

Duckworth and Yaphe (1971) fractionated a commercial agar, probably from *Gelidium cartilagineum* (L.) Gaill, and concluded that agar is composed of a spectrum of polymers. Neutral agarose, pyruvated agarose with little sulfation

150

and sulfated galactan containing no or little 3,6-anhydro-L-galactose or 4,6-O-(1-carboxyethylidene)-D-galactose are three extremes of structure in agar (Fig. 1).

Several studies on Gracilariaceae have pointed out that the chemical and rheological properties of agar are dependent on the species, season, location and environment (Hoyle, 1978a; Whyte & Englar, 1980; Patwary & van der Meer, 1983; Lahaye & Yaphe, 1988; Lahaye *et al.*, 1988; Chiles *et al.*, 1989). The quality of agar from the different life-history phases of *Gracilaria* has been also investigated (Hoyle, 1978b; Kim & Henríquez, 1979; Whyte *et al.*, 1981).

Chemical composition of agars from Chilean *Gelidium* has been studied in this laboratory during the last decade.

Recent studies (Santelices & Stewart, 1985; Santelices & Abbott, 1985) on the taxonomy of Gelidiaceae from Pacific South America have shown that only three species of the genus *Gelidium*, *G. lingulatum* Kützing, *G. rex* Santelices & Abbott and *G. chilense* (Montagne) Santelices & Montalva occur in Central Chile. *Gelidium lingulatum* is an intertidal species occurring with *G. chilense* (Santelices & Montalva, 1983).

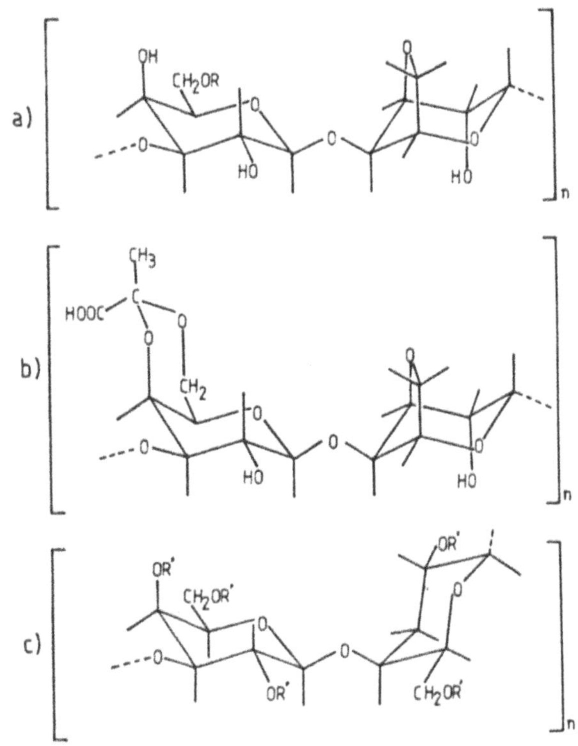

The chemical composition of agars extracted from *Gelidium filicinum* Bory and *G. lingulatum* collected near Valparaíso were analysed by Zanlungo (1979, 1980). According to Santelices (1989), *G. filicinum* is absent from Chile and seems to be restricted to warmer waters; the alga identified as *G. filicinum*, probably corresponding to *G. chilense*.

The soluble polysaccharides from cystocarpic and tetrasporic *G. lingulatum* and *G. rex* were previously studied in this laboratory (Matsuhiro & Urzúa, 1988, 1990). Seasonal effects on the yield and gel strength of *G. rex* agar have been reported previously (Matsuhiro & Urzúa, 1986).

In this work, results of the studies on cystocarpic, tetrasporic and vegetative *G. chilense* agars are compared with reported data of *Gelidium* agars. Structural analysis on the neutral fractions from tetrasporic *G. rex* and *G. chilense* agars are also presented.

Material and methods

Samples of *Gelidium chilense* were collected in Coquimbo Bay, Chile (29° 53' S, 71° 18' W) in December and were carefully sorted into cystocarpic, tetrasporic and vegetative fronds in the laboratory.

Extraction of agar was conducted as previously described (Matsuhiro & Urzúa, 1990). Gelation temperature and gel melting-point determinations of 1.0% agar solutions were conducted in duplicate, according to Whyte and Englar (1980). Gel strengths of 1.5% polysaccharide solutions were measured using a Marine Colloids Gel Tester (Model GT-2). The solvent systems used for paper chromatography were (A) pyridine-ethyl acetate-water (4:10:3), (B) n-butyl alcohol-

R = H or CH$_3$
R' = H or SO$_3^-$

Fig. 1. Structure of agar polysaccharides. (a) R = H: agarose, R = OCH$_3$: 6-O-methylagarose; (b) pyruvated agarose; (c) sulfated galactans.

water (4 : 1 : 2) and (C) ethyl acetate-formic acid-2-butanone-acetic acid-water (36 : 2 : 30 : 6 : 5). Gas-liquid chromatographic analysis was performed using dual 2.0 m × 2.0 mm stainless columns packed with (D) 3% ECNSS-M and (E) 3% SP-2340 using FID.

Fractionation of polysaccharides was done by anion-exchange chromatography on DEAE Sephadex as described by Matsuhiro and Urzúa (1988).

Analytical methods

Agar was hydrolysed with 2 M trifluoroacetic acid for 16 h in a boiling-water bath. The resulting solution was evaporated *in vacuo* and the acid was removed by co-distillation with water. The hydrolysate was studied by paper chromatography with systems A and B. An aliquot was reduced with NaBH$_4$ and acetylated with acetic anhydride in dry pyridine. The acetylated alditol mixture was analysed by gas-liquid chromatography with systems D and E. The percentage of 3,6-anhydrogalactose was determined by the previously-described procedure (Matsuhiro & Zanlungo, 1983); methyl 3,6-anhydro-α-D-galactopyranoside synthesized by the method of Lewis *et al.* (1963) was used as the standard sugar. The content of galactose and 6-*O*-methylgalactose was calculated by subtracting the anhydrogalactose content from the total sugar content determined by the anthrone method of Yaphe (1960). The amount of sulfate was determined following the technique of Dogson (1961). Pyruvic acid content was determined by the lactate dehydrogenase method of Duckworth and Yaphe (1970). Nitrogen was determined by elemental microanalysis in Laboratorio de Microanálisis, Universidad de Chile. ^{1}H NMR spectra were obtained with a Varian XL-100 spectrometer with 10% sample solutions in D$_2$O at 30 °C. Sodium 2,2-dimethyl-2-silapentane-5-sulfonate (DSS) was used as the internal standard.

Methylation analysis

Methylation followed the method of Hakomori (1964). Complete methylation, shown by the absence of hydroxyl absorption at 3600–3400 cm^{-1} in the IR spectrum, was achieved in a single treatment. The methylated polysaccharide was hydrolysed with 2 M trifluoroacetic acid at 90° for 16 h. The resulting methylated sugars were treated with NaBH$_4$, acetylated with acetic anhydride in dry pyridine and studied by gas-liquid chromatography with systems D and E.

Oxidative hydrolysis

An aliquot of the methylated polysaccharide was hydrolysed with bromine and sulfuric acid according to the method described by Cerezo (1973) and the hydrolysate was analysed by paper chromatography with system C. The hydrolysate was fractionated on a DEAE Sephadex A-25 column. Elution was with distilled water and 2% formic acid and monitored with phenol-sulfuric acid reagent (Dubois *et al.*, 1956). The fraction eluted with acid was evaporated *in vacuo*, dried and treated with 3% HCl in anhydrous methanol according to Anderson *et al.* (1973). The resulting syrup was acetylated with acetic anhydride in dry pyridine and studied by gas-liquid chromatography with column D. 2-*O*-methyl-4,5-di-*O*-acetyl-3,6-anhydrogalactonic acid methyl ester was used as the standard.

Partial hydrolysis

The polysaccharide was stirred with concentrated HCl for 15 min at room temperature. The resulting solution was poured into acetone, and the solid obtained was filtered, washed with acetone and dried. The dried material was dissolved in a minimum amount of water, neutralized with Zeo-Karb 215 and precipitated into acetone.

152

Results and discussion

Data for yields of agar from cystocarpic, tetrasporic and vegetative *Gelidium chilense* are presented in Fig. 2, together with values previously found in this laboratory for *G. lingulatum* and *G. rex* polysaccharides (Zanlungo, 1979, 1980; Matsuhiro & Urzúa, 1986, 1990). The yield of agars from Chilean *Gelidium* lies within those reported for northern Pacific Gelidiaceae: 35.9% from *G. japonicum* (Harvey) Okamura (Su & Young, 1977), 34.0% from *G. subcostatum* Okamura (Murakami, 1960) and 15.7% from *G. purpurascens* Gardner (Whyte & Englar, 1981).

It was reported (Matsuhiro & Urzúa, 1986) that the highest yield of agar from cystocarpic and vegetative *G. rex* was found in summer, whereas for tetrasporic thalli the highest yield was found in autumn. In Fig. 3 the yields of agar extracted from cystocarpic, tetrasporic and vegetative *G. rex* collected during June 1984–March 1985 are presented. It can be seen that the agar content is highest in December (early summer in the Southern Hemisphere). Onraët & Robertson (1987) found that the highest yields of agar from

sporophytic and gametophytic *Onikusa pristoides* (Turner) Akatsuka (Gelidiaceae) occur in late summer. For unsorted *Gelidium* species seasonal effects were also reported: the agar content of *G. cartilagineum* collected in Mexico is more than 30% higher in July than in January; in Spain, the agar content of *G. spinulosum* Kützing and *G. sesquipedale* (Clem.) Born. et Thur. peaks in July–August and again in November–December (Selby & Wynne, 1973).

Polysaccharides from cystocarpic, tetrasporic and vegetative *G. chilense* were hydrolysed with 2 M trifluoroacetic acid. By paper chromatography and gas-liquid chromatography analysis of the derived alditol acetates, the hydrolysates were shown to contain the acid-stable monosaccharides galactose and 6-*O*-methylgalactose. In Table 1, compositions of polysaccharides are presented. Neither xylose nor 4-*O*-methylgalactose were found in the hydrolysates. For unsorted *G. chilense* (as *G. filicinum*), Zanlungo (1979) reported the presence of minor amounts of 2-*O*-methylgalactose (0.7%), glucose (1.3%) and mannose (0.7%).

It is noteworthy that the three polysaccharides contain similar amounts of pyruvic acid. The content of this acid is indicative of the extent of replacement of D-galactose residues with 4,6–*O*-(1-carboxyethylidene)-D-galactose in agar-type polysaccharides. Agars from some species of *Gelidium* contain a small percentage of pyruvic acid [*G. amansii* (1.3%), *G. subcostatum* (1.2%), *G. cartilagineum* (1.29%)], whereas agars from *G. sesquipedale*, *G. japonicum* and *G. rex* are devoid of this acid (Young *et al.*, 1971; Percival, 1978; Matsuhiro & Urzúa, 1990).

Gelation and gel-melting temperatures of agars from different forms of *G. chilense*, *G. rex* and *G. lingulatum* were determined. Results are shown in Fig. 4. It can be seen that, within a species, the values for these parameters are similar.

Guiseley (1970) reported that increasing methoxyl content in agar samples from Gracilariaceae results in increasing gelation temperatures. For Chilean Gelidiaceae, the highest values in gelation temperatures were found in agars with

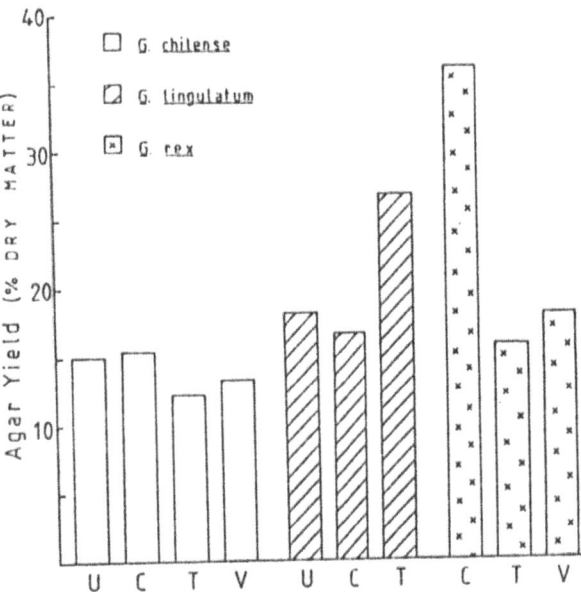

Fig. 2. Yield of agar extracted from unsorted (U), cystocarpic (C), tetrasporic (T) and vegetative (V) *Gelidium* species.

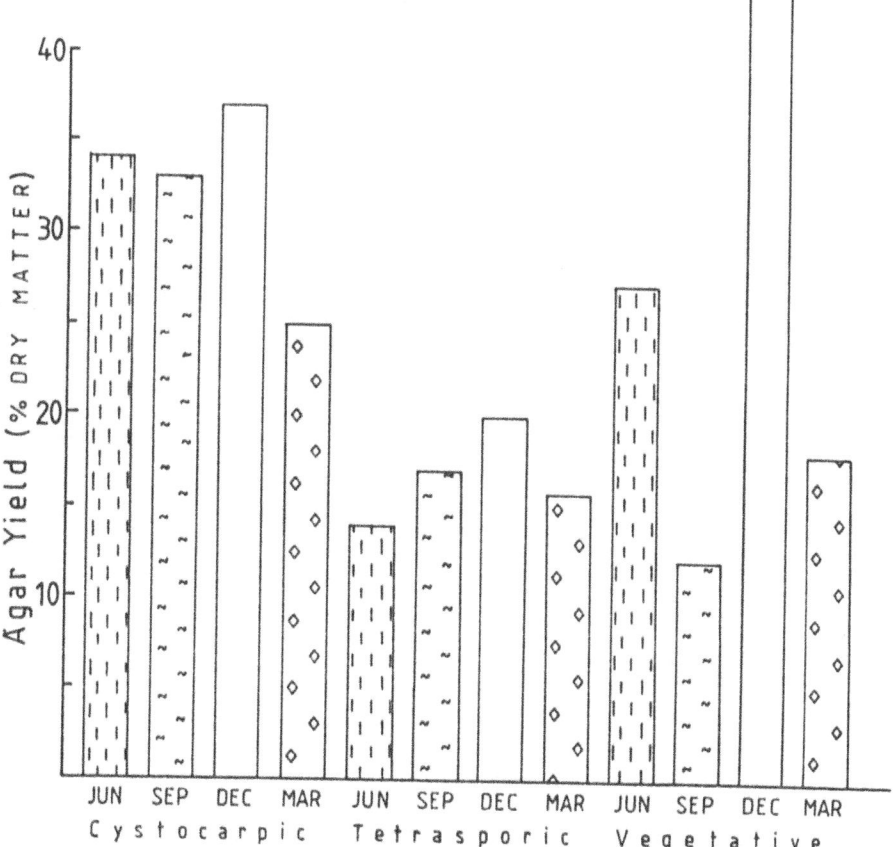

Fig. 3. Seasonal variation of agar yield from *Gelidium rex* separated into cystocarpic, tetrasporic and vegetative fronds.

lowest content of 6-*O*-methylgalactose. Cystocarpic *G. lingulatum* agar contained 1.4% of 6-*O*-methylgalactose and tetrasporic agar from the same species contained 1.3% of this monosaccharide.

Table 1. Composition of polysaccharides from cystocarpic, tetrasporic and vegetative fronds of *Gelidium chilense*.

	Percentage in		
	Cystocarpic	Tetrasporic	Vegetative
Galactose	56.0	40.8	42.9
6-*O*-Methylgalactose	6.2	5.7	6.2
3,6-Anhydrogalactose	30.7	36.3	32.5
Sulfate (SO_{3-})	2.1	1.3	2.8
Nitrogen	2.97	0.96	0.41
Pyruvic acid	0.54	0.42	0.48

Polysaccharides from different forms of *G. chilense* showed the highest gel-melting temperature values. This parameter is considered dependent on the molecular weight of the polysaccharide (Selby & Wynne, 1973).

Gel-strength determinations were conducted for agars from tetrasporic and vegetative *G. chilense*; values of 153 g cm^{-2} and 663 g cm^{-2} were obtained, respectively.

Agar from cystocarpic *G. chilense* contained a considerable amount of nitrogen, which may indicate the presence of carbohydrate-protein contaminants (Whyte & Englar, 1981), and it was not further studied.

Fractionation of tetrasporic and vegetative *G. chilense* polysaccharides was accomplished by anion-exchange chromatography on DEAE Sephadex. The yield and composition of the

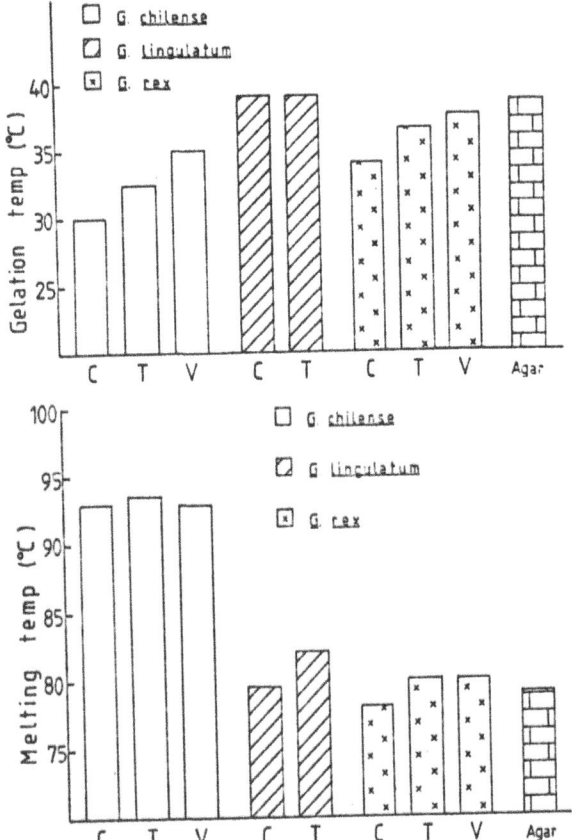

fractions are given in Table 2. The amounts of neutral fractions found in this work are lower than those published for cystocarpic and tetrasporic *G. lingulatum* and *G. rex* agars (Matsuhiro & Urzúa, 1988, 1990).

Structural studies were conducted on the neutral fractions from tetrasporic *G. rex* and *G. chilense* agars. The fraction devoid of sulfate, which was obtained by anion-exchange chromatography of the agar from tetrasporic *G. rex*, was methylated, hydrolysed with 3 M trifluoroacetic acid, and converted to partially-methylated alditol acetates. Gas-liquid chromatography analysis of the resulting derivatives showed a single peak which was identified as 1,3,5-tri-*O*-acetyl-2,4,6-tri-*O*-methylgalactitol by co-chromatography with an authentic sample. This result is indicative of the presence of 1–3 linked galactopyranosil residues. An aliquot of the permethylated fraction

Fig. 4. Gelation and gel melting temperatures of agars from cystocarpic (C), tetrasporic (T) and vegetative (V) *Gelidium chilense*, *G. lingulatum* and *G. rex*.

Table 2. Percentage yields and constituents of agar fractions obtained from tetrasporic and vegetative fronds of *Gelidium chilense* by chromatography on DEAE Sephadex.

	Eluant	Yield[a] %	Gals[b] %	3,6-An[c] %	SO₃ %
Tetrasporic					
Fraction 1	H₂O	19.8	48.4	36.5	0.0
Fraction 2	0.1 M KCl	16.1	47.1	31.9	5.3
Fraction 3	0.4 M KCl	7.6	51.2	23.5	6.2
Fraction 4	0.8 M KCl	20.7	45.7	23.7	9.8
Fraction 5	1.5 M KCl	7.7	38.3	–	15.2
Fraction 6	3.0 M KCl	1.3	20.1	–	18.6
Fraction 7	6.0 M Urea	10.7	28.9	17.5	14.3
Vegatative					
Fraction 1	H₂O	4.9	59.4	35.6	0.0
Fraction 2	0.1 M KCl	5.4	40.3	26.8	3.7
Fraction 3	0.4 M KCl	11.3	45.1	24.8	7.1
Fraction 4	0.8 M KCl	9.3	40.1	14.2	12.8
Fraction 5	1.5 M KCl	10.3	48.3	–	7.7
Fraction 6	6.0 M Urea	15.2	48.5	11.6	4.6

[a] Expressed as % of polysaccharide added to the column; [b] Gals = Galactose and 6-*O*-methylgalactose; [c] 3,6-An = 3,6-anhydrogalactose.

was studied by oxidative hydrolysis. The isolation of 2-*O*-methy-3,6-anhydrogalactonic acid, identified by gas-liquid chromatography as 2-*O*-methyl-3,6-anhydrogalactonic acid methyl ester and by co-chromatography with an authentic sample, indicates the presence of 3,6-anhydrogalactopyranosil residues linked 1–4 in the neutral fraction.

This fraction was depolymerized by partial hydrolysis and studied by ^{1}H NMR spectroscopy. Two pairs of doublets, centered at 4.96 ppm and 4.55 ppm assigned to the anomeric protons of α-L-3,6-anhydrogalactopyranosil and β-D-galactopyranosil residues were observed. A small signal, assigned to 6-*O*-methyl group was found at 3.65 ppm. From these results and the value reported (Matsuhiro & Urzúa, 1990) for molar ratio of galactose : 3,6-anhydrogalactose (1.10 : 1.00), an agarose structure, with small substitution by 6-*O*-methylagarose (Fig. 1), can be proposed for the neutral fraction.

The neutral fraction from the polysaccharide of tetrasporic *G. chilense* was depolymerized by partial hydrolysis. Its ^{1}H NMR spectrum is essentially similar to those of the partial hydrolysate of fraction 1 from tetrasporic *G. rex* and of partial hydrolysates of agars from *Gracilaria verrucosa* (Hudson) Papenfuss and *Gloiopeltis furcata* (Post. & Rupr.) J. Agardh (Izumi, 1973). The chemical analysis showed a molar ratio of 1.19 : 1.00 for galactose : 3,6-anhydrogalactose, close to the molar ratio (1.00 : 1.00) calculated for an ideal agarose structure.

From the results found in this work and preceding studies on agars from Chilean Gelidiaceae, it can be concluded that the hydrocolloids extracted with water at 95 °C are mixtures of related polymers, which range from neutral agarose to highly-sulfated galactans with low contents of 3,6-anhydrogalactose. Polysaccharides from different forms of *G. lingulatum* and *G. rex* contain mainly proportions of neutral polymers whereas those of *G. chilense* consist of a more random spectrum of polymers. *G. rex* is the most promising Chilean *Gelidium* species for agar production. Extraction of the alkali-treated seaweed is suggested for obtaining industrial-grade agar.

Acknowledgements

The financial support of the Dirección de Investigaciones Científicas y Tecnológicas of the Universidad de Santiago de Chile and the International Foundation for Science is gratefully acknowledged.

References

Anderson, N. S., J. C. S. Dolan & D. A. Rees, 1973. Carrageenans. Part VII. Polysaccharides from *Eucheuma spinosum* and *Eucheuma cottonii*. The covalent structure of ι-carrageenan. J. Chem. Soc.: 2173–2176.

Araki, C & K. Arai, 1957. Studies on the chemical constitution of agar-agar. XX. Isolation of a tetrasaccharide by enzymatic hydrolysis of agar-agar. Bull. Chem. Soc. Japan 30: 287–293.

Araki, C., K. Arai & S. Hirase, 1967. Studies on the chemical constituents of agar-agar. XXIII. Isolation of D-xylose, 6-*O*-methyl-D-galactose, 4-*O*-methyl-L-galactose and *O*-methylpentose. Bull. Chem. Soc. Japan 40: 959–962.

Cerezo, A. S., 1973. The carrageenan of *Gigartina skottsbergii* S. et G. Part III. Methylation analysis of the fraction precipitated with 0.3–0.4 M potassium chloride. Carbohydr. Res. 26: 335–340.

Chiles, T. C., K. T. Bird & F. E. Koehn, 1989. Influence of nitrogen availability on agar-polysaccharides from *Gracilaria verrucosa* strain G-16: structural analysis by NMR spectroscopy. J. appl. Phycol. 1: 53–58.

Dogson, K. S., 1961. Determination of inorganic sulphate in studies on the enzymic and non-enzymic hydrolysis of carbohydrate and other sulphate esters. Biochem. J. 78: 312–319.

Dubois, M., K. A. Gilles, J. K. Hamilton, P. A. Rebers & F. Smith, 1956. Colorimetric method for determination of sugars and related substances. Anal. Chem. 28: 350–356.

Duckworth, M. & W. Yaphe, 1970. Definitive assay of pyruvic acid in agar and other algal polysaccharides. Chem. Int. (London) 23: 747–748.

Duckworth, M. & W. Yaphe, 1971. The structure of agar. Part I. Fractionation of a complex mixture of polysaccharides. Carbohydr. Res. 16: 189–197.

Guiseley, K. B., 1970. The relationship between methoxyl content and gelling temperature of agarose. Carbohydr. Res. 13: 247–256.

Hakomori, S., 1964. A rapid permethylation of glycolipid and polysaccharides catalyzed by methylsulfinylcarbanion in dimethylsulfoxide. J. Biochem. 55: 205–208.

Hoyle, M. D., 1978a. Agar studies in two *Gracilaria* species (*G. bursapastoris* (Gmelin) Silva and *G. coronopifolia* J. Ag.) from Hawaii. Seasonal aspects. Bot. mar. 21: 247–252.

156

Hoyle, M. D., 1978b. Agar studies in two *Gracilaria* species (*G. bursapastoris* (Gmelin) Silva and *G. coronopifolia* J. Ag.) from Hawaii. I. Yield and gel strength in the gametophyte and tetrasporophyte generations. Bot. mar. 21: 343–345.

Izumi, K., 1971. Chemical heterogeneity of the agar from *Gelidium amansii*. Carbohydr. Res. 17: 227–230.

Izumi, K., 1973. Structural analysis of agar-type polysaccharides by NMR spectroscopy. Biochim. Biophys. Acta 320: 311–317.

Kim, D. H. & N. P. Henríquez, 1979. Yield and gel strength of agar from cystocarpic and tetrasporic plants of *Gracilaria verrucosa* (Florideophyceae). Proc. int. Seaweed Symp. 9: 257–262.

Lahaye, M., J. F. Revol, C. Rochas, J. McLachlan & W. Yaphe, 1988. The chemical structure of *Gracilaria crassissima* (P. et H. Crouan in Schramm et Mazé) P. et H. Crouan in Schramm et Mazé and *G. tikvahiae* McLachlan (Gigartinales, Rhodophyta) cell-wall polysaccharides. Bot. mar. 31: 491–501.

Lahaye, M. & W. Yaphe, 1988. Effects of seasons on the chemical structure and gel strength of *Gracilaria pseudoverrucosa* agar (Gracilariaceae, Rhodophyta). Carbohydr. Polym. 8: 285–301.

Lewis, B. A., F. Smith & A. M. Stephen, 1963. 2,5- and 3,6-anhydrosugars and their derivatives. Methods Carbohydr. Chem. 2: 172–188.

Matsuhiro, B. & C. C. Urzúa, 1988. Polysaccharides of *Gelidium* spp. In Maclean, J. L., L. B. Dizon & L. V. Hosillos (eds), The First Asian Fisheries Forum, The Asian Fisheries Society, Manila. pp. 165–168.

Matsuhiro, B. & C. C. Urzúa, 1988. Agarans from tetrasporic and cystocarpic *Gelidium lingulatum*. Bol. Soc. chil. Quím. 33: 135–140.

Matsuhiro, B. & C. C. Urzúa, 1990. Agars from *Gelidium rex*. Hydrobiologia, 204/205: 545–549.

Matsuhiro, B. & A. B. Zanlungo, 1983. Colorimetric determination of 3,6-anhydrogalactose in polysaccharides from red seaweeds. Carbohydr. Res. 118: 276–279.

McLachlan, J. & C. J. Bird, 1983. *Gracilaria* in the seaweed market: a prospectus. Symposium Internacional de Acuacultura, Coquimbo, Chile, Sept. 1983. pp. 133–157.

Mori, T., 1953. Seaweed polysaccharides. Adv. Carbohydr. Chem. 8: 315–350.

Murakami, S., 1960. Studies on the agar of Rhodophyceae III. The isolation and purification of agar. Sci. Rep. Saitama Univ. 3: 251–254.

Onraët, A. C. & B. L. Robertson, 1987. Seasonal variation in yield and properties of agar from sporophytic and gametophytic phases of *Onikusa pristioides* (Turner) Akatsuka (Gelidiaceae, Rhodophyta). Bot. mar. 30: 491–495.

Patway, M. & J. P. van der Meer, 1983. Genetics of *Gracilaria tikvahiae* (Rhodophycaea). IX: Some properties of agars extracted from morphological mutants. Bot. mar. 26: 295–299.

Percival, E., 1978. Do the polysaccharides of brown and red seaweeds ignore taxonomy? Syst. Assoc. Spec. 10: 47–62.

Santelices, B., 1989. Algas Marinas de Chile, Universidad Católica de Chile, Santiago, 310.

Santelices, B. & I. A. Abbott, 1985. *Gelidium rex* sp. nov. (Gelidiales, Rhodophyta) from central Chile. In Abbott I. A. & J. N. Norris (eds), Taxonomy of Economic Seaweeds. With Reference to Some Pacific and Caribbean Species. University of California, La Jolla, pp. 33–36.

Santelices, B. & S. Montalva, 1983. Taxonomic studies on Gelidiaceae (Rhodophyta) from central Chile. Phycologia 22: 185–196.

Santelices, B. & J. G. Stewart, 1985. Pacific species of *Gelidium* Lamouroux and other Gelidiales (Rhodophyta) with keys and descriptions to the common or economically important species. In Abott I. A. & J. N. Noris (eds), Taxonomy of Economics Seaweeds. With Reference to Some Pacific and Caribbean Species. University of California, La Jolla, pp. 17–31.

Selby, H. H. & W. H. Wynne, 1973. Agar. In Whistler, R. L. & J. N. BeMiller (eds), Industrial Gums. Academic Press, New York, pp. 29–48.

Su, H. & K. S. Young, 1977. An investigation of agars prepared from some Taiwan red seaweeds. J. Fish. Soc. Taiwan 6: 1–11.

Whyte, J. N. C. & J. R. Englar, 1980. Chemical composition and quality of agar in the morphotypes of *Gracilaria* from British Columbia. Bot. mar. 23: 277–283.

Whyte, J. N. C. & J. R. Englar, 1981. The agar component of the red seaweed *Gelidium purpurascens*. Phytochemistry 20: 237–240.

Whyte, J. N. C., J. R. Englar, R. G. Saunders & J. C. Lindsay, 1981. Seasonal variations in the biomass, quantity and quality of agar, from the reproductive and vegetative stages of *Gracilaria* (*verrucosa* type). Bot. mar. 24: 493–501.

Yaphe, W., 1960. Colorimetric determination of 3,6-anhydrogalactose and galactose in marine algal polysaccharide. Analyt. Chem. 32: 1327–1330.

Young, K., M. Duckworth & W. Yaphe, 1971. The structure of agar. Part III. Pyruvic acid, a common feature of agars from different agarophytes. Carbohydr. Res. 16: 446–448.

Zanlungo, A. B., 1979. Polisacáridos de algas chilenas. V. Composición del agar de *Gelidium filicinum*. Rev. latinoamer. Quím. 10: 149–151.

Zanlungo, A. B., 1980. Polysaccharides from Chilean seaweeds. Part IX. Composition of the agar from *Gelidium lingulatum*. Bot. mar. 23: 741–743.

Hydrobiologia **221**: 157–166, 1991.
J. A. Juanes, B. Santelices & J. L. McLachlan (eds), International Workshop on Gelidium.
© *1991 Kluwer Academic Publishers.*

Agar and agarose biotechnological applications

Rafael Armisén

Hispanagar S/A, Calle López Bravo, 1, Polígono Industrial de Villalonquejar, Apartado Postal 392, 09080 Burgos, Spain

Key words: agar, agarose, bacteriological agar, chromatography, electrophoresis

Abstract

Agar, a phycocolloid obtained commercially from species of *Gelidium* and *Gracilaria*, has been known for several centuries; its earliest industrial application was in the preparation of solid microbiological media. The numerous techniques available for the purification of agar affect the characteristics of bacterial-grade agar. The availability of agarose, that fraction of agar with the lowest possible charge, has enhanced the utilzation of this phycocolloid. The process of gelation of agarose is discussed and the applications of agarose gels in different types of chromatography are summarized. Agarose has many and diverse important applications in biotechnology. These uses, and newly-developed ones, can be expected to increase the demands for high-quality agarose in the rapidly expanding field of biotechnology.

Phycocolloids are water-soluble polysaccharides (hydrocolloids) that are extracted from seaweeds. These extracts are used as both human and animal foods in addition to having numerous industrial applications. The multitude of uses for these polysaccharides is based on their behaviour in aqueous solution. Numerous seaweed species yield phycocolloids of commercial importance (cf. McHugh, this volume). Agars and carrageenans are obtained from red seaweeds, the Rhodophyceae, and are galactans sulphated to varying degrees. Alginates are obtained from brown algae, Phaeophyceae, and the polymers are comprised of blocks of polyguluronic and polymanuronic units. Phycocolloids have little food value, and less than 10% of these seaweed extracts is assimilated by humans. These colloids are generally used in small quantities, less than 1% in most food preparations, where their applications are based on the physical properties imparted by these seaweed extracts to aqueous solutions.

Agar is the oldest utilized phycocolloid, and was discovered by Minoya Tarazaemon in 1658 in Japan, where it is called *kanten*, literally 'frozen sky'. This name figuratively describes the natural method of *kanten* production by freeze-thawing which has been used since its discovery in the 17th century. Agar was described in the West in 1859 by Payen who referred to it as gelose. Koch (1882) was the first to utilize agar in microbiology, and Parker and Leiking (1937) have documented the way in which agar came to be known to Koch. During the period that Koch was attempting to obtain axenic cultures of *Mycobacterium turbeculosis*, a country doctor from Saxony, Walter Hesse, came to learn of the new science of bacteriology. Hesse introduced Koch to the powerful gelling agent that has become the universal component of solid culture media for the cultivation of microorganisms. Hesse had learned about agar from his wife, Fanny Hesse, née Eilshemius, whose family had contact with the Dutch East

Indies. The term 'agar-agar' is a Malaysian word that initially referred to extracts from *Eucheuma*, which, interestingly, yields carrageenan, not agar.

Since the 17th century, agar has been produced in Japan in the form of strips and squares using traditional methods of freezing. Modern techniques of industrial freezing were introduced by Matsuoka in Glendale, California in 1922 to prepare agar for use in oriental cooking. In Europe, for use in microbiology, agar was purified by dialysis, redissolved, filtered, *etc.*, according to the German patents of Steinitzer (1912) and Merck (1913) or Lian Tjoa (1955a, b). Agar was purified by ion-exchange (Ionagar) in England. Production of agar in the United States was started just before the beginning of World War II as a strategic material according to U.S. Army specifications (1935). Bacteriological-grace agar manufactured by the American Agar Company of San Diego, California in the 1940's served as reference agar-material for the evaluation of the characteristics of other culture-media components, such as peptones. The American Agar Company ceased production in 1986, and the factory was dismantled in 1987.

The use of different seaweeds from various parts of the world allows industrialists to blend and produce bacteriological-grade agars with different properties such as ash content and gel strength. The most important properties of bacteriological-grade agar are: (1) good transparency in sol and gel forms; (2) consistent gel strength from lot-to-lot; (3) gelling temperature of $36 \pm 1.5\,°C$, and melting temperature of $87 \pm 1.5\,°C$ that is consistent from lot-to-lot (this temperature difference is hysteresis, which in agar is superior to that of other hydrocolloids); (4) low content of oligomers and proteins that cannot be utilized as a source of nutrients for micro-organisms; (5) low and regular content of electronegative groups that could cause differences in diffusion of electropositive molecules (antibiotics, nutrients, metabolites and so forth); (6) freedom from toxic substance (bacterial inhibitors); (7) freedom from hemolytic substances that might interfere with normal hemolytic reactions in culture media; (8) freedom from contamination by

thermophilic spores. It is also necessary to control other factors to assure proper performance of the agar in many different media formulations and with many different bacterial strains.

Agar comprises mixtures of galactans whose average sulphate content is between 1.5 and 6%, with lesser quantities of pyruvic and guluronic residues. These linear galactans are formed by alternating units of D- and L-galactopyranoses. Araki in 1937 reported on a method to fractionate agar into its main components, agarose and agaropectin. Araki's method was not generally appreciated until Hjerten (1961, 1962), in Uppsala, devised a simple technique for the production of agarose. More than 40 distinct methods have been published since then, and can be assigned to 13 basic groups (Table 1).

The secondary structure of agarose has a molecular weight of about 120 000 Daltons, determined by sedimentation procedures, represented in 400 agarobiose units linked together, or 800 hexose units connected linearly (Fig. 1). This linear and repetitive structure must be considered as an ideal simplification, as, in practice, it is impossible to obtain an agarose totally free from electronegative groups (sulphates, pyruvates, guluronates and, possibly, other carboxylic groups). The residual pyruvate groups are found in the form of 4,6-0-(1-carboxyethylidene)-D-galactose (Hirase, 1962).

Agaropectin has a similar structure, although with a much lower content of 3,6 anhydro-L-galactose. The major sulphate residues are L-galactose-6-sulphate and D-galactose-4-sulphate. The number of dimerous repetitions is around 75–100 and the molecular weight about 16 000 Daltons.

Araki (1956) defined agarose as:

The principal polysaccharide contained in agar-agar is a molecule in the form of neutral chains formed by residues of β-D-galactopyranoses, connected by way of C1–C3 with residues of α-L-galactopyranoses connected by way of C1 & C4. These two residues repeat themselves in alternating forms.

Table 1. The 13 basic methods for the preparation of agarose, listed in chronological order, and authors of the methods.

Method	Year	Author(s)
Acetylation	1937	Araki
Selective dissolution	1957	Glickman & Shubtosova
Ammonium quaternary precipitation	1962	Hjerten
Polyethylene glycol precipitation	1962	Russell *et al.*
Dimethylsulfoxide extraction	1966	Tagawa
Ionic exchange	1966	Zabin
Ammonium sulphate precipitation	1967	Azhitskii & Kobozev
Absorption over insoluble support	1969	Baterling
Acrinol precipitation	1970	Fuse & Goto
Chromotography	1970	Izumi
Electrophoresis	1971	Hjerten
Chitin or chitosan precipitation	1971	Allan & Johnson
Agarose precipitation (solved in urea buffer with ethanol or 2-methoxy ethanol)	1973	Patil & Kale

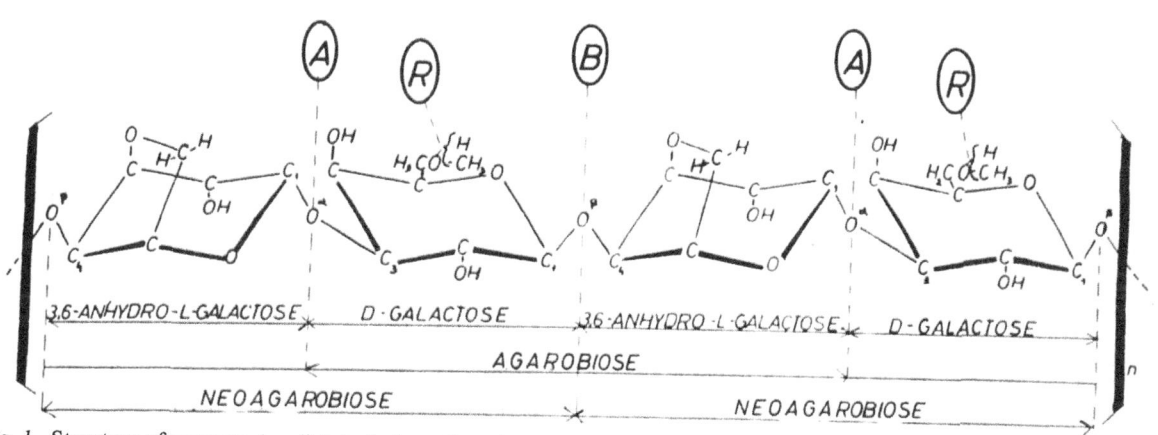

Fig. 1. Structure of agarose: A = link hydrolysed by acids and α-agarases; B = link easily hydrolysed by acids and β-agarases; R = substitution of − H by − CH$_3$ depends on the seaweed used (the gel point is increased by a higher content of − CH$_3$ groups); *n* = degree of polymerization which determines the gel strength.

Agarose was considered to be a neutral molecule, even though the best analytical methods could not establish chemical neutrality. Duckworth and Yaphe (1971) proposed as a definition:

Agarose is a mixture of agar molecules having the lowest charge content, and therefore the greatest gelling ability, fractionated from a whole complex of molecules, called agar, all differing in their extent of substitution with charged groups.

Good commercial agarose is considered to have less than 0.35% sulphates ('Pronarose D-3' has less than 0.10% sulphates); the content of pyruvate is likewise very low. There is no current technique available sufficiently sensitive to quantify the guluronic acid content of agarose.

The structure of agarose is extraordinarily resistant to enzymatic hydrolysis, and few bacteria are capable of producing α- or β-agarases. A few species, *Pseudomonas atlantica* and *Bacillus cereus*, can, however, enzymatically degrade agarose. The resistance of agar to hydrolysis is basic to its use in bacteriology. This, together with high gel strength, gelation in the absence of cations (unlike alginates and carrageenans) and large hystersis are the basic properties that have made agar indispensable in the preparation of solid culture media for microorganisms.

Drawing on my experience of nearly 30 years in the agar industry, I believe that the agaroses from species of *Gelidium* have greater stability against bacterial agarases than agaroses derived from species of *Gracilaria*. This is especially applicable to those species of the latter genus from warm waters, a fact that must be appreciated when storing seaweed resources prior to processing. The presence of methyl groups, which greatly affect gelling temperatures of the extract, must also be taken into account in species of *Gracilaria*. Guiseley (1970) showed a direct relationship between methoxyl content and gelling temperature of underivatized agaroses; an increase of 8% in methoxyl groups caused an increase of 12 °C in gelling temperature of a 1.5% agarose solution. Guiseley also detailed the synthetic methylation of agarose, and obtained lower gelling temperatures than in unmethylated agaroses. Methoxyl groups, in agaroses, occur primarily as 6-0-methyl-D-galactose and 4-0-methyl-L-galactose. Possibly synthetic methylation cannot distinguish between the carbon atoms that are naturally methylated, and this could be the reason that the gelling temperatures are lowered. Craigie and Jurgens (1989) have, in addition, established, for *Gracilaria tikvahiae* McLachlan, the branching of the 4-0-methyl-L-galactose from the polymeric line structure, and this is shown in Fig. 2.

The exothermic, reversible process of gelation of agarose is shown in Fig. 3; as random coils form double helices (tertiary structure) and finally the macroreticular network (Fig. 4), this determines the exclusion limits of the gel. The double helix has a cross-spiral distance of 1.90 nm and an intercaternary distance of 0.85 nm, as shown by X-ray diffraction, making the structure very compact (Fig. 5); the exclusion limits of the gel can be changed merely by changing its concentration (Fig. 6).

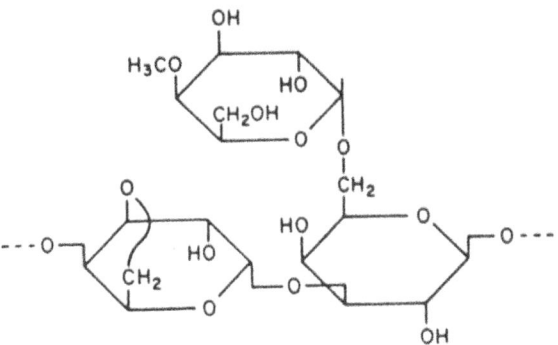

Fig. 2. Structure of branched neoagarobiose from *Gracilaria tikvahiae* (from Craigie & Jurgens, 1989).

Jimenez-Barbero *et al.* (1989) recently demonstrated a '3-fold left-handed single helix' with an 'advance' of 0.95 nm and a 'pitch' of 2.85 nm (Fig. 7). This structure of simple helices could be produced during gelling of the agarose.

The preparation of agarose 'beads' by Hjerten (1964) and Bengtsson and Philipson (1964) made agarose available for chromatographic separations (exclusion limits), and the ranges of these separations are shown in Fig. 8. Hjerten and Kunquan (1981) and Hjerten (1983, 1984) subsequently used higher concentrations of agarose beads (12% crossed-linked) than the usual 2, 4 and 6% levels that allowed chromatographic separations to proceed at pressures between 2×10^5 to 5×10^5 pascals. Hispanagar S/A currently produces agarose beads from 2–16%, both plain and cross-linked. It is possible to obtain a more rigid and irreversible gel bead by cross-linking with epichlorohydrin, divinylsulfone and so forth; it is also possible to sterilize packed columns of beads in the autoclave.

The ease with which beads can be derivatized permits attachment of proteins such as antigens or antibodies fixed by cyanogen bromide, p-nitrophenolcyanide, p-hydroxy-succinamide and sulfonyl halogens. The formation of aldehydic groups opens up possibilities for affinity chromatography and for the preparation of catalytic beds for ion-exchange.

The principal physical-chemical properties of agarose are:

1) Very high gel strength: Gel strength vs con-

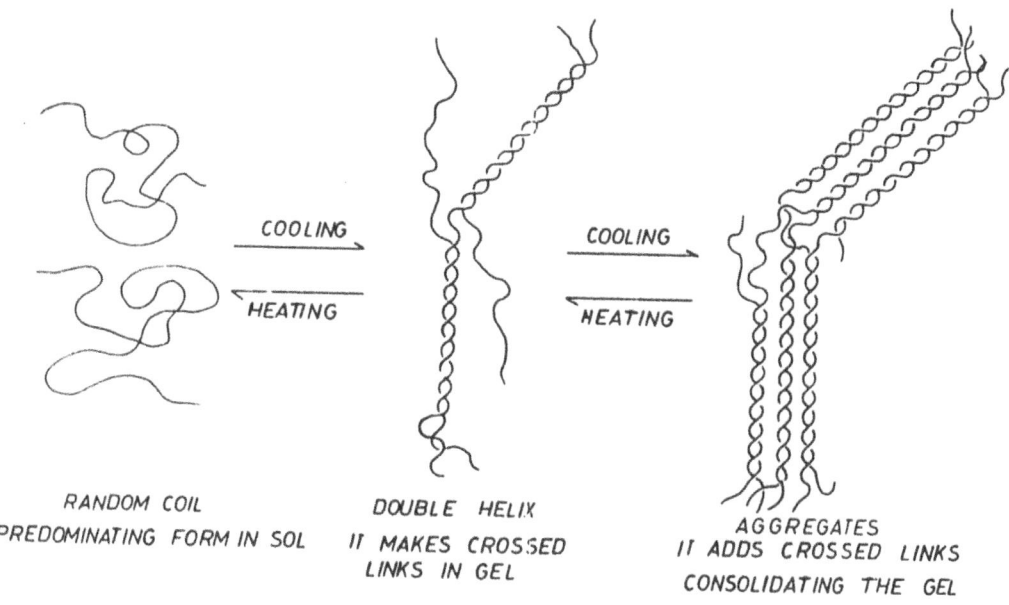

COOLING →

← HEATING

COOLING →

← HEATING

RANDOM COIL
PREDOMINATING FORM IN SOL

DOUBLE HELIX
II MAKES CROSSED
LINKS IN GEL

AGGREGATES
II ADDS CROSSED LINKS
CONSOLIDATING THE GEL

SOL ⇌ INCIPIENT GEL ⇌ ELASTIC CLEAR GEL ⇌ TURBID RIGID GEL ⇌ PHASES SEPARATION(SYNERESIS)

Fig. 3. Stages in the exothermic, reversible process of gelification in agarose.

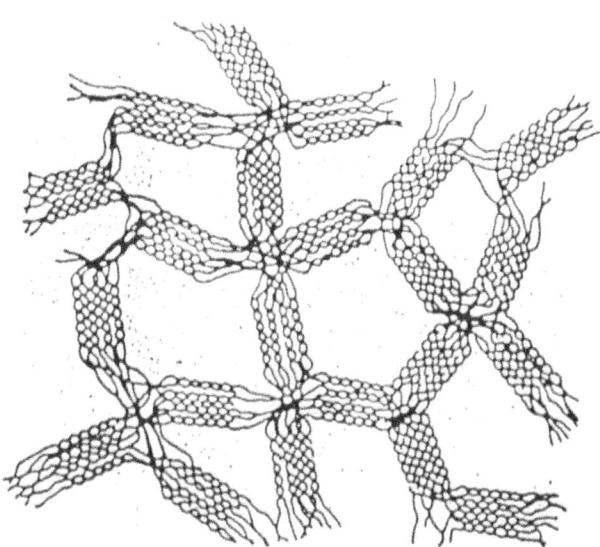

Fig. 4. Agarose gel network (based on Arnot *et al.*, 1974). The aggregates in agarose gels may contain 10 to 10 000 helices rather than the smaller numbers shown in this figure.

centration for different types of agar and agarose is illustrated in Fig. 6; 'Pronarose D-3' and 'D-5' agaroses have exceptionally high values. High gel strength is a consequence of high molecular and low sulphate content. There is, unfortunately, no correlation between gel strength and the average molecular weight of agarose, and gel strength for the same agarose increases with reduction in sulphate content. I believe that more basic research is necessary to understand the interrelationship between molecular weight and gel strength, though, in principle, this would appear obvious. We are working hard on this, but the matter is very complicated.

2) Hysteresis of gelation (Fig. 9): In agarose derivatized by synthesis, changes in gelling temperature (always a reduction) result in reductions of the melting temperature, and, in general, the value for gelling hysteresis is also reduced.

3) Presence of electronegative groups: It is easy to quantify sulphate ($\geq 0.35\%$) and pyruvate (usually 0.2% or less) groups; however, there is no technique sufficiently sensitive to measure the remaining guluronic acid groups. These groups can act as a block in electroendosmosis (EEO) measurements. As small variations in technique can result in important changes in numerical values of the results, EEO measurements must be

162

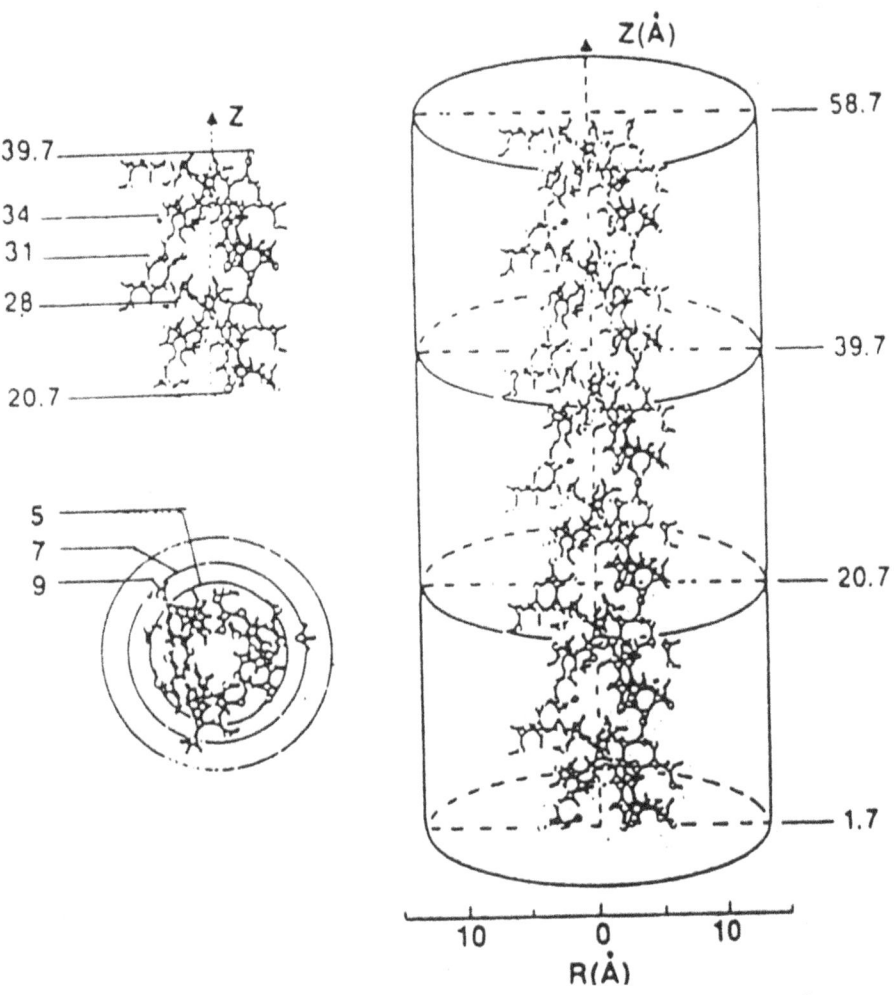

Fig. 5. Agarose double helix (from Corongiu *et al.*, 1983). The double helix fragment (three turns) of agarose is surrounded by 600 water molecules at 300° K. Sizes are given in Å.

standardized. An electrical current applied to an agarose gel causes a flow of soluble cations by electroendosmosis; this drags a neutral ion towards the cathode (Fig. 10). It is also necessary to standardize the protein (BSA) and the neutral molecule (Dextran T70), composition of the buffer, pH, intensity of the electrode current, type of wick and so forth. The relationship between electroendosmosis and the content of sulphates (in milliequivalents) and pyruvate per 100 g agarose is illustrated in Fig. 11.

The application of agarose in electrophoresis, initiated by Tiselius in Uppsala, began in earnest with the preparation of the first usable agarose by Hjerten (1962). The large mesh-size of the gel (quarternary structure), the ease of changing mesh size merely by altering its concentration, the elevated mechanical resistance of the gel and the lack of neurotoxicity associated with other separation substances, such as polyacrylamide, permitted the use of agarose in many different applications in the laboratory. In protein electrophoresis, the buffer system and pH are important factors because the electrical charge depends upon the degree of dissociation of the terminal amino acids of the protein. Also important are the buffer type and an agarose of adequate electroendosmosis.

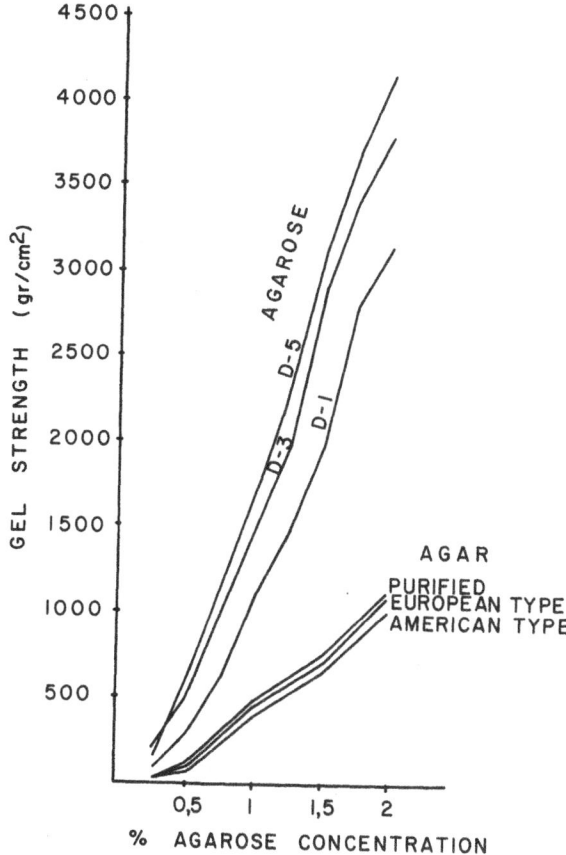

Fig. 6. Variation in gel strength as a function of concentration; D-1, D-3 and D-5 refer to 'Pronarose' agaroses.

Immunoelectrophoresis is a form of electrophoresis with extreme sensitivity. Separation becomes visible by the formation of an antigen-antibody precipitate, combining electrophoresis with immunoprecipitation; the influence of EEO on the position of the precipitating bands is shown in Fig. 12.

Separations of proteins, based on their isoelectric point, is done by electrofocussing. This technique requires an agarose with an EEO of 0. As previously stated, agarose totally free from electronegative charges cannot be prepared, although an EEO of 0 can be achieved using the following methods:

1) A number of electropositive groups are fixed on the agarose and these produce an equilibrium with the electronegative groups.

2) Addition of a neutral colloid of high viscosity to the agarose which fixes the free water, impeding the formation of large solvates of the cations.

In the early 1980's agarose, because of its large network, was used almost exclusively in the electrophoretic separation of large fragments of DNA and RNA, plasmids and chromosomes. The electrical charge used for transport is created solely by the phosphate groups, rendering this technique more independent of the buffer. In general, agarose of low EEO is used for the electrophoresis of nucleic acids, provided that the agarose is free from inhibitors of enzymes such as ligases, restriction enzymes, DN'ases and RN'ases. The agarose must also be tested for its ability to separate high and low molecular-weight nucleic acids, to detect ethidium bromide fixation and to meet many other requirements.

The electrophoresis of whole or large fragments

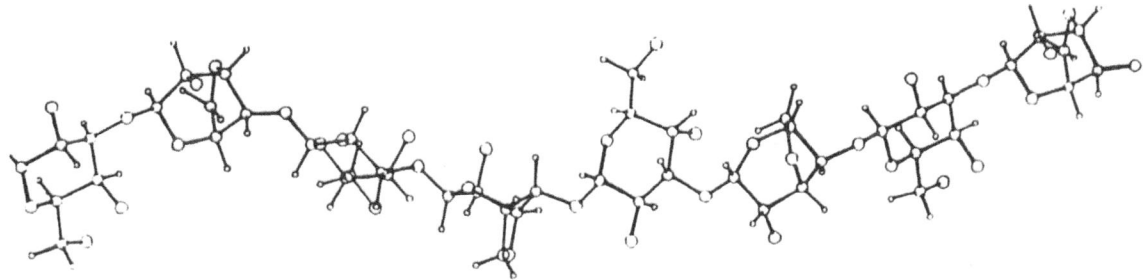

Fig. 7. Single chain of agarose (from Jimenez-Barbero *et al.*, 1989). This is the most stable single chain of agarose; it has a left-handed, threefold symmetry and a fibre repeat of 2.85 nm.

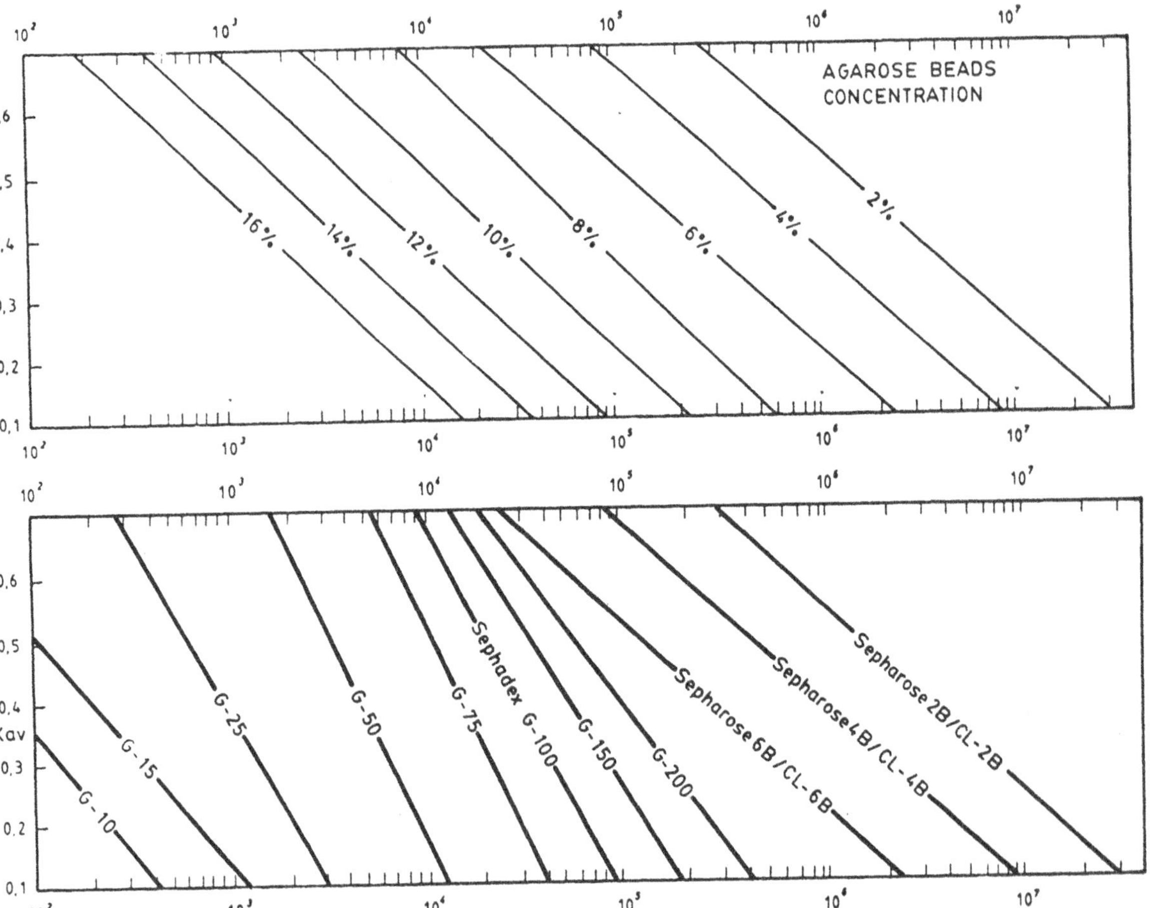

AGAROSE BEADS
CONCENTRATION

16% 14% 12% 10% 8% 6% 4% 2%

Kav

G-10 G-15 G-25 G-50 G-75 Sephadex G-100 G-150 G-200 Sepharose 6B/CL-6B Sepharose 4B/CL-4B Sepharose 2B/CL-2B

Fig. 8. Fractionational range for separation with 'Pronarose' agarose beads.

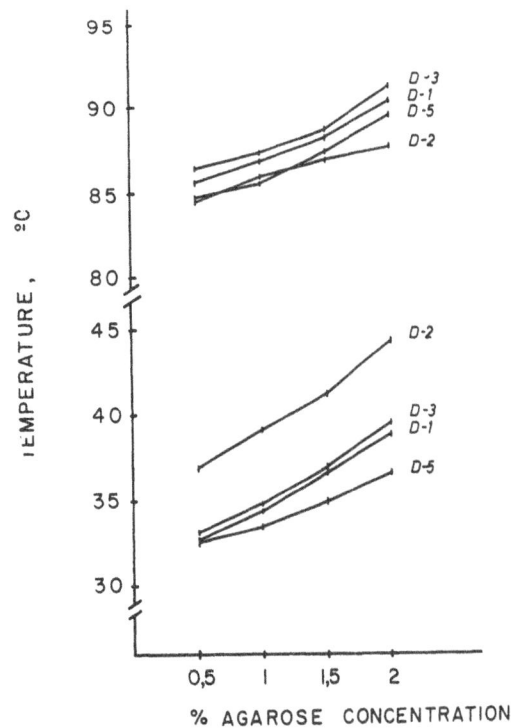

TEMPERATURE, °C

D-3
D-1
D-5
D-2

D-2

D-3
D-1

D-5

% AGAROSE CONCENTRATION

of chromosomes by means of pulse-field techniques will require the use of specially-designed agaroses. These agaroses, such as 'Pronarose D-3' and 'D-5', possess very high gel strengths not previously obtained on an industrial scale. Hispanagar S/A has produced these agaroses by selecting clones of special seaweeds, which have, at the same time, low levels of electronegative groups and a very low EEO. The high gel strength permits use at lower concentrations (0.5–0.7%), making good separations in a shorter amount of time with chromosomes, such as yeast chromosomes, containing high numbers of base-pairs. An important and interesting application of these agaroses is in the preparation of culture media for cells and difficult-to-grow autotrophic bacteria.

Fig. 9. Hysteresis of agarose, showing the gelling and melting temperatures of 'Pronarose D-1, D-2. D-3. D-5' agaroses as a function of concentration.

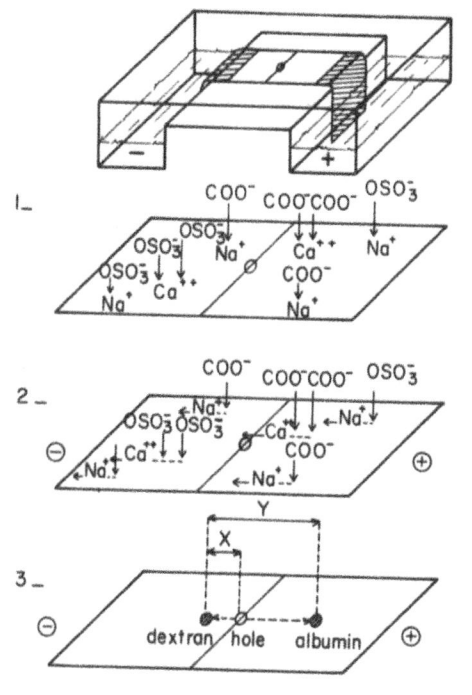

ELECTROENDOSMOSIS (EEO)-$Mr = \dfrac{X}{Y}$

Fig. 10. Graphic representation of electroendosmosis.

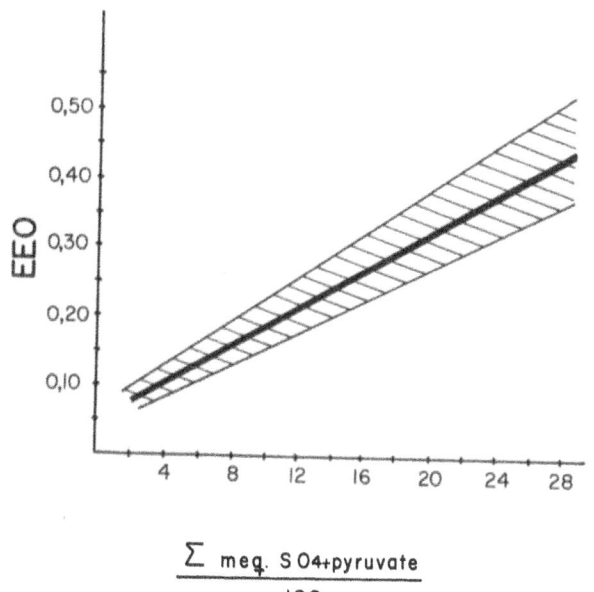

$$\dfrac{\Sigma \text{ meq. } SO_4 + \text{pyruvate}}{100 \text{ g.}}$$

Fig. 11. The relationship between electroendosmosis and the content of sulphate (milliequivalents) and pyruvate per 100 g of agarose.

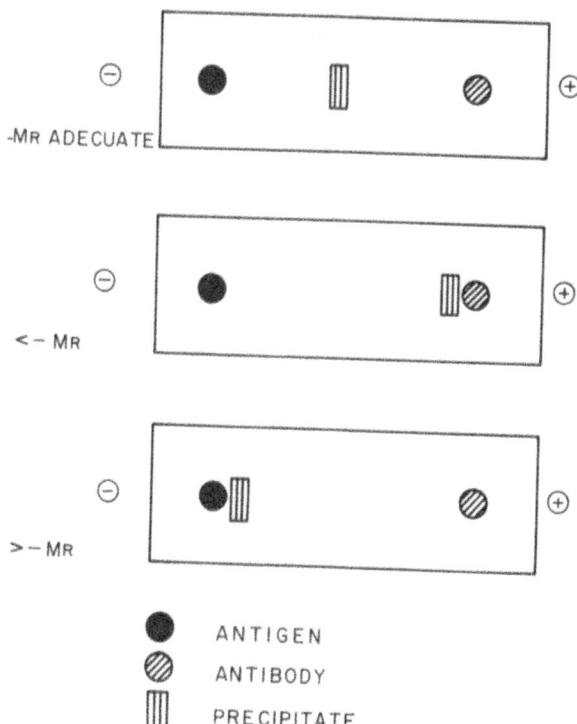

● ANTIGEN

▨ ANTIBODY

▥ PRECIPITATE

Fig. 12. Electroendosmosis showing the influence of the position of the precipitate in immunoelectrophoresis.

References

Allan, G. C. & P. G. Johnson, 1971. Marine polymers. Part I. A new procedure for the fractionation of agar. Carbohyd. Res. 17: 234–236.

Araki, C., 1937. Acetylation of agar like substance of *Gelidium amansii*. J. Chem. Soc. Jpn. 58: 1338–1350.

Araki, C., 1956. Structure of the agarose constituent of agar-agar. Bull. Chem. Soc. Jpn 29: 43–44.

Arnot, S., A. Fulmer, W. E. Scott, I. C. M. Deea, R. Morehouse & D. A. Rees, 1974. The agarose double helix and its function in agarose gel structure. J. mol. Biol. 90: 269–284.

Azhitskii, G. Y. & G. V. Kobozev, 1967. Use of ammonium sulfate to first eliminate agaropectin and then precipitate agarose. Lab. Delo 3: 143–145.

Baterling, S. J., 1969. A simple method for the preparation of agarose. Clin. Chem. 15: 1002–1005.

Bengtsson, S. J. & I. Philipson, 1964. Chromatography on animal viruses on pearl-condensed agar. Biochem. Biophys. Acta 79: 399–406.

Corongiu, G., S. L. Forlini & E. Clementi, 1983. Hydration of agarose double helix: a Montecarlo simulation. Int. J. Quant. Chem.: Quant. Biol. Symp. 10: 227–291.

Craigie, J. S. & A. Jurgens, 1989. Structure of agars from

166

Gracilaria tikvahiae, Rhodophyta: location of 4-0-methyl-L-galactose and sulphate. Carbohyd. Polym. 11: 265–278.

Duckworth, M. & W. Yaphe, 1971. The structure of agar. Part I. Fractionation of a complex mixture of polysaccharides. Carbohyd. Res. 16: 189–197.

Fuse, T. & F. Goto, 1971. Some properties of agarose and agaropectin isolated from various mucilaginous substances of red seaweeds. Agr. Biol. Chem. 35: 799–804.

Glickman, S. A. & I. G. Shubtosova, 1957. Physical chemistry of agar. Part II. Theory and practice of agar fractionation. Koll. Zhur. 16: 281–286.

Guiseley, K. B., 1970. The relationship between methoxyl content and gelling temperature of agar. Carbohyd. Res. 13: 247–256.

Hirase, S., 1962. Pyruvic acid as a constituent of agar-agar. Bull. Chem Soc. Jpn 41: 626–628.

Hjerten, S., 1961. Agarose as an anticonvection agent in zone electrophoresis. Biochim. Biophys. Acta 53: 514–517.

Hjerten, S., 1962. A new method for preparation of agarose for gel electrophoresis. Biochim. Biophys. Acta. 62: 445–449.

Hjerten, S., 1964. The preparation of agarose spheres for chromatography of molecules and particles. Biochim. Biophys. Acta. 79: 393–398.

Hjerten, S., 1971. Some new methods for the preparation of agarose. J. Chromogr. 61: 73–80.

Hjerten, S., 1983. High-performance liquid chromatography matrices of agarose. Protides Biol. Fluids. 30: 9–17.

Hjerten, S., 1984. Agarose gels in HPLC separation of biopolymers. Trends Analyt. Chem. 3: 89–90.

Hjerten, S. & Y. Kunquan, 1981. High-performance liquid chromatography of macromolecules on agarose and its derivatives. J. Chrom. 215: 317–322.

Izumi, K., 1970. A new method for fractionation of agar. Agr. Biol. Chem. 34: 1739–1740.

Jimenez-Barbero, J., C. Boufar-Roupe, C. Rochas & S. Perez, 1989. Modelling studies of solvent effects on the conformational stability of agarobiose and neoagarobiose and their relationship to agarose. Int. J. Macromol. 11: 265–272.

Koch, R., 1882. Die aetiologie der tuberculose. Berl. Klin. Wochensch. 15: 221–230.

Lian Tjoa, S., 1955a. Agar-agar. German Patent 925.969.

Lian Tjoa, S., 1955b. Purification and desalting of agar-agar. German Patent 925.970.

Merck, E., 1913. Verfahren zur herstellung eines klarloslichen agar-agar. German Patent 272.145.

Parker, H. A. & M. Leikind, 1937. The introduction of agar-agar into bacteriology. J. Bact. 37: 485–493.

Patil, N. B., Kale, N. R., 1973. A simple procedure for the preparation of agarose for gel electrophoresis. Ind. J. Biochem. Biophys. 10: 160–163.

Payen, M., 1859. Sur la gelose et les nids de salangane. C. R. Acad. Sci. Paris 1859: 521–532.

Russell, B., T. H. Mead & A. Polson, 1964. A method of preparing agarose. Biochim. Biophys. Acta. 86: 169–174.

Steinitzer, F., 1912. Process of purifying agar. German Patent 269.088.

Tagawa, S., 1966. Separation of agar-agar by dimethyl sulfoxide into agarose and agaropectin. J. Shiminoseki Fish. Univ. 14: 165–171.

U.S. Army Specifications, 1935. September 10 N° 4-1041.

Zabin, B., 1966. Agarose use of D.E.A.E. cellulose to remove the anionic polysaccharides from agar. U.S. Patent 3.423.396.

Hydrobiologia **221**: 167–179, 1991.
J. A. Juanes, B. Santelices & J. L. McLachlan (eds), International Workshop on Gelidium.
© 1991 *Kluwer Academic Publishers.*

Genetic alleviation of the self-fertilization complication when hybridizing monoecious *Gelidium vagum*

John P. van der Meer & Mohsin U. Patwary
National Research Council of Canada, Institute for Marine Biosciences, 1411 Oxford St., Halifax NS, B3H 3Z1, Canada

Key words: fertility, hybridization, mutants, reproduction, spermatia, sterility

Abstract

Fundamental genetic studies were initiated for the monoecious red alga *Gelidium vagum*. Color and sterility mutants were isolated and characterized to provide genetic tools, initially to identify hybrid plants when they occurred in crosses, and secondarily to eliminate self-fertilization altogether. When fertility phenotypes were scored, rapid onset of reproduction in culture was favored by long day-length, moderately high irradiance levels from fluorescent lights, warm temperature and the addition of Tris buffer to the medium. A recessive green mutant (designated *grn1*) was characterized and used in subsequent crosses to allow a clear distinction between non-hybrid (green) and hybrid (red) offspring. Additional color mutants and a variety of reproductive mutants were also isolated and characterized. Male-sterile mutants had phenotypes ranging from apparently normal plants to those that produced no spermatia. Female-sterile mutants also included a variety of phenotypes, some plants having post-fertilization malfunctions during the development of the carposporophyte. Only a fraction of the sterility mutations have been phenotypically or genetically characterized, but some are straightforwardly inherited as stable, nuclear, single-gene defects. From the genetic recombination pattern, one female-sterile mutant may be loosely linked (39 cMorgans) to the *grn1* marker gene. Male sterility very effectively eliminated selfing without affecting the production of carpospores in crosses, thereby overcoming one of the most serious genetic difficulties in working with this monoecious species.

Introduction

Some species in the genus *Gelidium* are among the finest agar-producing plants in the world, several of which are already extensively utilized in commerce. The wild resource is limited, however, and over-harvesting and pollution of some traditional collecting sites have led to reduced yields of wild plants, while the demand for high-quality agar and agarose has continued to climb (see other papers in this volume). This strong demand has encouraged interest in algal mariculture as an alternative source of supply, and culture experiments on *Gelidium* and other agar-producing species, particularly *Gracilaria*, are in progress in a number of centers around the world. At present, the general lack of detailed knowledge about *Gelidium* biology, and the comparatively slow growth of *Gelidium* species, make their mariculture a challenging undertaking, and selection of appropriate species and strains will likely be among the important factors determining commercial success or failure; for example, the reputed slow growth of *Gelidium* clearly does not apply to all species (see other reports in this workshop).

At the Atlantic Research Laboratory of the

National Research Council of Canada, there is a long-standing interest in the biology and genetics of red algae. Thus far, the genetic research has focused on a number of dioecious red algae, including species of *Champia, Chondrus, Deva-leraea, Gracilaria, Gelidium* and *Palmaria*, for which the two gametophytic sexes can be maintained in separate cultures prior to crossing. Genetic analyses can proceed in a straightforward manner, impeded only by whatever technical difficulties exist in managing the growth and reproduction of a particular species in laboratory culture.

In contrast to the above, some species of red algae are monoecious and present a considerably greater technical challenge for genetic studies, especially when the male and female reproductive structures cannot be separated from each other, by simple excision, to prevent self-fertilization. Several *Porphyra* species, for example, have intimately intermingled patches of spermatia and carpogonia, and it was not until recently that a method based on genetic markers (color mutations) was devised which allowed the identification of hybrid conchocelis and the selection of genetically recombinant fronds (Ohme *et al.*, 1986; Miura & Shin, 1990). We became interested in testing the feasibility of using reproductive mutants, for example male sterility, to eliminate self-fertilization in monoecious species, thereby greatly facilitating the production of hybrids in crosses. Demonstration of the utility of such mutants in a model monoecious species would provide encouragement for others to seek and use similar mutants in additional species whenever they might be useful.

Gelidium vagum Okamura, a Pacific Ocean species present on the west coast of Canada, is a monoecious alga that grows and reproduces reasonably well in culture (Renfrew, 1988; Renfrew *et al.*, 1989). We have chosen it as a model system for the genetic manipulation of a monoecious species and for more general genetic studies of this important agar-producing genus. *G. vagum* produces male and female gametes more or less simultaneously and in very close proximity to each other on specialized pinnules;

consequently, selfing can be expected to occur much more frequently than cross-fertilization under normal circumstances (this was confirmed experimentally through the use of a marker-gene conditioning mutant, green, frond color). On normal plants, hybrid offspring cannot be visually distinguished from sporelings produced through self-fertilization.

To facilitate genetic research on *G. vagum*, we decided first to obtain Mendelian color mutations (which provide marker-genes to detect F_1 hybrids), and then to attempt the isolation of male-sterile lines (which augment the production of F_1 hybrids by completely eliminating selfing). Because these genetic studies would depend on the accurate classification of sterile vs fertile plants, exploratory culture studies were necessary to determine growth conditions that promoted rapid fertility in laboratory-cultured plants. In this paper we summarize our progress, including both unpublished and previously published data.

Materials and methods

Wild-type stocks of *Gelidium vagum* were obtained as sub-cultures taken from isolates collected in the Strait of Georgia, British Columbia, Canada (Renfrew *et al.*, 1989). These stock cultures were maintained in small containers, without aeration, using an unbuffered seawater medium (Table 1). Stock cultures were kept at low biomass in 100 ml plastic containers (biomass concentration $1-2 \, g \, l^{-1}$ medium, wet wt), and the medium was renewed monthly. Temperature was maintained between 15 °C and 20 °C, with a photon flux density (PFD) of $35-40 \, \mu mol \, m^{-2} \, s^{-1}$ delivered from 40 W cool-white fluorescent lamps on a 12 : 12 L : D cycle.

To establish what culture conditions would be most conducive to the rapid induction of reproductive structures, sporelings and apical cuttings of mature plants were cultured under a range of temperatures, irradiances, photoperiods, and nutrient conditions. Carpospores were taken from two independently isolated, wild-type, gametophytic stocks, #7 and #16, to obtain

Table 1. The composition of seawater medium used to culture Gelidium vagum.

Component	Concentration in the Medium[a]
NaNO$_3$	2 mM
NaH$_2$PO$_4$	0.01 mM
Na$_2$SiO$_3$	0.2 mM
MnCl$_2$.4H$_2$O	7.0 μM
Zn Cl$_2$	0.8 μM
CoCl$_2$.6H$_2$O	0.02 μM
CuCl$_2$.2H$_2$O	0.0002 μM
Na$_2$ EDTA	11.7 mM
Fe EDTA	2 μM
Thiamine-HCl	500 μg·l^{-1}
Ca-pantothenate	100 μg·l^{-1}
Nicotinic Acid	100 μg·l^{-1}
p-amino benzoic acid	10 μg·l^{-1}
Biotin	1 μg·l^{-1}
Folic Acid	2 μg·l^{-1}
Thymine	3 μg·l^{-1}
B$_{12}$	1 μg·l^{-1}
(Tris)[b]	0.1 μg·l^{-1}

[a] See McLachlan (1973) for original formulation and preparation.

[b] Added only to Tris-buffered medium.

tetrasporophyte sporelings, and tetraspores were taken from tetrasporophytes closely related to stocks #7 and #16, respectively, to obtain gametophytes for testing. Tetraspores were also taken from a third tetrasporophyte, #132. Spores were collected over a 16-h period, then dispensed into experimental dishes by pipette. For sporelings, reproduction was scored as the percentage of plants that became fertile without regard to the number of reproductive structures formed. A second set of experiments used small apical cuttings (2–3 mm) taken from the mature parent plants, including a gametophyte derived from tetrasporophyte stock #132 to determine if these responded to environmental factors in the same manner as sporelings. The parent plants were maintained under uniform, defined conditions (20 °C, 16:8 L:D, 50 μmol m^{-2} s^{-1} PFD) for at least one week prior to the taking of the cuttings. For the experiments, 10 apical tips were placed in each dish with the medium renewed

weekly. Due to the small number of apical tips per dish, reproduction was measured as the average number of reproductive pinnules per tip rather than the percentage of tips becoming fertile.

In all experiments other than those testing culture medium, plants were grown in half-strength Tris medium (Table 1). Temperature and photoperiod experiments were conducted on a single collection of spores from each isolate, cultured simultaneously in small growth chambers having different combinations of four temperatures (10, 15, 20, 25 °C) and three photoperiods (16:8, 12:12, and 8:16 L:D), lacking only the combination 8:16 at 25 °C. The PFD for these experiments was 50–55 μmol m^{-2} s^{-1}. After the first week, the number of sporelings in each dish was reduced to around 50 ($\pm$ 15). Sporeling cultures were maintained under the specified conditions for at least 6 wk. Two dishes of sporelings were used for each strain at each condition.

To determine the effect of irradiance on the development of reproductive structures, dishes of sporelings were placed under PFDs of 30, 50, 100, 150 and 200 μmol m^{-2} s^{-1}. This experiment was conducted at 20 °C and a 16:8 L:D photoperiod. Owing to technical problems during the course of these trials, no useful data were obtained for young tetrasporophytes at PFDs of 30 and 50 μmol m^{-2} s^{-1}, and the data for the gametophytes were obtained in a separate experiment.

Fertility was tested in sterile seawater without added nutrients, and in seawater culture medium, both with and without Tris buffer, at three nutrient concentrations (100%, 50% and 25%). Unbuffered medium was diluted with sterile seawater, whereas Tris-buffered medium was diluted with sterile seawater containing Tris at the same concentration as found in the medium. For these experiments, the constant conditions were: 20 °C, 16:8 photoperiod and 100 μmol m^{-2} sec^{-1} PFD (in the mid-range of irradiances found to be suitable for laboratory cultures).

Mutagenesis was performed by applying a seawater solution containing either ethyl methanesulfonate (0.2 M for 45 min at 20 °C) or N-(methyl-N'-nitro-N-nitrosoguanidine (25 μg ml^{-1} for 30 min at 23 °C) to dishes of germinating tetraspores. Following treatment, sporelings were screened for visible mutations and about 4500, derived from the NNG treatment, were put into individual containers to screen for sterility mutations (van der Meer, 1990a).

Crosses to determine the transmission of color mutants were made by taking gametangial fronds from mutant and wild-type isolates, or from two different mutant isolates, and placing these together in a disposable plastic dish containing about 30 mL of medium. Crossing dishes were placed on trays and agitated 2–3 times a week (usually overnight) on a rotary shaker.

Detection and genetic characterization of reproductive mutants were achieved in three phases. In the first phase, all self-fertile plants, constituting the bulk of the original isolates, were discarded; in the second phase, the putative reproductive mutants were co-cultured with a self-fertile green mutant (25 grn1), to determine whether they were defective in their male or their female functions; in the final phase, the heritability of some of the putative sterility mutants was tested (van der Meer, 1990a, b).

For SEM examination, fronds were fixed for several hours at 4 °C in seawater containing 5% gluteraldehyde. The samples were then thoroughly rinsed in distilled water by gradually adding distilled water to the fixative. Next, they were frozen in liquid nitrogen, freeze-dried, and gold-coated on stubs. They were stored in desiccators before and after SEM observation and photography.

Specimens for anatomical observation were treated overnight in algal preservative (4% formaldehyde and 5% glycerine in distilled water), rinsed in distilled water, imbedded in agar blocks and sectioned on a freezing microtome. Sections were stained with 1% aniline blue, washed in distilled water and mounted on slides. A drop of 70% 'Karo' syrup was added to the sections before placing a coverglass over the specimen. Slides prepared in this way were left overnight under mild pressure and then made semi-permanent by spreading clear nail polish around the periphery of the coverglass.

Results

Culture experiments

For both gametophytic and sporophytic sporelings, reproductive structures formed rapidly at 20 °C and 25 °C (Fig. 1A & B). At 20 °C a number of sporelings became reproductive within two weeks, when they were only 2–3 mm in length. During the 6-wk course of this experiment, sporelings did not become fertile at 10 °C and 15 °C. Photoperiod also influenced reproductive maturity, especially of gametophytic sporelings, with a larger percentage becoming fertile under long-day conditions (Fig. 1B). PFDs between 30 and 200 μmol m^{-2} s^{-1} appeared to have little effect on the induction of fertility in sporelings (Fig. 1E). Plants become quite pale at the higher irradiance values. Tris buffer and nutrient concentration in the medium both had pronounced effects on the fertility of sporelings (Fig. 2A & B). Reproduction was most favored in Tris-containing medium with nutrients diluted 50%. In the absence of Tris, full-strength medium promoted the most fertility.

Mature plants behaved much like the sporelings but appeared to have more latitude in becoming fertile. As for sporelings, temperature and photoperiod were strong influences on the development of reproductive structures, which formed on pinnules of some excised tips within

Fig. 1. The effect of light and temperature on the rapid induction of reproduction in *Gelidium vagum*, (A & C, 12-h day length; B, D, E & F, 16 h day length; A, B & E response for sporelings, C, D & F response for mature plants). Figs. 1E & F exhibit the response of sporelings and mature plants to various levels of irradiance. In Fig. 1E, tetrasporophyte sporeling values for 30 and 50 μmol m^{-2} s^{-1} are missing owing to technical problems during the experiment. For sporeling experiments, individual bars in the graph clusters are in the order (left to right) tetrasporophyte sporelings from line #7 and #16 followed by gametophytes from line #7, #16 and #132. For experiments of mature plants the order is tetrasporophyte tips from #7, #16 and #132, followed by gametophyte tips from lines #7, #16 and #132. For experiments using mature plant tips (both Figs. 1 and 2) standard deviations were variable but generally not more than 60% of the mean.

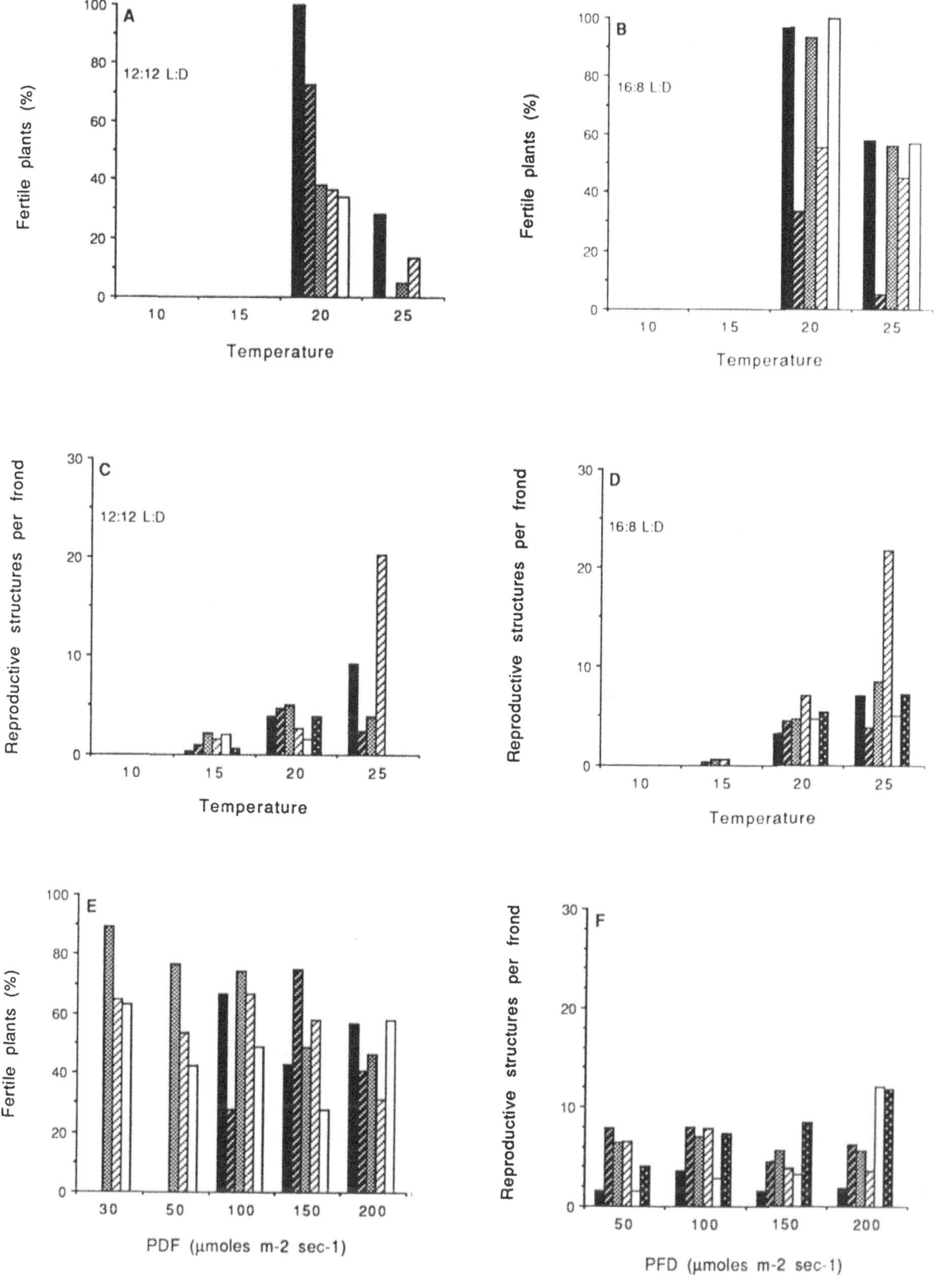

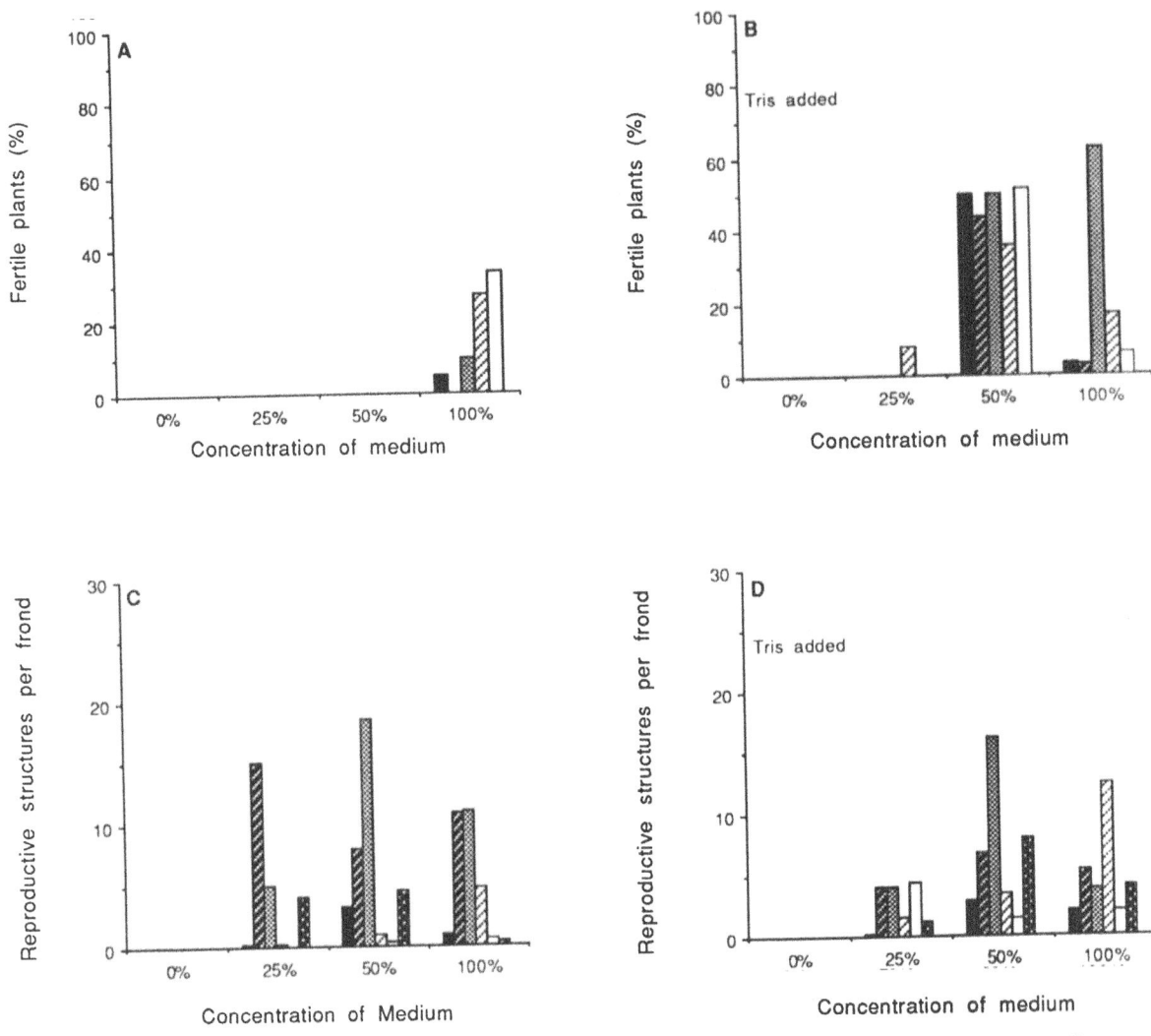

Fig. 2. The effect of medium composition on reproduction in *Gelidium vagum*, (A & B response for sporelings, C & D response for mature plants). The order of individual bars is the same as for Fig. 1.

one week under favorable conditions. Sporophytic and gametophytic fronds became most fertile at 20 and 25 °C with 12:12 and 16:8 photoperiods (Fig. 1C & D). Almost no plants became fertile under short-day conditions (not graphed). As with sporelings, irradiance did not have an effect on mature plant tips over the PFD range tested, but no really extreme levels were included (Fig. 1F). Nutrient concentration (apart from starvation in unenriched seawater) did not have a consistent effect on fertility. Although it is our experience that gametophytes have greatly-enhanced fertility (greater number of reproductive

structures per frond tip) in the presence of Tris buffer, in these particular trials the difference was not statistically significant ($p = 0.1$ in 't' test). Tetrasporophytes became fertile quite readily in unbuffered medium (Fig. 2C & D).

Color mutants

From the small population of germinating spores treated with EMS, only a few color mutants were observed, most of these as sectors on otherwise normal-looking plants. The most robust, and the first of the mutants to become reproductive, was a green frond on an otherwise red plant. The green and red components of this plant were left together and spermatia from the red portion was transferred to the green frond through gentle brushing. After several days, red spore clusters appeared in some of the cystocarps developing on the green frond. Interestingly, these cystocarps frequently contained both red and green spores (*cf.* van der Meer, 1987), directly confirming that, in *Gelidium vagum*, the gonimoblast tissue found in individual cystocarps may be derived from two or more zygotes. In subsequent weeks, the red spores developed into red tetrasporophytes, which, upon reproductive maturity, yielded a 1 red:1 green segregation of gametophytes; thus, this first mutation was transmitted in a classical 'Mendelian' fashion, as a recessive nuclear gene. It was designated *grn1* and used extensively in subsequent crosses as a marker-gene to detect hybrids.

Because it would be useful to have more than one marker-gene in the genetic 'tool box', the remaining color mutations from the EMS treatment, and several of those obtained in the subsequent mutagenesis treatment with NNG, were put into crosses, in hope of identifying recessive mutations that could complement *grn1*, and dominant mutations that could be detected visually as a marker-gene in the diploid phase. Although this part of the research was not afforded a high priority, a considerable number of the color mutants was genetically characterized.

Group 1: These mutants exhibited the same nuclear gene-transmission pattern as the *grn1* marker in crosses with normal plants, with the hybrid tetrasporophytes yielding statistically acceptable (chi square) 1:1 ratios of mutant to wild-type sporelings. Included in this group were the green mutants #27, #41, #73, #76, #77, and #78. Surprisingly, none of the non-green color mutations that were tested exhibited a Mendelian transmission pattern.

Crosses are underway to test complementation between the nuclear green mutants to determine the number of cistrons represented. The mutants #25 (*grn1*), #27, #41, and #77 failed to

complement each other in a crossing matrix and thus all appear to be alleles of *grn1*. Mutants #76, and #78 have complemented with #25 *grn1*, with red spores forming on both parents in these crosses. These results define a second cistron, *grn2* for the mutation in #78, and it remains to be seen, from crosses between #67 and #78, whether these two are alleles or represent different cistrons.

Group 2: The color mutants in this group exhibited the maternal transmission pattern found to be characteristic of non-nuclear mutations in other members of the Florideophyceae; thus these likely have defects in genes encoded in plastid DNA (van der Meer, 1990c). In crosses to #25 *grn1*, red spore clusters formed in some cystocarps of the *grn1* plant, but never on the plant being tested (*i.e.* the spores on the mutant plants always had the maternal color). Hybrid plants from red spores on the #25 *grn1* plant produced F_1 progenies in which the *grn1* marker segregated as expected, but the mutation from the male parent did not reappear. Counted within this group are 30 green and bright-green mutants, 5 yellow and yellow-brown mutants, and 7 pink mutants.

Reproductive mutants

Many reproductive mutants were isolated after treatment of germinating spores with NNG. Some of the results from the initial genetic characterization of these mutants have already been published (van der Meer, 1990a, b); however, for the sake of completeness here, these data are included with new data in Table 2.

Among the apparently male-sterile mutants, isolates #59 and #66 were found to be poorly fertile rather than completely male sterile (van der Meer, 1990a). Mutant #97 appeared to be the same sort of poorly-fertile mutation because the skew towards fertile plants in the F_1 progeny was pronounced; however, unlike #59 and #66, this isolate has not produced sporelings through self-fertilization. A large group of mutants (#64,

Table 2. The inheritance of reproductive mutants of *Gelidium vagum.*

Cross	Phenotype of F_1 Gametophytes	
	Red plants	Green plants (grnl)
Putative male-sterile mutants		
66 ms × 25 grnl[V]	20 fertile	25 fertile
59 ms × 25 grnl[V]	15 fertile : 8 sterile	17 fertile : 10 sterile
97 ms × 25 grnl	20 fertile : 5 sterile	16 fertile : 9 sterile
64 ms × 25 grnl	10 fertile : 16 sterile	12 fertile : 13 sterile
89 ms × 25 grnl[T]	11 fertile : 14 sterile	10 fertile : 15 sterile
90 ms × 25 grnl	12 fertile : 10 sterile	12 fertile : 11 sterile
90 grnl ms F_1 × 7 wt	16 fertile : 9 sterile	11 fertile : 13 sterile
92 ms × 25 grnl[T]	12 fertile : 13 sterile	14 fertile : 11 sterile
94 ms × 25 grnl[V]	15 fertile : 10 sterile	11 fertile : 14 sterile
100 ms × 25 grnl[T]	12 fertile : 13 sterile	11 fertile : 14 sterile
117 ms × 25 grnl	9 fertile : 11 sterile	12 fertile : 8 sterile
148 ms × 25 grnl	15 fertile : 10 sterile	13 fertile : 12 sterile
192 ms × 25 grnl	11 fertile : 15 sterile	10 fertile : 13 sterile
204 ms × 25 grnl	10 fertile : 10 sterile	14 fertile : 6 sterile
210 ms × 25 grnl	9 fertile : 11 sterile	8 fertile : 12 sterile
216 ms × 25 grnl	12 fertile : 13 sterile	13 fertile : 12 sterile
99 ms × 25 grnl[T]	7 fertile : 18 sterile	6 fertile : 19 sterile
211 ms × 25 grnl	7 fertile : 17 sterile	9 fertile : 16 sterile
188 ms × 25 grnl	12 fertile : 13 dwf st	20 fertile : 5 gwf st
Putative female sterile mutants		
25 grnl × 52 fs[V]	24 fertile	25 fertile
25 grnl × 63 fs	20 fertile	20 fertile
25 grnl × 225 fs[T]	25 fertile	25 fertile
25 grnl × 91 fs	19 fertile : 11 sterile	17 fertile : 12 sterile
25 grnl × 101 fs	16 fertile : 9 sterile	15 fertile : 10 sterile
25 grnl × 250 fs	16 fertile : 4 sterile	13 fertile : 7 sterile
25 grnl × 96 fs	8 fertile : 12 sterile	11 fertile : 9 sterile
25 grnl × 297 fs[V]	10 fertile : 8 sterile	8 fertile : 10 sterile
25 grnl × 261 fs	16 fertile : 23 sterile	27 fertile : 13 sterile
261 grnl fs F_1 × 7 wt	11 fertile : 9 sterile	8 fertile : 12 sterile

[V] Presented at XIIIth International Seaweed Symposium, Vancouver 1989.

[T] Presented at First International Marine Biotechnology Conference, Tokyo 1989.

#89, #90, #92, #94, #100, #117, #148, #192, #204, #210, and #216) yielded a Mendelian ratio of 1 fertile : 1 sterile in the F_1 gametophytic progenies, and thus appear to have stable, nuclear, single-gene mutations. In mutant #99, the segregation ratio of fertile-to-sterile plants was clearly 1 : 3, and arises from the segregation of two sterility mutations present simultaneously in this isolate (van der Meer, 1990b). The male sterility of #99 masks a second mutation (abortive cystocarps) expressed only after fertilization). Mutant #211 exhibited a segre-

gation ratio similar to that of #99 and may also result from a double defect; however, a second sterility phenotype was not identified, and further characterization would be necessary to establish this point. During the genetic characterization of male-sterile isolate #188, it became apparent that the sterility was completely associated with the segregation of an altered, somewhat-dwarfed growth-habit. The large skew for this morphology in the green plants of the F_1 gametophytic progeny must have resulted from sampling.

Mutant #52 initially appeared to be female sterile, but, on re-testing, was found to undergo rare selfing, producing abnormal, dwarfed tetrasporophytes (van der Meer, 1990a). The next two mutants listed in Table 2 (#63 and #225) are curious in that they yielded only normally-fertile offspring in crosses, while remaining entirely self-sterile, despite producing many reproductive branches under conditions that should allow selfing. Mutants #91, #101 and #250 had a majority of fertile plants in the F_1 gametophytic progeny, for both the red and green subgroups; however, with relatively small numbers, the results are statistically within the acceptable fluctuations expected for a 1 : 1 ratio (except the red group for mutant #250, which exceeds the 1% chi-square limit). While it seems likely that most such mutants would eventually prove to be poorly-fertile isolates, these three mutants have not selfed in culture, and thus may actually have gene defects conferring full sterility. Only further testing could resolve the matter. Three of the female-sterile mutants, #96, #261 and #297, gave a good Mendelian ratio of 1 fertile : 1 sterile plant in crosses, indicating that they are caused by stable, single-factor defects located on the nuclear chromosomes. One of these, #261, had a segregation where parental phenotypes (red sterile and green fertile) far exceeded those with recombinant phenotypes (red fertile and green sterile) suggesting a linkage of about 28 cMorgans between *grn 1* and the sterility mutation in #261. It appears that this distribution may have been partially a fluke attributable to small sample-size, as results from further testing showed less evidence of linkage, with the combined results giving a linkage value of only 39 cMorgans.

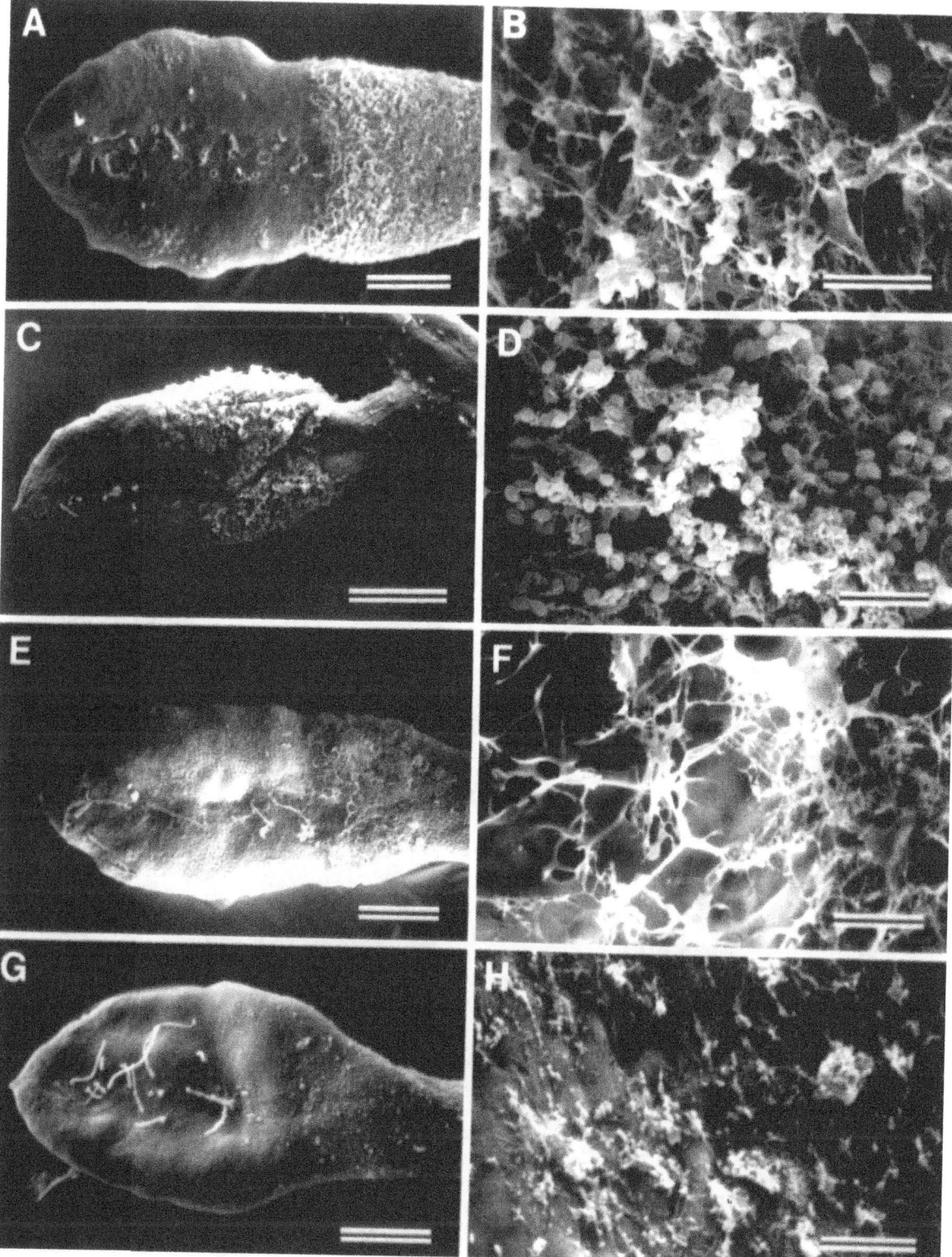

Fig. 3. Phenotypes of male-sterile mutants as revealed by scanning electron microscopy, with normal plants for comparison. Fig. 3 A & B = wild type; C & D = mutant #310; E & F = mutant #159; G & H = mutant #136.

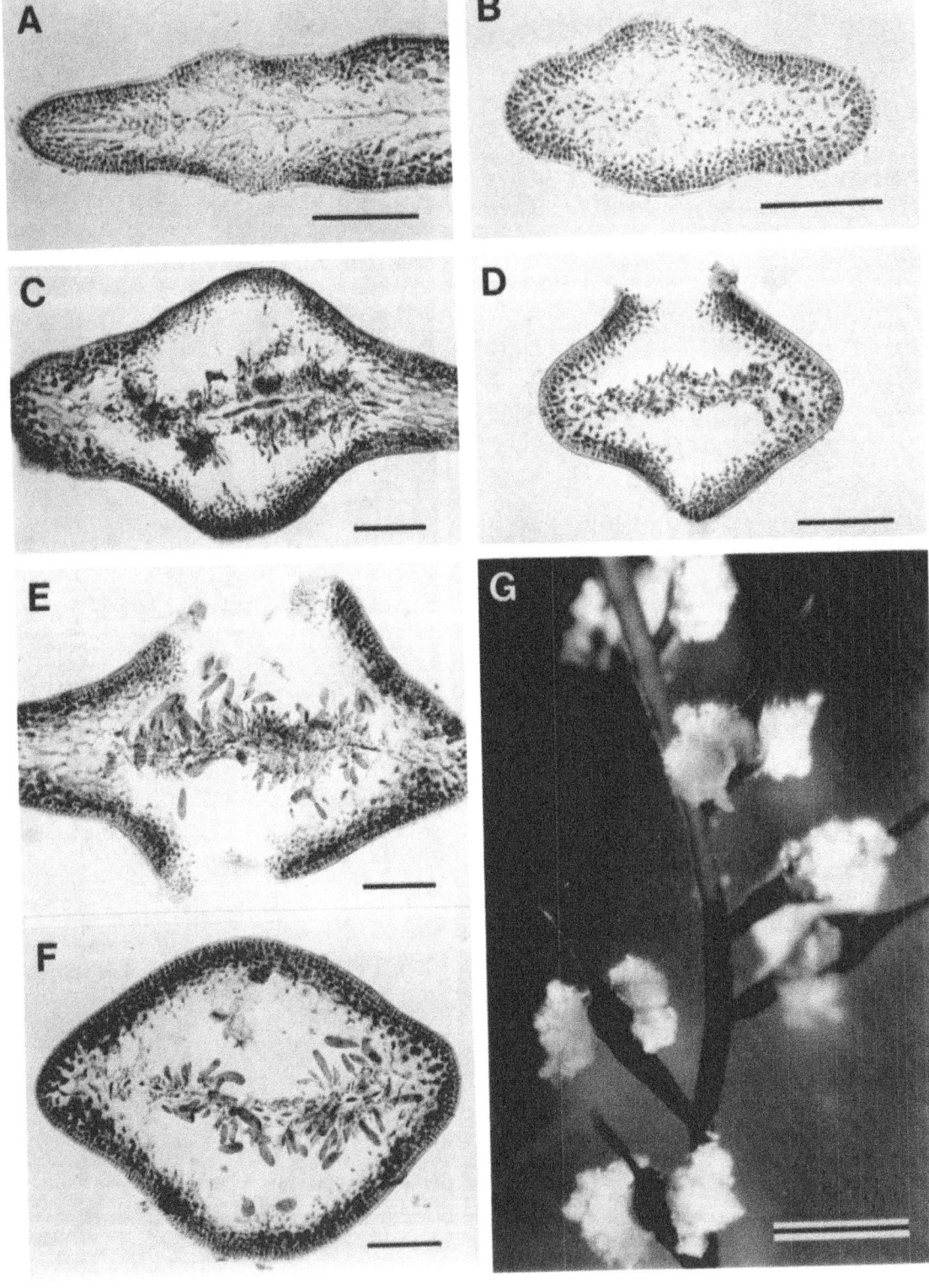

Phenotypic characterization of the sterile mutants has only just begun, but differences are clearly evident for those that have been examined. Male-sterile mutant #310 (Fig. 3C & D) makes abundant spermatia and cannot be phenotypically distinguished from wild-type (Fig. 3A & B) even by SEM examination. Mutant #159 produces surface-eruptions, superficially resembling normal spermatangial zones (Fig. 3E), but no spermatia can be detected (Fig. 3F). No spermatangial zone is visually evident on male-sterile mutant #136 (Fig. 3G), and observations of the surface by SEM reveal only a smooth vegetative surface (Fig. 3H).

Female-sterile plants also have variable phenotypes. All that have been examined by SEM thus far produced trichogynes (data not shown), but it is not possible to tell whether these are functional. Mutant 305 undergoes extremely limited development of the cystocarp, forming small ostioles, but with little swelling of the fronds. In sectioned material of this mutant, only the axial filament (Fig. 4A) dividing the reproductive area into upper and lower locules and limited development of nutritive filaments around axial and periaxial cells (Fig. 4B) is seen. Gonimoblast tissue was not recognized and probably does not form. There is far more development of the cystocarp and internal tissues for mutant 297 (Fig. 4C & D). The reproductive area is divided into two locules, with extensive development of nutritive and other filaments, but differentiation is disrupted before the production of recognizable carposporangia which are readily apparent in wild-type (Fig. 4E). Another example of developmental problems arising late in the development of the cystocarp is seen for mutant 152 (Fig. 4F) which forms normal-looking cystocarps but releases masses of whitish, elongate, inviable cells rather than carpospores.

Discussion

The culture experiments undertaken in connection with this study were initiated solely to obtain some guidelines for our work on the selection and characterization of sterile plants, because we discovered, soon after beginning the work, that inappropriate choice of culture conditions would make classification of fertility mutants nearly impossible. Results obtained in these trials were sufficiently consistent to indicate general conditions promoting the rapid onset of reproduction in G. vagum in culture, namely: moderate fluorescent lighting, long photoperiod, warm temperature and the presence of Tris buffer in the growth medium being used; however, the results were also sufficiently variable to point out that more extensive experimentation would be necessary to establish the effects of these variables on G. vagum. These culture results cannot be extrapolated to what happens under natural conditions in the field. There are indications of differences in the fertility of various strains, in the behavior of the two phases and in the responses of sporelings and mature plants, but the significance of these differences cannot be established from our limited data; furthermore, when these apparently most favorable environmental conditions were applied to progeny of crosses, there were always some offspring that remained vegetative, or took a long time to become fertile. This lack of uniform response under largely defined conditions is often observed for algal lines newly introduced into culture, and indicates that there are additional unidentified, probably genetic, factors that influence the fertility of these plants.

Mutagenesis of G. vagum was readily achieved with both EMS (only a small number of plants treated) and NNG, with the latter appearing to be much more effective, bearing in mind, however, that only a single treatment was evaluated for each chemical. The frequency of mutants obtained

◀ *Fig. 4.* Reproductive anatomy of female-sterile mutants. Fig. 4 A & B = reproductive zone of mutant #305 in longitudinal and cross-section; Fig. 4 C & D = reproductive zone of mutant #297 in longitudinal and cross section; Fig. 4 E = reproductive zone of wild type in longitudinal section; Fig. 4 F = mutant #152 releasing elongate, nearly colorless cells produced in the cystocarp. Ontogeny of these abnormal cells was not determined.

with NNG was higher than any values previously obtained for EMS treatments of *Gracilaria* and other species (van der Meer, 1979; Kehoe & van der Meer, 1990). The majority of *G. vagum* color mutations induced by the NNG treatment were transmitted in a non-nuclear rather than Mendelian fashion, an outcome that has not been observed for mutagenesis treatments of other red algal species. With results from only a single treatment, it is not possible to tell if this represents a typical or a unusual outcome for NNG mutagenesis of *Gelidium*. The effectiveness of NNG mutagenesis, and the balance of nuclear to non-nuclear mutations, may be dependent on the time of day the treatment is applied (*i.e.* stage of the replication cycle for nuclear and plastid DNA).

One of the nuclear color-mutations (*grn1*) was obtained and characterized early in the study, and its value as a color marker for the rest of the investigation would be hard to exaggerate. It was almost essential for characterizing the putative sterile mutations, allowing us to determine relatively quickly, and clearly, whether the defective plants retained any made or female reproductive capability. It also allowed us to characterize other color mutations more efficiently as hybrid plants could be identified, regardless of whether the mutant being tested had Mendelian or non-Mendelian transmission. It is now clear that color markers will always play a central role in genetic studies of *G. vagum*, even after sterility mutations become incorporated into the crossing lines.

The additional cistron, *grn2*, identified in the partial complementation analysis of the green Mendelian mutants, should be a useful addition to the 'genetic toolbox' for *G. vagum*. It will now be possible to put complementing mutations in the two parents of a cross. It might initially seem odd that no Mendelian color mutations other than green were isolated; however, this is understandable, considering the preponderance of non-nuclear mutations obtained in the NNG treatment, and the small number of non-green mutants characterized. It will be left for future studies to identify additional marker mutations, including one or more dominant ones.

It is noteworthy that all the non-nuclear mutations characterized for *G. vagum* exhibited strictly maternal inheritance in crosses (*i.e.* there was no evidence of biparental inheritance or transmission of the mutation through spermatia). This is consistent with results obtained for similar mutations in other species of the Florideophyceae, and appears to reflect the transmission behavior of plastids, and consequently, of plastid-encoded genes (van der Meer, 1990c).

Of the plants tested for fertility, about 5% had mutations that rendered them sterile or nearly so. It thus appears that a substantial number of nuclear genes are involved in the differentiation and function of the reproductive structures. In view of the fact that this NNG mutagenesis treatment produced a modest yield of Mendelian pigmentation mutants, it is possible that the frequency of nuclear reproductive mutations obtained in this experiment may actually be a low estimate. The high yield of sterility mutations should be encouraging for anyone interested in developing male sterile strains in other monoecious species.

Initial genetic and phenotypic characterization of the large number of reproductive mutants has allowed us to group many of them into general categories: vegetative, completely sterile, male-sterile, female-sterile, carposporophyte-lethal and a few 'exotic' reproductive mutations (van der Meer, 1990b). This having been done, further study must now be directed toward consolidating the results for a selected group of the more interesting mutants. There is no compelling reason to fully characterize all of the mutants as only one or two reliable male-sterile mutations are needed as hybridization tools. Far too many sterility mutants were obtained to examine them all in detail or even to maintain the stocks for an exended period of time; nevertheless, a good collection will be preserved in storage against the possibility that some will be useful for future studies, for example, on the genetic control of reproductive development.

This experimentation on sterility mutants of *G. vagum*, while only a model study, has clearly demonstrated the feasibility of using mutations to eliminate self-fertilization in monoecious species. This is not only important for our future genetic studies on this species, but also has important implications for future selective breeding and growth of any monoecious species used in mariculture. Male-sterility genes similar to the ones described here could be used to facilitate crosses in genetic-improvement programs. It might also be possible to use male sterility to prevent fertilization of fronds during mariculture which might lead to improved growth rates. In *Porphyra*, complete or partial male sterility might be useful to delay the erosion of cultivated fronds by preventing fertilization of reproductively mature fronds, thereby extending the time during which good-quality fronds can be harvested.

Acknowledgements

We gratefully acknowledge the technical assistance of J. Osborne and the staff of the IMB Aquaculture Research Station. In the early stages of our interest in *Gelidium*. Dr. R. Foreman kindly provided subcultures of several *Gelidium* species in his collection. Dr. D. Renfrew provided several wild-type isolates of *G. vagum*, some of which were used as stocks for these studies.

Published as NRC No. 31969.

References

Kehoe, D. M. & J. P. van der Meer, 1990. Genetics of *Champia parvula* (Rhodymeniales, Rhodophyta): Induction, characterization and mapping of mutants. Bot. mar. 33: 393–399.

McLachlan, J., 1973. Growth medium-marine. In Stein J. R., (ed.), Handbook of Phycological Methods. Cambridge University Press, London: 25–51.

Miura, A. & J. A. Shin, 1990. Crossbreeding in cultivars of *Porphyra yezoensis* (Bangiales, Rhodophyta). In Miyachi S., I. Karube & Y. Ishida (eds) Current Topics in Marine Biotechnology. The Japanese Society for Marine Biotechnology, Tokyo: 209–212.

Ohme, M., Y. Kunifuji & A. Miura, 1986. Cross experiments of the color mutants in *Porphyra yezoensis* Ueda. Jpn. J. Phycol. 34: 101–106.

Renfrew, D. E., 1988. Gelidiales (Rhodophyta, Red Algae) in British Columbia and Northern Washington: Taxonomy, Morphology, Development and Life History. Ph.D. Thesis, Dept. of Botany, University of British Columbia, Vancouver, Canada.

Renfrew, D. E., P. W. Gabrielson & R. F. Scagel, 1989. The marine algae of British Columbia, northern Washington and southeast Alaska: division Rhodophyta (red algae), class Rhodophyceae, order Gelidiales. Can. J. Bot. 67: 3295–3314.

van der Meer, J. P., 1979. Genetics of *Gracilaria* sp. (Rhodophyceae, Gigartinales). V. Isolation and characterization of mutant strains. Phycologia 18: 47–54.

van der Meer, J. P., 1987. Using genetic markers in phycological research. Proc. int. Seaweed Symp. 12: 49–56.

van der Meer, J. P., 1990a. Isolation and genetic characterization of self-sterile mutants in a monoecious red alga *Gelidium vagum*. Proc. int. Seaweed Symp. 13: 389–395.

van der Meer, J. P., 1990b. Self-sterile mutants of *Gelidium vagum*. In Miyachi S., I. Karube & Y. Ishida (eds). Current Topics in Marine Biotechnology. The Japanese Society for Marine Biotechnology, Tokyo: 205–208.

van der Meer, J. P., 1990c. Genetics. In Sheath R. G. & K. M. Cole (eds) Biology of the Red Algae, Cambridge University Press, New York: 103–121.

Hydrobiologia **221**: 181–194, 1991.
J. A. Juanes, B. Santelices & J. L. McLachlan (eds), International Workshop on Gelidium.
© 1991 *Kluwer Academic Publishers.*

Actual, potential and speculative applications of seaweed cellular biotechnology: some specific comments on *Gelidium*

G. Garcia-Reina, J. L. Gómez-Pinchetti, D. R. Robledo & P. Sosa
Marine Plant Biotechnology Laboratory, University of Las Palmas, Box 550, Las Palmas, Canary Islands, Spain

Key words: callus, cell culture, domestication, protoplast, tissue culture

Abstract

Cellular biotechnology is a promising application in the propagation and selection of superior strains of seaweeds. Although axenic cultures, organogenetic tissue cultures, vegetative micro-propogation, callus induction and high yields of agar from calli have been described for several species of *Gelidium*, a number of basic problems remain to be solved. These include standardized methods for obtaining axenic cultures, identification of requirements for organic nutrients, PGR's, cellular disorganization and reorganization, somaclonal variation and somatic incompatibilities. Future progress in seaweed biotechnology will depend on the resolution of many of these problems.

Introduction

Cellular biotechnology was introduced into the phycological field at the beginning of the 1980's (Gibor, 1980), and is considered important in future development of seaweed cultivation. The four major techniques in seaweed cellular biotechnology, tissue-, callus-, cell- and the protoplast-culture, offer means in understanding seaweed domestication (propagation and selection) germplasm storage and for production of biomass. We first discuss seaweed cellular biotechnology as highly effective techniques, and then consider some still unresolved questions, which limit their potential and actual applications.

Seaweed biotechnology

Vegetative micropropagation

The possibility of using small thallus fragments (tissue culture) to establish cultures, instead of spores, exploits the *organogenetic potential* of explants. Chen and Taylor (1978) described the regeneration of whole plants from decorticated explants of *Chondrus crispus* Stackh. Waaland (1982) and Sylvester and Waaland (1983) obtained plants following inoculation of finely-chopped fragments of the thallus of *Gigartina* into braided rope. Their data indicated that the best strategy for commerical purposes is to maximize the amount of propagules per donor plant. During the early 1980's, similarly successful experiments were done with *Gelidium nudifrons* Gardner and *G. robustum* (Gardner) Hollenberg & Abbott (Polne-Fuller, 1988).

In China, Luqin *et al.* (1988), described a technique, using small explants of *Gelidium pacificum* Okamura, that permitted the firm attachment of small explants to ropes through the development of adventitious rhizoids and the regeneration of new plantlets. Garcia-Reina *et al.* (1988a) reported on organogenetic development from wounded explants of *Gelidium versicolor* (Gmel.) Lamouroux, and Robaina *et al.* (1990a) discussed the effects of solid culture medium and its osmolality on organogenetic potential.

If the best commercial strategy for vegetative micropropagation is to maximize propagules per donor-plant ratios, the logical strategy would be to reduce the plant to single, somatic cells, using these as inocula. 'Somatic spores' implies totipotentiality, as pointed out by Saga *et al.* (1978). Single somatic cells, obtained by enzymatic digestion of the thalli of *Porphyra* and *Ulvaria*, have been inoculated onto ropes and, from these, attached plants have developed (Polne-Fuller *et al.*, 1984; Kapraun, 1987; Kapraun & Sherman, 1989). This technique has been tested in commercial *Porphyra* farms (Mumford, 1987).

Somatic-cell isolation in seaweeds is not a difficult task as it has been achieved by partial enzymatic digestion of thalli (in *Porphyra*: Tang, 1982; Chen, 1986; Wang *et al.*, 1986, 1987a, 1987b; Tait *et al.*, 1990; Wang & Yan, 1990; *Sphacelaria*: Ducreux *et al.*, 1988), by cell-wall regeneration from protoplasts (*Porphyra*: Chen, 1989) or spheroplasts (*Undaria*: Kapraun & Sherman, 1989), by mechanical disruption of thalli (*Prasiola*: Schiff *et al.*, 1972; Bingham & Schiff, 1973; *Griffithsia*: Duffield *et al.*, 1972; *Porphyra*, Tait *et al.*, 1990), or of frozen thalli (*Porphyra*: Zhao & Zhang, 1984), by mechanical disruption of calli (*Laminaria*: Saga *et al.*, 1978; McLean & Connolly, 1989; *Alaria*: Saga *et al.*, 1978; *Undaria*: Zhang, 1982) or by combinations of mechanical and enzymatic treatments of both thallus and callus (*Pterocladia*: Liu & Gordon, 1987).

There are advantages to vegetative micropropagation over conventional propagation from spores. It is unnecessary to know or control the life-history of the species; 'seedlings' can be produced in quantity, both when they are needed and from a small number of donor plants, requiring cheap storage and handling.

These advantages are directly related to propagation, while the techniques are applicable to selection. Mono-phase crops (*i.e.* monoculture of *Gelidium* sporophytes or of male or female gametophytes) can be produced. Mutant plants (natural or induced) can be propagated rapidly, even if sterile, and even if the selected phenotype has a non-genetic basis (epigenetic) but is stable.

Selection

Tissue culture of seaweeds is based on natural genetic variability of the somatic cells; for example, *Gracilaria* species (van der Meer, 1986) have been shown to have a high degree of somatic recombination. Genetic diversity among cells of the thallus, normally masked by more common non-variant cells, is exploited, as described for higher plant tissue culture (D'Amato, 1978). This 'island effect' (retention of pigmentation, growth, *etc.*, by some cells or cell aggregates) is a common phenomenon in seaweed tissue culture (Fries, 1980; Polne-Fuller *et al.*, 1986; Garcia-Reina *et al.*, unpubl. data).

Polne-Fuller and Gibor (1986) have described the isolation and growth to callus on solid medium of a few living cells from *Porphyra* tissue cultures contaminated by *Pythium*. The callus was enzymatically digested to cells and small clumps of cells and co-cultured with *Pythium*, where healthy, normal callus developed. Although no plant regeneration was reported, this simple experiment shows how quickly and effectively tissue-culture techniques can yield results by selection.

Selected strains of *Sargassum* and *Enteromorpha* (showing differences in temperature tolerance, orphology and vitamin auxotrophy) have been established through tissue culture (Polne-Fuller *et al.*, 1986), apparently owing to intercellular variability among explants. Tissue culture (*i.e.* callus culture) is not, however, considered the best technique for cultivar improvement. The application of callus culture for selection is based on a process referred to as somaclonal variation by Larkin and Scowcroft (1981). This term covers genomic, chromosomic, genetic and even epigenetic variations associated with or induced by the disorganized callus state, and is also expressed in the reorganized plants (Karp, 1989). Callus is associated with a high degree of genetic (or epigenetic) variability. Fang *et al.* (1983) have described genomic alteration in the callus cells of *Laminaria*, and Lee (1986), varying levels of polyploidy.

Callus culture has many advantages over conventional selection procedures. Within a small space (*e.g.* Petri dish incubated in growth chamber), large numbers of 'plants' (assuming cell = plant), having natural or induced genetic variabilities, can be screened. The stability of selected phenotypes can be determined in a short time, as the following generation is the next mitotic division, not the new adult sporophytic phase as in *Laminaria*. Any selected phenotype can be easily propagated to large amounts (using cell culture), even if it is an epigenetic but stable character.

In seaweeds, obtaining haploid cultures through cell culture is easy, whereas with higher plants the only way to obtain haploid cultures is by anther or pollen culture (Han & Hongyuan, 1986). For the latter group of plants, diploid cells can proliferate and few haploid cells remain viable, so that the efficiency is low. In seaweeds, haploid cultures depend only on the type of explant (gametophyte) chosen. Thus, exploitation of recessive information, haploid somatic recombination and gametoclonal variation (somaclonal variation in haploid cells, Evans *et al.*, 1984), can be readily employed as can the production of fully-homozygous cells and plants by colchicine treatment or by spontaneous chromosome doubling. All such characteristics (haploid culture, gametoclonal variation, chromosome doubling) have been demonstrated and utilized for *Laminaria*-improvement programs (Fang, 1984; Wu & Lin, 1988). Reports of regeneration of variant plants (altered morphology and life-history: monospore producing) from isolated cells or protoplasts of *Porphyra* (Tang, 1982; Fujita & Migita, 1985; Chen, 1987) and *Monostroma*

gametophytes, with different developmental patterns and apogamy (Saga & Kudo, 1989), may be explained by differential regenerative potential or different variant cell strains pre-existing in the haploid thallus.

Selection through callus culture has some drawbacks, as the techniques allow selection only of cellular-based characteristics. It is impossible to screen for stipe-length or frond-thickness at the cellular level or for sulphate content in the cell walls of *Gelidium*. Monoclonal antibodies (Vreeland *et al.*, 1987) can be useful markers. There are also possibilities of selecting variant cells with characteristics expressed only at the cellular level and not in fully-differentiated plants.

The regeneration of plants from calli can yield surprises, some of which could be useful, even in the absence of directionally selective pressures. Several reports indicate inadvertent selection resulting from seaweed somaclonal variation. Yan (1984) described plants, reorganized from calli of *Laminaria* and *Undaria*, with more rapid growth and tolerance of high temperatures for a longer period than the normal sporophytes. Garcia-Reina *et al.* (1988b) selected two types of callus of *Laurencia* obtained from the same plant and similar in appearance (pigmentation, texture). These differed markedly in organogenetic potential (number of plantlets/callus) and growth-rate of the regenerated plants.

A major advantage of selection using cellular biotechnologies is in protoplast culture. This allows for somatic hybridization (through the breakdown of sexual barriers, through interspecific or intergeneric fusion of somatic cells, and exclusively maternal extrachromosomal inheritance). There is also improved efficiency of *genetic transformation* by the removal of physical barriers, and host-vector recognition specificities through elimination of the cell wall.

Protoplast isolation and regeneration of whole plants have been successful (Table 1). The same techniques as applied to higher plant protoplasts induce somatic hybridization in seaweeds. These include electrofusion (*Enteromorpha*: Saga *et al.*, 1986; *Porphyra*: Fujita & Saito, 1989) and PEG (*Ulva* × *Monostroma*: Zhang, 1982; *Enteromor-*

184

pha, Saga *et al.*, 1986; *Porphyra*: Fujita & Migita, 1987; Fujita & Saito, 1989; *Gracilaria*: Cheney, 1989). True somatic hybrids have been obtained in microalgae (*Chlamydomonas*: Matagne *et al.*, 1979; *Dunaliella* × *Porphyridium*: Lee & Tan,

1988). Recently, Fujita & Migita (1987), with wild × mutant *Porphyra* and Kapraun (1987) *Enteromorpha*, in anatomically-simple species, and Cheney (1989), with anatomically more complex species (*Gracilaria tikvahiae* McLachlan ×

Table 1. Seaweeds from which protoplasts have been isolated, types of isolation (Iso; M = mechanical, F = enzymes obtained from phycophages, C = commercial enzymes, B = enzymes obtained from bacteria, A = enzymes obtained from amoeba) and types of further development (Dev; R = cell wall regeneration, R = without cell wall regeneration, R + = plant regeneration, Ca = callus formation, Ca + = callus and plants).

	Iso.	Dev.	
Chlorophyta			
Bryopsis plumosa	M	R +	Tatewaki & Nagata 1970
Enteromorpha linza	C	R +	Fujita & Migita 1985
Enteromorpha linza	C	R	Saga 1984
E. intestinalis	C	R +	Millner *et al.* 1979
E. intestinalis	C	R +	Saga *et al.* 1986
Monostroma angicava	C + F	R +	Zhang 1982
Monostroma angicava	C	R	Saga & Kudo 1989
Monostroma nitidum	C	R +	Fujita & Migita 1985
Monostroma zoostericola	C	R +, Ca	Saga *et al.* 1986
Ulva conglobuta	C + F	R +, Ca	Reddy *et al.* 1989
Ulva fasciata	C + F	R +	Reddy *et al.* 1989
Ulva linza	C + F	R +	Zhang 1982
Ulva pertusa	C	R +	Fugita & Migita 1985
Ulva pertusa	C	R +	Saga 1984
Ulva pertusa	C	R +, Ca	Fujimura & Kajiwara 1989
Ulva pertusa	C + F	R +	Reddy *et al.* 1989
Ulvaria oxysperma	C	R +	Kapraun & Sherman 1989
Uronema gigas	F		Gabriel 1970
Phaeophyta			
Fucus distichus	C + F	R +	Kloareg *et al.* 1988
Laminaria	F		Saga & Sakai 1984
Laminaria digitata	F		Butler & Evans 1988
Laminaria digitata	C + F	R	Butler *et al.* 1989
Laminaria japonica	F	R –	Saga & Sakai 1984
Laminaria saccharina	F		Butler & Evans 1988
Laminaria saccharina	C + F	R	Butler *et al.* 1989
Macrocystis pyrifera	C + F		Saga *et al.* 1986
Macrocystis pyrifera	C + F	R, Ca	Kloareg *et al.* 1989
Macrocystis pyrifera	M + F	R –	Davison & Polne-Fuller 1990
Sargassum echinocarpum	C + F		Fisher & Gibor 1987
Sargassum muticum	F	R	Neushul 1984
Sargassum muticum	C + F		Saga *et al.* 1986
Sargassum muticum	C + F		Fisher & Gibor 1987
Sargassum muticum	A		Polne-Fuller & Gibor 1987c
Sargassum polyphyllum	C + F		Fisher & Gibor 1987
Sphacelaria	C + F	R +	Ducreux & Kloareg 1988
Undaria pinnatifida	B	R –	Fujita & Migita 1985
Undaria pinnatifida	F	R	Tokuda & Kawashima 1988

Table 1. (Continued)

	Iso.	Dev.	
Rhodophyta			
Chondrus crispus	B	R –	LeGall *et al.* 1989
Chondrus crispus	B		Smith & Bidwell 1989
Gracilaria lemaneiformis	C	Ca	Cheney *et al.* 1986
Gracilaria lemaneiformis	C		Björk *et al.* 1990
Gracilaria secundata	C		Björk *et al.* 1990
Gracilaria tenuistipitata	C		Björk *et al.* 1990
Gracilaria tikvahiae	C	Ca	Cheney *et al.* 1986
Gracilaria verrucosa	C		Björk *et al.* 1990
Palmaria palmata	C	R	Liu 1989
Porphyra	F	R +	Liu *et al.* 1984
Porphyra	C + B		Fujita & Saito 1989
Porphyra haitanensis	C + F		Wang *et al.* 1986
Porphyra leucosticta	C + F	Ca, R +	Chen 1987
Porphyra linearis	C + F	R +, Ca	Chen *et al.* 1988
Porphyra maculosa	C + F		Waaland & Dickson 1987
Porphyra nereocystis	C + F	Ca	Waaland & Dickson 1987
Porphyra perforata		R +	Saga *et al.* 1986
Porphyra perforata	C + F	R +, Ca +	Polne & Gibor 1984
Porphyra perforata	C + F		Waaland & Dickson 1987
Porphyra pseudolanceolata	C + F		Waaland & Dickson 1987
Porphyra suborbiculata	F	R	Tang 1982
Porphyra yezoensis	F	R +	Araki *et al.* 1987
Porphyra yezoensis	F	R +, Ca +	Polne *et al.* 1984
Porphyra yezoensis	F	R +, Ca +	Polne & Gibor 1984
Porphyra yezoensis	B	R –	Saga & Sakai 1984
Porphyra yezoensis	C + F	R +	Fujita & Migita 1985
Porphyra yezoensis	F	R +	Saga *et al.* 1986
Porphyra yezoensis		R +	Dai 1987

G. chilensis Bird, McLachlan & Oliveira), have described the recovery of somatic hybrids. These results suggest that the production of 'agarophytic corn' (*Gelidium* × *Zea mays*), of 'agarageenan' (*Gelidium* × *Eucheuma*) or other imaginative hybrids, can be realized.

There are a few somewhat speculative and distantly related reports on genetic engineering in seaweeds. Thes include *Agrobacterium tumefaciens*, used genetically to engineer algae, at least in *Chlamydomonas* (Ausich, 1983); the possible existence of *Agrobacterium*-like bacteria that promote the induction and growth of seaweed tumours, galls and calli (Cantacuzene, 1930; Apt, 1988; Tsekos, 1982; Garcia-Reina *et al.*, 1988a); DNA-transferring parasitic seaweeds, possibly

usable as vectors (Goff & Coleman, 1984); chromosomal location and molecular biological regulatory studies of alginate-related genes in *Pseudomonas*, with characteristics similar to those in *Fucus* alginates (Deretic *et al.*, 1987); development of cloning vectors in photosynthetic microorganisms (Chauvat *et al.*, 1988).

Germplasm storage

Tissues and callus cultures can be highly efficient techniques for germplasm storage in selection programs or 'seed' storage for propagation as they avoid expensive whole-plant management. The longest, 5-year-old culture of seaweed calli

(Fang, 1983; Polne-Fuller & Gibor, 1987b), and culture of somatic cells (Chen, 1989) parallels the immortality shown by higher plant-cell cultures (Gautheret, 1985). Cryopreservation will require even less handling, and van der Meer and Simpson (1984) have described techniques for *Gracilaria* tissue.

Production of biomass or metabolites

The cultivation of tissues, calli or cells in photo-bioreactors, bioreactors or immobilized bioreactors are alternative techniques for direct production of biomass or of economically-significant metabolites. The best example (Misawa, 1977) is described in Japanese patent No. 74-101561 obtained by Nakamura in 1984 (Kureha Chem. Co. Ltd). This patent protects the right to produce agar from calli of *Gelidium amansii* Lamouroux and *G. subcostatum* Okamura (among other agarophytes). Callus growth increased 11.3 times in 20 days and, from 100 g of dried callus, 75 g of agar was obtained.

In phycocolloid production in cell culture, Cheney *et al.* (1987) did not find a significant difference between iota-type carrageenan extracted from tissue culture of *Agardhiella subulata* (C. Ag.) Kraft & Wynne compared with extractive from field-collected material. Tait *et al.* (1990) obtained comparative spectra for polysaccharides produced by *Porphyra* cell-cultures and the native plants, but, as the authors concluded, 'it should be demonstrated whether specific growth conditions can affect the structure and properties of the polysaccharides produced by cell cultures'... or by disorganized cells.

Bryhni (1978) found highly-significant quantitative and structural differences between polysaccharides extracted from organized strains of *Ulva mutabilis* Føyn and those from a strain of disorganized aggegates of undifferentiated cells. Liu *et al.* (1989) obtained 50% agar yield from dried callus of *Pterocladia capillacea* (Gmel.) Born. & Thur., 25% less than Nakamura's patent, but the agarose molecule contained less sulfate and fewer methyl groups. The callus of *Laurencia*

consists of small photoautotrophic, pseudo-meristematic cells, filled with starch-like granules, and with cell walls at least twice the width of normal cells (Garcia-Reina *et al.*, unpub. data). These data indicate that phycocolloid production, with high quality and distinct characteristics, is possible through cell and callus culture.

In addition to phycocolloid production, Chen (1989) has suggested the possibility of producing *Porphyra* for human consumption through cell culture. Fujimura and Kajiwara (1989) produced a bioflavor (released into the culture media) from *Ulva pertusa* Kjellman cell-suspension cultures.

Some critical considerations on applicability

All the applications described above are possible. The question is whether application on a commercial scale will be practical. *Laminaria* selection by monoploid cell-culture has yielded industrial results, and propagation by somatic cell-inoculation of morphologically simple species could be achieved commercially in a few years. There remain, however, many unresolved basic questions, limiting types of applications and species to which these processes can be applied.

Axenic cultures

Owing to seasonal variation of contaminants, differential interspecific sensitivities to biocides and the widespread phenomenon of endophytism in seaweeds and other algae, few generalizations can be made on obtaining axenic cultures. Evaluation of axenic cultures do not include tests for viruses, mycoplasma, rickettsia and so forth. Even if judged axenic, contaminants can appear in the cultures after months or years (Fries, 1963; Lee, 1986; Cheney, 1986; Tait *et al.*, 1990), and even with morphologically simple species, there are contradictory results of the efficacy of sterilization treatments (Bonneau, 1976).

Callus, callus control and PGR's

Gall, tumour, cancer, tumour-like, callus and callus-like are all terms used to describe proliferative, disorganized growth in seaweeds (Apt, 1988). The term 'callus-like' has been widely-used to qualify filamentous, crustose, hyperplasic and other abnormal outgrowth of the explant. If 'callus-like' is not true callus, as some authors claim, the literature on seaweed callus culture can be dramatically reduced. A consensus of what a callus in seaweed is has not been achieved.

Callus control implies callus induction, growth (without the explant) and regeneration. Although auxins and cytokinins exist in seaweeds, their effects when applied *in vitro* are not clear. Based on the literature of cell and callus culture of agarophytes (*Gelidium* and *Pterocladia*), three types of responses have been described for the effects of these hormones: 'success', although not compared with hormone-free media (Nakamura, 1974; Gusev *et al.*, 1987); 'no effect' (Polne-Fuller & Gibor, 1987a; Garcia-Reina *et al.*, 1988a); 'not clear' (Liu & Gordon, 1987).

Ranking of calligenic potential (Table 2) can be done (Gusev *et al.*, 1987; Garcia-Reina *et al.*, 1988a; Polne Fuller & Gibor, 1987a), indicating genotypic differences in disorganization (Fang *et al.*, 1979). 'Wounding-response' seems to be the key factor in initiating callus growth in seaweeds.

True callus culture (excised from the explant) has been described for a few species (*Gelidium* and *Gracilaria*: Nakamura, 1974; *Laurencia*: Garcia-Reina *et al.*, 1988b; *Grateloupia*: Robaina *et al.*, 1990b; *Phyllophora*: Gusev *et al.*, 1987; *Sargassum, Cystoseira, Macrocystis, Ulva, Enteromorpha, Eucheuma* and *Porphyra*: Polne-Fuller & Gibor, 1987a, 1987b; *Ecklonia*: Lawlor *et al.*, 1989; *Pterocladia*: Liu & Gordon, 1987). Only the Nakamura (1974) patent describes (true) callus culture of *Gelidium amansii* and *G. subcostatum*. Calli from *Gelidium vagum* Okamura (Gusev *et al.*, 1987), *G. nudifrons*, *G. robustum* (Polne-Fuller & Gibor, 1987a) and *G. versicolor* (Garcia-Reina *et al.*, 1988a) have not been described, either excised from the explant or with organogenetic potential.

Table 2. Ranking of 'calligenic potential' expressed as percent of tissue explants which developed callus (data from Gusev *et al.* 1987; Kawashima & Tokuda, 1989; Garcia-Reina, unpubl.; Polne-Fuller & Gibor, 1987a).

Chlorophyta

88.0%	*Enteromorpha intestinalis*
86.0%	*Ulva augusta*

Phaeophyta

70.0%	*Ecklonia cava*
29.0%	*Macrocystis pyrifera (gametophyte)*
20.0%	*Laminaria sinclairii (gametophyte)*
17.0%	*Sargassum muticum*
17.0%	*Pelvetia fastigiata*
10.0%	*Sargassum hystrix*
10.0%	*Cystoseira osmundacea*
9.2%	*Laminaria sinclairii (sporophyte)*
9.0%	*Sargassum fluitans*
7.8%	*Macrocystis pyrifera (sporophyte)*

Rodophyta

87.0%	*Porphyra lanceolata*
84.0%	*Porphyra perforata*
81.0%	*Porphyra nereocystis*
75.0%	*Smithora naiadum*
33.0%	*Laurencia sp*
18.0%	*Phyllophora nervosa*
16.0%	*Furcellaria fastigiata*
15.0%	*Gelidium vagum*
10.0%	*Gracilaria ferox*
7.0%	*Eucheuma alvarezii*
4.0%	*Gracilaria verrucosa*
2.0%	*Ceramium kondoi*
1.0%	*Gigartina exasperata*
0.9%	*Eucheuma uncinatum*
0.6%	*Gelidium robustum*
0.5%	*Gracilaria papenfusii*
0.3%	*Gelidium versicolor*

Neither the physiological status of the explant (differential nutritional levels yielding differences in calligenic potential (Fries, 1980) seasonal differences in endogenous hormones (Mooney & van Staden, 1984; Featonby-Smith & van Staden, 1984) nor seasonally-different calligenic potentials, (Polne-Fuller & Gibor, 1987b) could be key factors for success. With so many unknown variables, it is difficult to project a productive future in the short term for the application of callus culture to seaweeds.

Culture media

As organic requirements for axenic or aseptic seaweed cultures are largely unknown, the more common culture media are undefined (enriched seawater). The addition of organic complexes (coconut milk, algal and yeast extracts, *etc.*) to increase low growth rates (indicating suboptimal conditions) of seaweed tissue (Fries, 1973; 1984) and callus cultures (Nakamura, 1974; Saga & Sakai, 1983) has resulted in both stimulatory effects and no effects (Robaina, 1988; Lawlor *et al.*, 1988). Nutrient depletion could be the reason for callus and 'callus-like' inductions or reversions (Pedersén, 1968; Fries, 1980; Bradley & Cheney, 1986; Polne-Fuller & Gibor, 1986; Lawlor *et al.*, 1989). Sugars have been reported to be unnecessary or even inhibitory to tissue (*Fucus spiralis*: Fries, 1984), callus (*Ecklonia radiata*: Lawlor *et al.*, 1989) and cell culture (*Porphyra umbilicalis*: Tait *et al.*, 1990) growth or organogenesis. Our studies on the interrelationship between the osmolality of the culture medium and the effect of osmotically active sugar supplementation has shown that, as in higher plants (van Rensburg & Vcelar, 1989), the effects of carbohydrates in seaweed tissue and callus culture seem osmotic or metabolic or both (Robaina *et al.*, 1990a; 1990b). Solidity of culture media promotes disorganization (Robaina *et al.*, 1990a; 1990b).

Heterotrophic growth is required for biomass production in bioreactors, but heterotrophic frowth of seaweed tissue remains to be demonstrated (Fries, 1973; Robaina, 1988; Lawlor *et al.*, 1989; Robaina *et al.*, 1990b). Although a callus of *Gelidium* has been obtained with the addition of sucrose (Nakamura, 1974), 20 g l^{-1} mannitol (Gusev *et al.*, 1987), or in the absence of sugars in the culture media (Polne-Fuller & Gibor, 1987a; Garcia-Reina *et al.*, 1988a), callus production has not occurred in darkness. However, as seaweed tissues, (Fries, 1977; 1980), cell (Saga *et al.*, 1978; Polne-Fuller *et al.*, 1987; Tait *et al.*, 1990) and callus cultures (Polne-Fuller *et al.*, 1986; Polne-Fuller & Gibor, 1987b; Garcia-Reina *et al.*, 1988a; Lawlor *et al.*, 1989) are phototrophic, including *Gelidium* species, photobioreactors remain a possibility for biomass production.

True cell culture

Porphyra is the only genus for which continuous cell cultures have been established (Chen, 1989; Tait *et al.*, 1990). Other reports on 'cell culture' refer to dynamic steps between protoplasts, pseudoprotoplasts or enzymatically/mechanically-isolated cells and their development to callus or to plants. The application of cell culture of other species for selection and for the production of biomass or metabolites requires confirmation of true cell culture.

Viable protoplasts

Protoplast isolation is not a problem in some seaweeds (Table 1), whereas the isolation of *large* numbers of *highly-viable* protoplasts is. Trial and error assays with cocktails of commercial (Table 3), phycophage-extracted (Table 4) or microorganism-extracted (Table 5) enzymes are common approaches to obtain protoplasts. Basic knowledge of composition and structure of cell walls of seaweeds and enzymatic activity of cell-wall digesting enzymes are scarce. Until more information becomes available, possible benefits from protoplast culture as applied to *Gelidium*, for example, are likely limited.

Somatic incompatibility

Somatic incompatibility needs to be overcome to realize the potential for somatic hybridization. In somatic incompatibility chromosomes, chloroplasts and mitochondria from one of the fused protoplasts degenerate, as noted in heterokaryons of *Zygnema* × *Porphyridium*, *Zygnema* × *Mougeotia* (Ohiwa, 1978; 1980; 1981) and *Daucus* × *Chlamydomonas* (Fowke *et al.*, 1979). Other undesirable processes could result from the

Table 3. Commercial enzymes commonly used for the digestion of seaweed cell walls.

Abalone acetone powder	Sigma A7514
Agarase	Sigma A6162
Cellulase	Sigma C2415
Cellulisin	Sigma C2274
Laminarinase	Sigma L7758
Limpet I	Sigma L1251
Limpet II	Sigma L0630
Limpet III	Sigma L8755
Hemycelulase	Sigma H2125
Papain	Sigma P3125
Pectinase	Sigma P2401
Pectolyase	Sigma P3026
Protease XXIV	Sigma P8163
Lysozyme	Sigma L6876
Cellulase onozuka RS	Yakult
Cellulase onozuka R-10	Yakult
Macerozyme R-10	Yakult
Cellulase onozuka R-10	Serva
Maceroyme R-200	Serva
Cellulysin	Calbiochem 219466
Macerase	Calbiochem 441201
Laminarinase	BDH 39120 2G
Driselase	Kyowa hakko
Pronase E	Merck 7433

Table 4. Phycophages from which have been isolated seaweed cell wall digestive enzymes.

Aplysia dactylomela	Gómez-Pinchetti *et al.* 1989
Aplysia depilans	Boyen *et al.* 1990
Aplysia punctata	Kloareg & Quatrano 1987a, 1987b
Aplysia vaccaria	Kloareg *et al.* 1989
Aplysia kurodai	Tokuda & Kawashima 1988
Crassostrea gigas	Onishi *et al.* 1985
Diadema antillarum	Gómez-Pinchetti *et al.* 1989, Lewis 1964
Diadema setosum	Benitez & Macaranas 1979
Dolabella auricula	Nisizawa *et al.* 1968
Haliotis cracherodii	Dai 1987
Haliotis rufescens	Dai 1987
Haliotis tuberculata	Ducreux & Kloareg 1988
Haliotis corrugata	Nakada & Sweeny 1967
Katherina tunicata	Kloareg & Quatrano 1987a
Littorina sp	Elyakova & Favorov 1974
Littorina striata	Gomez-Pinchetti *et al.* 1989
Littorina littorea	Chen 1986
Littorina brevicula	Onishi *et al.* 1985
Lunella cornata	Zhu 1982
Melagraphia aethiops	Liu & Gordon 1987
Monodonta labio	Zhu 1982
Mytilus edulis	Onishi *et al.* 191985
Nordotis discus	Onishi *et al.* 1985
Patella vulgata	Ducreux *et al.* 1988
Purpura clavigera	Zhu 1982
Strongylocentrotus intermedius	Saga & Sakai 1984
Strongylocentrotus purpuratus	Neushul 1984
Tegula funebralis	Galli & Giese 1959
Turbo coronatus	Tang 1982
Turbo sp	Liu *et al.* 1984
Turbo cornutus	Muramatsu *et al.* 1977

formation of different types of chimaera in plants regenerated from the heterokaryon. This occurred in the first seaweed somatic hybrid; irregularly-variegated chimeral thalli from the fusion of wild (red) and mutant (green) *Porphyra* protoplasts occurred (Fujita & Migita, 1987). Cheney (1989) claimed to verify true hybridization in green and red plants regenerated from the fused green mutant (*Gracilaria tikvahiae*) and red (*G. chilensis*) protoplasts by isoenzyme analysis. Although suggestive of true hybridization, this technique cannot distinguish chimaeric plants from true somatic-hybrid plants.

As with somatic hybridization in microalgae (Matagne *et al.*, 1979), the first recovery of a stable seaweed heterokaryon (Kapraun, 1987) has been achieved using haploid cells, although these were zoospores, not somatic protoplasts. If the reduction of somatic incompatibility from haploid state of the fusants can be confirmed, potential applications of seaweed somatic hybridization can be reexamined with some likelihood of success.

Genetic engineering

Few data are available on seaweed gene-identification, location and cloning, vectors, effectiveness of DNA injection techniques or possible expression of foreign genes. The genetic engineering of seaweeds is thus largely speculative at this time. When, for instance, the genes codifying agar synthesis are transferred, the beginning of phyco-colloid-producing bacterial biotechnology will be a real possibility.

190

Table 5. Microorganisms from which has been isolated seaweed cell wall digestive enzymes.

Achromobacter	Quatrano & Caldwell 1978
Aeromonas sp. F-25	Araki *et al.* 1987
Alcaligenes	Quatrano & Caldwell 1978
Alteromonas	Quatrano & Caldwell 1978
Amoeba (Am I-7)	Polne-Fuller & Gibor 1987c
Arthrobacter	Quatrano & Caldwell 1978
Cytophaga flevensis	van der Meulen 1975
Flavobacterium	Quatrano & Caldwell 1978
Pseudomonas atlantica	Yaphe 1957, Morrice *et al.* 1983
Pseudomonas sp. strain P-1	Fujita & Migita 1987
Pseudomonas carrageenovora	Smith & Bidwell 1989
Vibrio sp. AP-2	Araki *et al.* 1987
Vibrio sp. AX-4	Araki *et al.* 1987

References

Apt, K. E., 1988. Etiology and development of hyperplasia induced by *Streblonema* sp. (Phaeophyta) on members of the Laminariales (Phaeophyta). J. Phycol. 24: 28–34.

Araki, T., T. Aoki & M. Kitamikado, 1987. Preparation and regeneration of protoplasts from wild-type of *Porphyra yezoensis* and green variant of *P. ternera*. Bull. Jpn. Soc. Sci. Fish. 53: 1623–1627.

Ausich, R. L., 1983. Method for introducing foreign genes into green algae utilizing T-DNA of agrobacterium. European patent application no. 83306603.8. Publication no. 0108580. Applied for by Standard Oil Company, 200 East Randolph Drive, Chicago, IL 60601.

Benítez, L. V. & J. M. Macaranas, 1979. Partial purification of a carrageenase from the tropical sea urchin *Diadema setosum*. Proc. Int. Seaweed Symp. 9: 353–359.

Bingham, S. E. & J. A. Shiff, 1973. Conditions for attachment of single cells released from mechanically disrupted thalli of *Prasiola stipitata*. Biol. Bull. 154: 425.

Björk, M., P. Ekman, A. Wallin & M. Pedersén, 1990. Effects of growth rate and other factors on protoplast yield from four species of the red seaweed *Gracilaria* (Rhodophyta). Bot. mar. 33: 433–439.

Bonneau, E. R., 1976. Variation in growth and morphology of *Ulva* in axenic culture. Thesis, Connecticut. University, Gorton CT, 155 p.

Boyen, C., B. Kloareg, M. Polne-Fuller & A. Gibor, 1990. Preparation of alginate lyases from marine molluscs for protoplast isolation in brown algae. Phycologia 29: 173–181.

Bradley, P. M. & D. P. Cheney, 1986. Morphogenetic variation in tissue cultures of a red seaweed. Plant Physiol. Suppl. 80: 129.

Bryhni, E., 1978. Quantitative differences between poly-saccharide compositions in normal differentiated *Ulva mutabilis* and the undifferentiated mutant lumpy. Phycologia 17: 119–124.

Butler, D. M. & L. V. Evans, 1988. Isolation of protoplasts from *Laminaria*. Br. Phycol. J. 23: 284.

Butler, D. M., K. Ostgaard, C. Boyen, L. V. Evans, A. Jensen & B. Kloareg, 1989. Isolation conditions for high yields of protoplasts from *Laminaria saccharina* and *L. digitata* (Phaeophyceae). J. exp. Bot. 40: 1237–1246.

Cantacuzene, A., 1930. Contribution à l'étude des tumeurs bactériennes chez les Algues marines. Thèse présentée à la Faculté des Sciences de l'Université de Paris, Paris, 97 p.

Chauvat, F., J. Labarre, F. Ferino, P. Thuriaux & P. Fromageot, 1988. Gene transfer to the cyanobacterium *Synechocystis* PCC 6803. In Stadler T., J. Mollion, M.-C. Verdus, Y. Karamanos & M. H. Christiaen (eds), Algal Biotechnology. Elsevier. pp. 89–99.

Chen, L. C.-M. & A. R. A. Taylor, 1978. Medullary tissue culture of the red alga *Chondrus crispus*. Can. J. Bot. 56: 883–886.

Chen, L. C.-M., 1986. Cell development of *Porphyra leucosticta* in culture. Bot. mar. 30: 399–403.

Chen, L. C.-M., 1987. Protoplast morphogenesis of *Porphyra leucosticta* in culture. Bot. mar. 30: 399–403.

Chen, L. C.-M., M. F. Hong & J. S. Craigie, 1988. Protoplasts development from *Porphyra linearis* – an edible marine red alga. In Puite K. J., J. J. M. Dons, H. J. Huizing, A. J. Kool, M. Koornneef & F. A. Krens (eds), Progress in Plant Protoplast Research. Kluwer Academic Publishers. 123–124.

Chen, L. C.-M., 1989. Cell suspension culture from *Porphyra linearis* (Rhodophyta) a multicellular marine red alga. J. appl. Phycol. 1: 153–159.

Cheney, D. P., 1986. Genetic engineering in seaweeds: applications and current status. Nova Hedwigia 81: 22–29.

Cheney, D. P., E. Mar, N. Saga & J. van der Meer, 1986. Protoplast isolation and cell division in the agar-producing seaweed *Gracilaria* (Rhodophyta). J. Phycol. 22: 238–243.

Cheney, D. P., A. H. Luistro & P. M. Bradley, 1987. Carrageenan analysis of tissue cultures and whole plants of *Agardhiella subulata*. Hydrobiologia 151/152: 161–166.

Cheney, D. P., 1989. Interspecific protoplast fusion and somatic hybridization in the agarophyte *Gracilaria*. Abstracts XIII Int. Seaweed Symp. Vancouver, p. 17–69.

Dai, J., 1987. The effects of digestive enzymes of five species of marine shellfish on the isolation of *Porphyra yezoensis* cells. Trans. Oceanol. Limnol. 1: 84–88.

D'Amato, F., 1978. Chromosome number variation in cultured cells and regenerated plants. In Thorpe T. A. (ed.), Frontiers of Plant Tissue Culture Calgary, 288–296.

Davison, I. R. & M. Polne-Fuller, 1990. Photosynthesis in protoplasts of *Macrocystis pyrifera* (Phaeophyta). J. Phycol. 26: 384–387.

Deretic, V., J. F. Gill & A. M. Chakrabarty, 1987. Alginate biosynthesis: a model system for gene regulation and function in *Pseudomonas*. Biotechnology 5: 469–477.

Ducreux, G., P. Maillard & D. Sihachard, 1988. Patterns of differentiation and regeneration in single cells and protoplasts of *Sphacelaria* (Phaeophyceae). In Stadler T., J. Mollion, M.-C. Verdus, Y. Karamanos & M. H. Christiaen (eds), Algal Biotechnology. Elsevier, 129–137.

Ducreux, G. & B. Kloareg, 1988. Plant regeneration from protoplasts of *Sphacelaria* (Phaeophyceae). Planta 174: 25–29.

Duffield, E. C. S., S. D. Waaland & R. Cleland, 1972. Morphogenesis in the red alga *Griffithsia pacifica*: regeneration from single cells. Planta 105: 185–195.

Elyakova, L. A. & V. V. Favorov, 1974. Isolation and certain properties of alginate lyase VI from the mollusk *Littorina* sp. Biochimica et Biophysica Acta 358: 341–354.

Evans, D. A., W. R. Sharp & H. P. Medina-Felho, 1984. Somaclonal and gametoclonal variation. Am. J. Bot. 71: 759–774.

Fang, T. C., T. Chi-Hsun, O. Yu-Lin, T. Chin-Chin & C. Ten-Chin, 1979. Some genetic observations on the monoploid breeding of *Laminaria japonica*. Oceanic Selections 2: 1–12.

Fang, T. C., 1983. A summary of the genetic studies of *Laminaria japonica* in China. In Tseng C. K. (ed.) Proceedings of the joint China-USA. Phycology Symposium. Science Press. Beijing, 123–136.

Fang, T. Ċ., Z. Yan & Z. Wang, 1983. Some preliminary observations on tissue cultures in *Laminaria japonica* and *Undaria pinnatifida*. Kezue Tougbao Sci. Bull. Jpn. 28: 247–249.

Fang, T. C., 1984. Some genetic features revealed from culturing the haploid cells of kelps. Hydrobiologia 116/117: 317–318.

Featonby-Smith, B. C. & J. van Staden, 1984. Identification and seasonal variation of endogenous cytokinins in *Ecklonia maxima*. Bot. mar. 27: 527–531.

Fisher, D. D. & A. Gibor, 1987. Production of protoplasts from the brown alga, *Sargassum muticum* (Yendo) Fensholt (Phaeophyta). Phycologia 26: 488–495.

Fowke, L. C., P. M. Gresshoff & M. J. Marchant, 1979. Transfer of organelles of the alga *Chlamydomonas reinhardii* into carrot cells by protoplast fusion. Planta 144: 341–347.

Fries, L., 1963. On the cultivation of axenic red algae. Physiol. Plant. 16: 695–708.

Fries, L., 1973. Requirements for organic substances in seaweeds. Bot. mar. 26: 19–31.

Fries, L., 1977. Growth regulating effects of phenylacetic acid and p-hydroxyphenylacetic acid on *Fucus spiralis* in axenic culture. Phycologia 16: 451–455.

Fries, L., 1980. Axenic tissue cultures from the sporophytes of *Laminaria digitata* and *L. hyperborea*. J. Phycol. 16: 475–477.

Fries, L., 1984. D-vitamins and their precursors as growth regulators in axenically cultivated marine macroalgae. J. Phycol. 20: 62–66.

Fujimura, T. & T. Kajiwara, 1989. Production of bioflavor by regeneration from protoplasts of *Ulva pertusa*. Abstracts XIII Int. Seaweed Symp. Vancouver, 17–68.

Fujita, Y. & S. Migita, 1985. Isolation and culture of protoplasts from some seaweeds. Bull. Fac. Fish. 57: 39–45.

Fujita, Y. & S. Migita, 1987. Fusion of protoplasts from thalli of two different color types in *Porphyra yezoensis* Ueda and development of fusion products. Jpn. J. Phycol. 35: 201–208.

Fujita, Y. & M. Saito, 1989. Protoplast isolation and fusion in *Porphyra*. Abstracts XIII Int. Seaweed Symp. Vancouver, 17–68.

Gabriel, M., 1970. Formation, growth, and regeneration of protoplasts of the green alga, *Uronema gigas*. Protoplasma 70: 135–138.

Galli, D. R. & A. C. Giese, 1959. Carbohydrate digestion in a herbivorous snail, *Tegula funebralis*. J. exp. Zool. 140: 415–439.

García-Reina, G., R. Robaina & A. Luque, 1988a. Attempts to establish axenic cultures and photoautotrophic growth of *Gelidium versicolor Gracilaria ferox* and *Laurencia* sp. In Stadler T., J. Mollion, M.-C. Verdus, Y. Karamanos & M. H. Christiaen (eds), Algal Biotechnology. Elsevier. 111–118.

García-Reina, G., R. Romero & A. Luque, 1988b. Regeneration of thalliclones from *Laurencia*. In Pais M. S. S., F. Mavituna & J. M. Novais (eds), Plant Cell Biotechnology. NATO ASI Series. Springer-Verlag. 81–86.

Gautheret, R. J., 1985. History of plant tissue and cell culture: a personal account. In Vasil I. K. (ed.), Cell Culture and Somatic Cell Genetics of Plants. Academic Press. 2–60.

Goff, L. J. & A. W. Coleman, 1984. Transfer of nuclei from a parasite to its host. Proc. natl. Acad. Sci. USA 81: 5420–5424.

Gómez-Pinchetti, J. L., R. Robaina & G. García-Reina, 1989. Quantification of the agarolytic activity from marine herbivores. Its potential use for protoplast isolation. Abstracts XIII Int. Seaweed Symp. Vancouver, 17–97.

Gusev, M. W., A. H. Tambiev, N. N. Kirikova, N. N. Shelyastina & R. R. Aslanyan, 1987. Callus formation in seven species of agarophyte marine algae. Mar. Biol. 95: 593–597.

Han, H. & Y. Hongyuan, 1986. Haploids of Higher Plants *in vitro*. Springer Verlag. 211 pp.

Kapraun, D. F., 1987. Marine biologist clones seaweed cells. Gen. Engineering News 7: 42–43.

Kapraun, D. F. & S. G. Sherman, 1989. Strain selection and cell isolation of *Ulvaria oxysperma* (Chlorophyta) for net cultivation. Hydrobiologia 179: 53–60.

Karp, A., 1989. Can genetic instability be controlled in plant tissue cultures? Newsletter IAPTC, 58: 2–11.

Kawashima, Y. & H. Tokuda, 1989. Callus formation of *Eklonia cava*. Abstracts XIIIth Int. Seaweed Symp. Vancouver, 17–27.

Kloareg, B. & R. S. Quatrano, 1987a. Enzymatic removal of the cell walls from zygotes of *Fucus distichus* (L.) Powell (Phaeophyta). Hydrobiologia 151/152: 123–129.

Kloareg, B. & R. S. Quatrano, 1987b. Isolation of protoplasts from zygotes of *Fucus distichus* (L.) Powell. Plant Science 50: 189–194.

Kloareg, B., D. Kropf, R. S. Quatrano, C. Boyen & V. Vreeland, 1988. Protoplast production, cell wall regeneration and photopolarization in zygotes of *Fucus distichus* (L.) Powell (Phaeophyta). In Stadler T., J. Mollion, M.-C. Verdus, Y. Karamanos & M. H. Christiaen (eds), Algal Biotechnology. Elsevier. 119–128.

Kloareg, B., M. Polne-Fuller & A. Gibor, 1989. Mass production of viable protoplasts from *Macrocystis pyrifera*. Plant Science 62: 105–112.

Larkin, P. J. & W. R. Scowcroft, 1981. Somaclonal variation – a novel source of variability from cell cultures for plant improvement. Theor. appl. Genetics 60: 197–214.

Lawlor, H. J., J. A. McComb & M. A. Borowitzka, 1988. The development of filamentous and callus-like growth in axenic cultures of *Ecklonia radiata*. Proc. Int. Seaweed Symp. 12: 139–149.

Lawlor, H. J., J. A. McComb & M. A. Borowitzka, 1989. Tissue culture of *Ecklonia radiata* (Phaeophyceae, Laminariales): effects on growth of light, organic carbon source and vitamins. J. appl. Phycol. 1: 105–112.

Lee, T. F., 1986. Callus development from *Laminaria saccharina* sporophytes derived from gametophytes of aposporous origin. Abstracts for the meeting of the American Society of Limnology and Oceanography. Phycologial Society of America. 75.

Lee, Y.-K. & H. M. Tan, 1988. Interphylum protoplast fusion and genetic recombination of the algae *Porphyridium cruentum* and *Dunaliella* spp. J. gen. Microbiol. 134: 635–641.

LeGall, Y., J. P. Braud & B. Kloareg, 1989. Mass production of protoplasts from *Chondrus crispus* Stackhouse. Abstracts XIIIth Int. Seaweed Symp. Vancouver, 17–32.

Lewis, J. B., 1964. Feeding and digestion in the tropical sea urchin *Diadema antillarum* Philippi. Can. J. Zool. 42: 549–557.

Liu, W.-X., Y.-L. Tang, X.-W. Liu & T. C. Fang, 1984. Studies on the preparation and properties of sea snail enzymes. Hydrobiologia 116/117: 319–320.

Liu, Q. Y., 1989. The ultrastructure of cell wall regenerated by isolated *Palmaria palmata* protoplasts. Abstracts XIIIth Int. Seaweed Symp. Vancouver, 17–32.

Liu, X.-W. & M. E. Gordon, 1987. Tissue and cell culture of New Zealand *Pterocladia* and *Porphyra* sp. Hydrobiologia 151/152: 147–154.

Liu, X.-W., C. Rochas & M. E. Gordon, 1989. Callus culture of cell wall polysacharides. Abstract XIIIth. Int. Seaweed Symposium. Vancouver, 17–32.

Luqin, P., Z. Fei, G. Ma, J. Zhou & Y. Zhu, 1988. A preliminary study on the raising of seedlings of *Gelidium pacificum* by regeneration of thallus fragments. J. Zhejiang Coll. Fisheries 7: 100–105.

Matagne, R. F., R. Deltour & L. Ledoux, 1979. Somatic fusion between cell wall mutants of *Chlamydomonas reinhardi*. Nature 278: 344–346.

McLean, R. O. & D. J. Connolly, 1989. Callus and suspension culture in the Laminariales. Abstracts XIIIth Int. Seaweed Symp. Vancouver, 17–98.

Millner, P. A., M. E. Callow & L. V. Evans, 1979. Preparation of protoplasts from the green alga *Enteromorpha intestinalis* (L.) Link. Planta 147: 174–177.

Misawa, M., 1977. Production of natural substances by plant cell cultures described in Japanese patents. In Barz W. (ed.), Plant Tissue Cultures and its Biotechnological Applications. Springer-Verlag. 49–101.

Mooney, P. A. & J. van Staden, 1984. Lunar periodicity of the levels of endogenous cytoquinins in *Sargassum heterophyllum*. Bot. mar. 27: 467–472.

Morrice, L. M., M. McLean, F. B. Williamson & W. F. Long, 1983. β-Agarases from *Pseudomonas atlantica*. Hydrobiologia 116/117: 576–579.

Mumford, T., 1987. Commercialization strategy for nori cultivation in Puget Sound, Washington. In Bird K. & P. H. Benson (eds), Seaweed Cultivation for Renewable Resources. Elsevier. 351–372.

Muramatsu, T., S. Hirose & M. Katayose, 1977. Isolation and properties of alginate lyase from the mid-gut gland of wreath shell *Turbo cornutus*. Agric. Biol. Chem. 41: 1939–1946.

Nakada, H. I. & P. C. Sweeny, 1967. Alginic acid degradation by eliminases from abalone hepatopancreas. J. Biol. Chem. 242: 845–851.

Nakamura, T., 1974. (See Misawa, 1977).

Neushul, M., 1984. Marine algal tissue culture. Topical technical report for Gas Research Institute. Neushul, M. (ed.) Gas Research Institute 84/0076, Goleta CA. 46 p.

Nisizawa, K., S. Fujibayashi & Y. Kashiwabara, 1968. Alginate lyases in the hepatopancreas of a marine mollusc, *Dolabella auricula* Solander. J. Biochem. 64: 25–37.

Ohiwa, T., 1978. Behavior of cultured fusion products from *Zygnema* and *Spirogyra* protoplasts. Protoplasma 97: 185–200.

Ohiwa, T., 1980. Fine structure of degenerating nuclei in intergeneric fusion products from *Zygnema taceae* protoplasts. Protoplasma 102: 77–95.

Ohiwa, T., 1981. Intergeneric fusion of *Zygnema taceae* protoplasts. Bot. Mag. Tokyo 94: 261–271.

Onishi, T., M. Suzuki & R. Kikuchi, 1985. The distribution of polysaccharide hydrolase activity in gastropods and bivalves. Bull. Jpn. Soc. Sci. Fish. 51: 301–308.

Pedersén, M., 1968. *Ectocarpus fasciculatus*: marine brown algae requiring kinetin. Nature 218: 5143.

Polne-Fuller, M., M. Biniaminov & A. Gibor, 1984. Vegetative propagation of *Porphyra perforata*. Hydrobiologia 116/117: 308–313.

Polne-Fuller, M., N. Saga & A. Gibor, 1986. Algal cell, callus, and tissue cultures and selection of algal strains. Nova Hedwigia 83: 30–36.

Polne-Fuller, M. & A. Gibor, 1986. Calluses, cells, and protoplasts in studies towards genetic improvement of seaweeds. Aquaculture 57: 117–123.

Polne-Fuller, M. & A. Gibor, 1987a. Tissue culture of

seaweeds. In Bird K. & P. H. Benson (eds), Seaweed culti-vation for renewable resources. Elsevier. 219–239.

Polne-Fuller, M. & A. Gibor, 1987b. Calluses and callus-like growth in seaweeds: Induction and culture. Hydrobiologia 151/152: 131–137.

Polne-Fuller, M. & A. Gibor, 1987c. Microrganisms as digestors of seaweed cell walls. Hydrobiologia 151/152: 405–409.

Polne-Fuller, M., 1988. The past, present, and future, of tissue culture and biotechnology of seaweeds. In Stadler T., J. Mollion, M.-C. Verdus, Y. Karamanos & M. H. Christiaen (eds), Algal Biotechnology. Elsevier. 17–31.

Quatrano, R. S. & B. A. Caldwell, 1978. Isolation of a unique marine bacterium capable of growth on a wide variety of polysaccharides from macroalgae. Appl. envir. Microbiol. 36: 979–981.

Reddy, C. R. K., S. Migita & Y. Fujita, 1989. Protoplast isolation and regeneration of three species of Ulva in axenic culture. Bot. mar. 32: 783–490.

Robaina, R., 1988. Biotecnología del cultivo in vitro de algas rojas de interés industrial. Thesis. Universidad de Las Palmas de Gran Canaria. 236 pp.

Robaina, R., G. García-Reina & A. Luque, 1990a. The effects of the physical characteristic of the culture medium on the development of red seaweeds in tissue culture. Hydro-biologia (In press).

Robaina, R., P. García, G. García-Reina & A. Luque, 1990b. Morphogenetic effect of glycerol on tissue cultures of the red seaweed Grateloupia doryphora. J. appl. Phycol. 2: 137–143.

Saga, N., T. Uchida & Y. Sakai, 1978. Clone Laminaria from single isolated cell. Bull. Japan. Soc. Sci. Fish. 44: 87.

Saga, N. & Y. Sakai, 1983. Axenic tissue culture and callus formation of the marine brown alga Laminaria angustata. Bull. Jpn. Soc. Sci. Fish. 49: 1561–1563.

Saga, N., 1984. Isolation of protoplasts from edible seaweeds. Bot. Mag. Tokyo 97: 423–427.

Saga, N. & Y. Sakai, 1984. Isolation of protoplasts from Laminaria and Porphyra. Bull. Japan. Soc. Sci. Fish. 50: 1085.

Saga, N., M. Polne-Fuller & A. Gibor, 1986. Protoplasts from seaweeds: production and fusion. Nova Hedwigia 83: 37–43.

Saga, N. & T. Kudo, 1989. Isolation and culture of proto-plasts from the marine green alga Monostroma angicava. J. appl. Phycol. 1: 25–30.

Schiff, J. A., R. S. Quatrano, G. C. Harris, M. Legg & R. Stanley, 1972. Development of single cells released from mechanically disrupted thalli of Prasiola stipitata. Biol. Bull. 143: 476.

Smith, R. G. & R. G. S. Bidwell, 1989. Inorganic carbon uptake by photosynthetically active protoplasts of the red macroalga Chondrus crispus. Mar. Biol. 102: 1–4.

Sylvester, A. W. & J. R. Waaland, 1983. Cloning the red alga Gigartina exasperata for culture on artificial substrates. Aquaculture 31: 305–318.

Tait, M. I., A. M. Milne, D. Grant, J. A. Somers, J. Staples, W. F. Long, F. B. Williamson & S. B. Wilson, 1990. Por-phyra cell cultures: isolation and polysaccharide produc-tion. J. appl. Phycol. 2: 63–70.

Tang, Y., 1982. Isolation and cultivation of vegetative cells and protoplasts of Porphyra suborbiculata Kjellm. J. Shandong Coll. Oceanol. 12: 49–50 (English abstract).

Tatewaki, M. & K. Nagata, 1970. Surviving protoplasts in vitro and their development in Bryopsis. J. Phycol. 6: 401–403.

Tokuda, H. & Y. Kawashima, 1988. Protoplast isolation and culture of brown alga Undaria pinnatifida. In Stadler T., J. Mollion, M.-C. Verdus, Y. Karamanos & M. H. Christiaen (eds), Algal Biotechnology. Elsevier. 151–157.

Tsekos, I., 1982. Tumour-like growths induced by bacteria in the thallus of a red alga, Gigartina teedii (Roth) Lamour. Ann. Bot. 49: 123–126.

van der Meer, J. P. & F. J. Simpson, 1984. Cryopreservation of Gracilaria tikvahiae (Rhodophyta) and other macro-phytic marine algae. Phycologia. 23: 195–202.

van der Meer, J. P., 1986. Genetic contributions to research on seaweeds. Prog. in Phycol. Res. 4: 1–38.

van der Meulen, H. J., 1975. The enzymatic hydrolysis of agar by Cytophaga flevensis sp. nov. Thesis, University of Groningen, Netherlands, 80 pp.

van Rensburg, J. G. J. & B. M. Vcelar, 1989. The effect of the sucrose concentration on the initiation and growth of adventitious buds from leaf tissue of Lachenalia. S. Afr. J. Bot. 55: 117–121.

Vreeland, V., E. Zablackis, B. Doboszewski & W. M. Laetsch, 1987. Molecular markers for marine algal poly-saccharides. Hydrobiologia 151/152: 155–159.

Waaland, J. R., 1982. Cloning marine algae for mariculture. J. World. Maricul. Soc. 14: 404–414.

Waaland, J. R. & L. G. Dickson, 1987. Preparation of Porphyra protoplasts. J. Phycol. 23: 13.

Wang, S., Z. Xu, G. Wang & Z. Xia, 1986. Ultrastructural study on protoplasts of Porphyra haitanensis (Bangio-phyceae, Rhodophyta). Mar. Sci. 10: 21–24 (Chinese with English abstract).

Wang, S., X. Zhang, Z. Zhidong & Y. Sun, 1986. A study on the cultivation of the vegetative cells and protoplasts of Porphyra haitanensis. Oceanol. Limnol. Sinica 17: 217–221.

Wang, S., Y. Sun, A. Lu & G. Wang, 1987a. Early stage differentiation of thallus cells of Porphyra haitanensis. Chin. J. Oceanol. Limnol. 5: 217–224.

Wang, S., G. Wang, Y. Sun & A. Lu, 1987b. Isolation and cultivation of the vegetative cells of Porphyra haitanensis. Chin. J. Oceanol. Limnol. 5: 333–339.

Wang, S. & X. Yan, 1990. Observations on the monospore-like cells in the somatic cell culture of Porphyra haitanensis. Oceanol. Limniol. Sinica 21: 166–169.

Wu, C. Y. & G. Lin, 1988. Progress in the genetics and breeding of economic seaweed in China. Hydrobiologia 151/152: 57–61.

194

Yan, Z. M., 1984. Studies on tissue culture of *Laminaria japonica* and *Undaria pinnatifida*. Hydrobiologia 116: 314–316.

Yaphe, W., 1957. The use of agarase from *Pseudomonas atlantica* in the identification of agar in marine algae (Rhodophyceae). Can. J. Microbiol. 3: 987–993.

Zhang, D., 1982. Study on protoplasts preparation, culture and fusion of somatic cells from two species of green algae *Ulva linza* and *Monostroma angicava* Kjellm. J. Shandong College of Oceanol. 13: 57–65.

Zhao, H. & X. Zhang, 1984. On the cultivation of the isolated vegetative cells of *Porphyra yezoensis* Ueda. J. Fish. China 8: 197–202.

Zhu, R., 1982. A comparative study of marine algae cell wall decomposition by the digestive enzymes isolated from three species of marine gastropods. Biochem. Biophys. 2: 43–45.

Hydrobiologia **221**: 195–196, 1991.
J. A. Juanes, B. Santelices & J. L. McLachlan (eds), International Workshop on Gelidium.
© 1991 *Kluwer Academic Publishers.*

Concluding remarks

Key words: *Gelidium, Pterocladia,* stock management, cultivation, agars, genetics

The main purpose of this International Workshop on *Gelidium* was to bring together experts on the biology, chemistry and applications of species of *Gelidium* and their derivatives. It was anticipated that such a multidisciplinary approach would result in a comprehensive overview of the present state of our knowledge of this algal group, and equally importantly, it would reveal significant gaps in the scientific and technological information available for these species. This, in turn, would suggest requirements for new developments to advance the biology and utilization of these resources.

While no formal lecture was delivered on the state of taxonomy of *Gelidium* and related genera, the importance of this subject emerged in several of the round-table discussions. Clarification of *Gelidium* taxonomy must precede diversified usage of these algae, as different species produce different quantities, qualities and kinds of agars. The inclusion in this volume of recent advances describing intrageneric differences in cystocarp structure, in both *Gelidium* and *Pterocladia*, is intended to fulfill the information gap that became apparent during the workshop. The results of this study were anticipated in the discussions. These suggested heterogeneity with respect to the basic taxonomic character segregating *Pterocladia* from *Gelidium*; consequently, changes in the delimitation of both genera are anticipated.

Those reports dealing with *Gelidium* as a resource indeed confirmed the major importance of this group of algae, despite the comparatively low contribution made by these species to total algal stocks. Several contributions stressed, in addition, the sociological importance of this algal group, as significant numbers of harvesters depend partially or totally on *Gelidium* resources for their livelihood. This is not only a Third World reality, but it also affects small populations in a number of developed countries.

Data available suggest gradual decrease in harvestable stocks of *Gelidium* in several areas (*e.g.* Japan, Portugal) and point to a strong need for biologically-based management programmes. Such programmes have traditionally considered the ecological effects of harvesting on *Gelidium* populations; recent studies have also included evaluations of harvesting effects on other organisms inhabiting the *Gelidium* beds. Biologically-based resource management programmes require extensive field-testing of ecological concepts applied to these resources. This type of work is expensive and time-consuming; however, such resource management not only preserves the beds but can also increase production. Several *Gelidium* and *Pterocladia* beds from Portugal, described in the round-table discussions, were shown to exhibit 100% production increments through management.

In spite of recent advances in knowledge resulting from field-ecological investigations of *Gelidium*, several conceptual areas require additional development. Different beds of some species apparently respond differently to the effects of similar ecological factors; furthermore, patterns of recruitment, growth and reproduction for some species (*e.g. G. sesquipedale*) seem to change from place to place. These findings beg caution against the use of results gathered in one area, transferred and blindly applied to another area; in addition, it is evident that many commercial beds still require basic ecological studies prior to management. Demographic and experimental field ecology studies emerged as general requirements to understand community organization of these resources. There has apparently been no demographic study completed for any population of *Gelidium*, while experimental field studies elucidating biotic interactions within these communities are only starting.

The contributions dealing with physiological-ecology and cultivation suggest that it is technologically possible to cultivate species of *Gelidium*, although it has not yet become economically feasible. At least seven species (*G. amansii*, *G. coulteri*, *G. latifolium*, *G. pusillum*, *G. purpurascens*, *G. robustum*, *G. vagum*) have been maintained for extended periods under experimental cultivation, either in laboratory-size containers, tanks or pilot farms in the sea. At least one species is being propagated following inoculation of artificial substratum with thallus fragments. The optimization of all of these cultivation systems and the possibility of integrating them with other types of marine organisms into polyculture hold the promise of rendering these systems commercially viable.

Attempts to manipulate red algae genetically have been rather limited, and species of *Gelidium* are no exception. Present studies are oriented to the production of colour mutants, to mimic developments attained for species of *Gracilaria*. Hopefully these efforts will lead to morphological mutants or to polyploids with increased growth rates or productivities comparable with wild types.

The need to know population patterns of genotypic variability emerged during this workshop as one of the more relevant areas for future research on *Gelidium*. Population genetics have been badly neglected in algal studies, and advances in this area should enhance our understanding of eco-physiological responses, patterns of biomass and agar production, artificial selection of desirable strains and individual and population responses under different cultivation systems.

Although traditional genetic techniques still have much to offer and have been under-utilized in *Gelidium* studies, tissue and cell culture, and protoplast fusion also hold considerable promise for production of improved clones of algae. Present studies with species of *Gelidium* and future investigations on these organisms are expected to complement the present efforts in classical genetics.

The scientific contributions of *Gelidium*-derived agar summarized our present knowledge of the chemical structure of this family of compounds, present applications and the prospects for future uses. It became evident from the discussions that the galactans produced by *Gelidium* are more varied than what industry and scientists have normally recognized as agars. Therefore, if the primary goal is to understand the biology of production of galactans in seaweeds, caution must be taken not to limit considerations to those compounds now defined as agar; furthermore, investigators must be aware that the abundance and physical-chemical properties of these molecules can be significantly modified by the biological state of the seaweed at the time of harvest, by field manipulations and by subsequent processing treatments. All of these variables must be described in much greater detail than is presently being done, if we are to understand the effects of environmental factors and processing technologies on the abundance and qualities of galactans produces by *Gelidium*. It must be re-emphasized, strongly, that considerably more critical research is required to describe and understand environmental effects on quantities and properties of agars produced by species of *Gelidium*.

In general, The International Workshop on *Gelidium* has provided a useful platform for assessing our present body of knowledge on gelidiod organisms, and it is anticipated that this will stimulate new research and developments in different aspects of basic science and utilization. Given the biological and the social-economic importance of this algal group, meetings of this, or of a similar nature, should be held at regular intervals in the future.

B. SANTELICES

Hydrobiologia **221**: 197–203, 1991

Subject and taxonomic index

Page numbers are the first page of a paper in which the entry is discussed.

The manufacturer's authorised representative in the EU is Springer
Nature Customer Service Centre GmbH, Europaplatz 3, 69115 Heidelberg,
Germany. If you have any concerns regarding our products, please
contact ProductSafety@springernature.com

Printed and bound by CPI Group (UK) Ltd, Croydon, CR0 4YY

23/04/2026

02095657-0005